T0399379

NANOCELLULOSE MATERIALS

MICRO AND NANO TECHNOLOGIES SERIES

NANOCELLULOSE MATERIALS
Fabrication and Industrial Applications

Edited by

RAMESH ORAON

Department of Nanoscience and Technology, Central University of Jharkhand, Brambe, Ranchi, Jharkhand, India

DEEPAK RAWTANI

School of Pharmacy, National Forensic Sciences University (Ministry of Home Affairs, GOI), Gandhinagar, Gujarat, India

PARDEEP SINGH

Department of Environmental Studies, PGDAV College, University of Delhi, New Delhi, India

CHAUDHERY MUSTANSAR HUSSAIN

Department of Chemistry and Environmental Science, New Jersey Institute of Technology, Newark, New Jersey, USA

ELSEVIER

Elsevier
Radarweg 29, PO Box 211, 1000 AE Amsterdam, Netherlands
The Boulevard, Langford Lane, Kidlington, Oxford OX5 1GB, United Kingdom
50 Hampshire Street, 5th Floor, Cambridge, MA 02139, United States

Notices

Knowledge and best practice in this field are constantly changing. As new research and experience broaden our understanding, changes in research methods, professional practices, or medical treatment may become necessary.

Practitioners and researchers must always rely on their own experience and knowledge in evaluating and using any information, methods, compounds, or experiments described herein. In using such information or methods they should be mindful of their own safety and the safety of others, including parties for whom they have a professional responsibility.

To the fullest extent of the law, neither the Publisher nor the authors, contributors, or editors, assume any liability for any injury and/or damage to persons or property as a matter of products liability, negligence or otherwise, or from any use or operation of any methods, products, instructions, or ideas contained in the material herein.

British Library Cataloguing-in-Publication Data
A catalogue record for this book is available from the British Library

Library of Congress Cataloging-in-Publication Data
A catalog record for this book is available from the Library of Congress

ISBN: 978-0-12-823963-6

For Information on all Elsevier publications visit our website at
https://www.elsevier.com/books-and-journals

Publisher: Matthew Deans
Acquisitions Editor: Simon Holt
Editorial Project Manager: Gabriela D. Capille
Production Project Manager: Surya Narayanan Jayachandran
Cover Designer: Miles Hitchen

Typeset by Aptara, New Delhi, India

Working together
to grow libraries in
developing countries

www.elsevier.com • www.bookaid.org

Contents

Contributors

Shivakalyani Adepu
Cellulose and composites group, Department of Materials Science and Metallurgical Engineering, IIT Hyderabad, Telangana, India

Sevgi Aslıyüce
Department of Chemistry, Hacettepe University, Beytepe, Ankara, Turkey

Suddhasatwa Basu
CSIR-Institute of Minerals and Materials Technology, Bhubaneswar, Odisha, India

Birendra Bharti
Department of Water Engineering and Management, Central University of Jharkhand, Jharkhand, India

Dian Burhani
Research Center for Biomaterial – Indonesian Institute of Sciences (LIPI), Cibinong, Bogor, Indonesia

Tulsi Chandak
School of Engineering and Technology, National Forensic Sciences University (Ministry of Home affairs, GOI), Gandhinagar, Gujarat, India

Mansi Chugh
School of Engineering and Technology, National Forensic Sciences University (Ministry of Home affairs, GOI), Gandhinagar, Gujarat, India

Rafael Belasque Canedo da Silva
Departamento de Engenharia de Alimentos, Faculdade de Zootecnia e Engenharia de Alimentos, Universidade de São Paulo, Rua Duque de Caxias Norte 225, Pirassununga, SP, Brazil; Postgraduate Programme in Materials Science and Engineering, University of São Paulo, USP/FZEA, Av. Duque de Caxias Norte, 225, Pirassununga, Brazil

Natalia Cristina da Silva
Departamento de Engenharia de Alimentos, Faculdade de Zootecnia e Engenharia de Alimentos, Universidade de São Paulo, Rua Duque de Caxias Norte 225, Pirassununga, SP, Brazil; Postgraduate Programme in Materials Science and Engineering, University of São Paulo, USP/FZEA, Av. Duque de Caxias Norte, 225, Pirassununga, Brazil

Adil Denizli
Department of Chemistry, Hacettepe University, Beytepe, Ankara, Turkey

Ankita Gupta
Institute of Environment & Sustainable Development Banaras Hindu University Varanasi, Uttar Pradesh, India

Mani Pujitha Illa
Cellulose and composites group, Department of Materials Science and Metallurgical Engineering, IIT Hyderabad, Telangana, India

Sapna Jain
Department of Chemistry, School of Engineering, University of Petroleum and Energy Studies, Dehradun, Uttarakhand, India

Shruti Jha
Sardar Vallabhbhai National Institute of Technology, Surat, Gujarat, India

Swaminathan Jiji
DRDO-BU Center for Life Sciences, Bharathiar University Campus, Coimbatore, Tamil Nadu, India

Krishna Kadirvelu
DRDO-BU Center for Life Sciences, Bharathiar University Campus, Coimbatore, Tamil Nadu, India

Mudrika Khandelwal
Cellulose and composites group, Department of Materials Science and Metallurgical Engineering, IIT Hyderabad, Telangana, India

Sanjeev Kumar
Department of Chemistry, School of Engineering, University of Petroleum and Energy Studies, Dehradun, Uttarakhand, India

Vibhanshu Kumar
Department of Water Engineering and Management, Central University of Jharkhand, Jharkhand, India

Himanshu Kumar
Department of Water Engineering and Management, Central University of Jharkhand, Jharkhand, India

Bhawna Yadav Lamba
Department of Chemistry, School of Engineering, University of Petroleum and Energy Studies, Dehradun, Uttarakhand, India

Kannan Maharajan
Biology Institute, Qilu University of Technology (Shandong Academy of Sciences), Jinan, Shandong Province, China

Santosh Kr. Mishra
Department of Production Engineering, National Institute of Technology Tiruchirappalli, Tamil Nadu, India

Mamata Mohapatra
CSIR-Institute of Minerals and Materials Technology, Bhubaneswar, Odisha, India; Academy of Scientific & Innovative Research, New Delhi, India

Aarohi Natu
School of Pharmacy, National Forensic Sciences University (Ministry of Home affairs, GOI), Gandhinagar, Gujarat, India

Nadia Obrownick Okamoto-Schalch
Departamento de Engenharia de Alimentos, Faculdade de Zootecnia e Engenharia de Alimentos, Universidade de São Paulo, Rua Duque de Caxias Norte 225, Pirassununga, SP, Brazil; Postgraduate Programme in Materials Science and Engineering, University of São Paulo, USP/FZEA, Av. Duque de Caxias Norte, 225, Pirassununga, Brazil

Ramesh Oraon
Department of Nanoscience and Technology, Central University of Jharkhand, Brambe, Ranchi, Jharkhand, India

Arun. K. Padhy
Department of Chemistry, Central University of Jharkhand, Brambe, Ranchi, Jharkhand, India

Bansari Parikh
School of Pharmacy, National Forensic Sciences University (Ministry of Home affairs, GOI), Gandhinagar, Gujarat, India

Garvita Parikh
School of Pharmacy, National Forensic Sciences University (Ministry of Home affairs, GOI), Gandhinagar, Gujarat, India

Işık Perçin
Department of Biology, Hacettepe University, Beytepe, Ankara, Turkey

Prajesh Prajapati
School of Pharmacy, National Forensic Sciences University, Gandhinagar, Gujarat, India

Benjamin Raj
CSIR–Institute of Minerals and Materials Technology, Bhubaneswar, Odisha, India

Deepak Rawtani
School of Pharmacy, National Forensic Sciences University (Ministry of Home Affairs, GOI), Gandhinagar, Gujarat, India

Yulianti Sampora
Research Center for Chemistry – Indonesian Institute of Sciences (LIPI), Kawasan PUSPIPTEK, Serpong South Tangerang, Indonesia

Syed Saquib
Institute of Environment & Sustainable Development Banaras Hindu University Varanasi, Uttar Pradesh, India

Athanasia Amanda Septevani
Research Center for Chemistry – Indonesian Institute of Sciences (LIPI), Kawasan PUSPIPTEK, Serpong South Tangerang, Indonesia

Archana Singh
School of Engineering and Technology, National Forensic Sciences University (Ministry of Home Affairs, GOI), Gandhinagar, Gujarat, India

Gurudatta Singh
Institute of Environment & Sustainable Development Banaras Hindu University Varanasi, Uttar Pradesh, India

Purushottam Kumar Singh
Department of Mechanical Engineering, BIT Sindri, Jharkhand, India

P. R Sreeraj
Department of Production Engineering, National Institute of Technology Tiruchirappalli, Tamil Nadu, India

Swati
Department of Botany, Institute of Science, Banaras Hindu University Varanasi, Uttar Pradesh, India

Emel Tamahkar
Department of Bioengineering, Bursa Technical University, Bursa, Turkey

Anamika Shalini Tirkey
Department of Geoinformatics, School of Natural Resource Management, Central University of Jharkhand, Brambe, Ranchi, Jharkhand, India

Aykut Arif Topçu
Department of Medical Services and Techniques, Aksaray University, Aksaray, Turkey

Milena Martelli Tosi
Departamento de Engenharia de Alimentos, Faculdade de Zootecnia e Engenharia de Alimentos, Universidade de São Paulo, Rua Duque de Caxias Norte 225, Pirassununga, SP, Brazil; Postgraduate Programme in Materials Science and Engineering, University of São Paulo, USP/FZEA, Av. Duque de Caxias Norte, 225, Pirassununga, Brazil

Riddhi Trivedi
School of Pharmacy, National Forensic Sciences University, Gandhinagar, Gujarat, India

Shashikant Shivaji Vhatkar
Department of Nanoscience and Technology, Central University of Jharkhand, Brambe, Ranchi, Jharkhand, India

CHAPTER 1

Bacterial cellulose nanofibers for separation, drug delivery, wound dressing, and tissue engineering applications

Emel Tamahkar[a], Aykut Arif Topçu[b], Işık Perçin[c], Sevgi Aslıyüce[d], Adil Denizli[d]
[a]Department of Bioengineering, Bursa Technical University, Bursa, Turkey
[b]Department of Medical Services and Techniques, Aksaray University, Aksaray, Turkey
[c]Department of Biology, Hacettepe University, Beytepe, Ankara, Turkey
[d]Department of Chemistry, Hacettepe University, Beytepe, Ankara, Turkey

1.1 Introduction

Bacterial cellulose (BC) nanofibers can be produced by bacteria, plants, and algae. Among them the most utilized source is *Acetobacter xylinum* due to the production efficiency. Although BC has similar chemical structure with plant cellulose, BC attracted tremendous interest with its high purity devoid of hemicellulose and lignin thus, avoiding the purification cost [1]. BC presents many advantages over plant cellulose regarding the production expenses. Cost-efficient bacterial culture medium is utilized in the preparation of BC. Also the synthesis is simple and requires no special equipment and any harsh chemicals [2]. BC is synthesized as a pellicle composed of unoriented fibers of cellulose chains resulting in 3D nanofibrous network. These nanofibers are approximately 50–100 nm in diameter. BC is formed via beta glycosidic bonds, and glucan chains are linked through intermolecular and intramolecular hydrogen bonds [3].

BC is one of the mostly used biomaterial for academic, industrial, and pharmaceutical areas designed as drug delivery vehicle, wound dressing, separation medium, and tissue engineering scaffold. BC presents excellent potential for these areas with its high water absorption capacity, high porosity, high purity, and high mechanical strength. Also BC can be fabricated in any size and shape due to its *in-situ* moldability properties suggesting great applicability for diverse applications [4]. Thus, the aim of this chapter is focused on the recent developments of BC in these application areas.

1.2 Support as separation medium

As BC nanofibers exhibit nanoporous structure, biocompatibility, high water-holding capability, and chemical stability, they have big potential to be used for separation and removal of many compounds. BC nanofibers are attractive materials to be prepared as a separation media for the removal of environmental pollutants and purification of biologically important molecules. BC nanofibers are resistant to different environmental conditions and have high surface area with nanofibrous network [5]. Separation processes based on many intermolecular interactions need to be selective and specific to target molecules. Thus, modifications of BC using in separation applications have great importance for preparation of effective adsorbents [6]. BC nanofibers with their unique properties can be functionalized by many different chemical methods to show good bio-affinity properties. BC nanofibers also can be prepared by combining natural and synthetic polymers, thus changing surface and bulk properties.

Protein-specific BC-based nanofibers were prepared by Denizli and coworkers. Göktürk et al. designed human serum albumin imprinted composite BC nanofibers for proteomic applications. Molecular imprinting is a promising technique to obtain selective binding sites formed via polymerization around the template molecule with further extraction of the template from the polymeric material. Molecularly imprinted polymers (MIP) show great potential with high selectivity. As albumin is a high-abundance protein, they fabricated albumin imprinted composite BC nanofibers via metal ion coordination interactions using N-methacryloyl-(L)-histidine methyl ester (MAH) and Cu(II) ions (Fig. 1.1A). They tested selectivity of albumin imprinted BC by using human transferrin and myoglobin as competitive agents. Furthermore, the depletion of human serum albumin from artificial human plasma was performed by albumin-imprinted

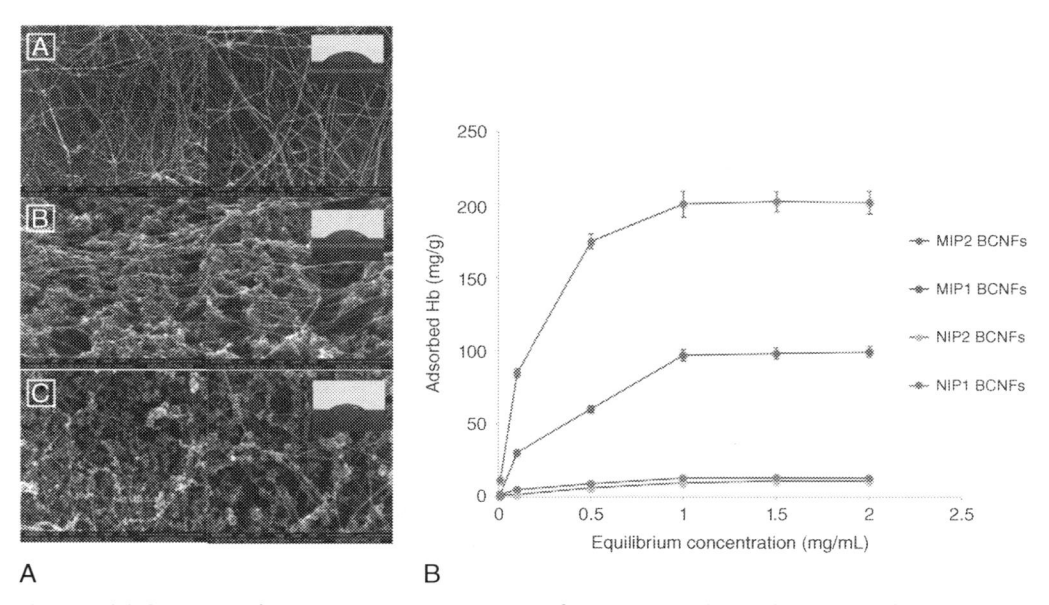

Fig. 1.1 (A) Scanning electron microscope images of BC (top), molecularly imprinted BC (middle) and nonimprinted BC (bottom). (B) The adsorption capacity of hemoglobin onto MIP-BC and NIP-BC. (A) Adapted with permission from [7], (B) Adapted with permission from [10].

BC nanofibers and efficiency of these BC nanofibers was shown with SDSPAGE and 2D gel electrophoresis analyses [7]. Tamahkar et al. aimed to deplete human serum albumin from human serum to provide easier study of low-abundance proteins, which have potential value for clinical diagnosis, for proteomic applications. For this purpose, they produced cibacron blue F3GA (CB) bound BC nanofibers. Specific surface area of CB-bound BC nanofibers was obtained as 914 m^2/g and albumin binding capacity of CB-bound BC nanofibers was found as 1800 mg/g. Elution of albumin from CB-bound BC nanofibers was achieved by 1 M NaCl and reusability was shown. In addition, albumin was purified from human serum successfully by using CB-bound BC nanofibers [8]. Saylan et al. developed lysozyme-imprinted BC (Lyz–MIP/BC) nanofibers for recognition of lysozyme. They applied surface molecular imprinting method by imprinting lysozyme on the surface of BC nanofibers in the presence of metal ions. The maximum lysozyme adsorption capacity of lysozyme-imprinted BC nanofibers obtained was 71.1 mg/g. Bovine serum albumin and cytochrome c were chosen as competitive proteins in this study and lysozyme-imprinted BC nanofibers showed high selectivity for lysozyme against these proteins. Reusability and high characteristic properties of lysozyme-imprinted BC nanofibers were also shown by authors [9]. Bakhshpour et al. fabricated a hemoglobin-imprinted film onto BC nanofibers. They preferred surface imprinting method by taking advantage of metal ion coordination interactions using MAH and copper ions. Hemoglobin was selectively purified from hemolysate using hemoglobin surface–imprinted BC nanofibers. Authors indicated structural and geometrical complementarity between recognition sites and template hemoglobin molecules by performing selectivity studies using cytochrome c, lysozyme, bovine serum albumin, and myoglobin. Maximum hemoglobin binding capacity of hemoglobin-imprinted BC nanofibers was 208.39 mg/g (Fig. 1.1B) [10]. Tamahkar et al. designed surface-imprinted BC nanofibers for cytochrome c purification. They used metal ion coordination interactions using copper and MAH as a functional monomer orienting cytochrome c molecules. After determining optimum experimental conditions they have obtained high recognition capacity. Authors also extracted cytochrome c from rat liver extract and eluted cytochrome c from cytochrome c–imprinted BC nanofibers successfully. Thus, they presented a novel, cheap, and simple procedure for efficient purification of cytochrome c [11]. Another study from Denizli and coworkers was reported for BC nanofibers as a protein adsorbent. 4-vinylimidazole (VIm) and a household monomer, MAH, were used as metal-chelating monomers. These monomers were complexed with Cu(II) and Ni(II) metal ions for carrying out metal ion coordination interactions with proteins. They have chosen hemoglobin as a model protein and the maximum hemoglobin binding capacity of the modified BC nanofibers was found as 47.40 mg/g [12].

The removal of some toxic substances from environment was also achieved by using BC nanofibers. Derazshamshir et al. used N-methacryloyl-(L)-amido phenylalanine as a functional monomer for the preparation of phenol-imprinted BC nanofibers by surface imprinting approach. They have applied surface-imprinted BC (BC-MIP) nanofibers to real wastewater samples from Ergene basin in Turkey and evaluated phenol removal efficiency of BC-MIP nanofibers. 2-chlorophenol, 4-chlorophenol and 2,4-dichlorophenol were used as computing agents and phenol removal efficiency was obtained as 97% [13].

Hou et al. presented functional BC membranes for water/oil separation for a solution to oil spill accidents that cause serious environmental pollution. They modified BC nanofibers by hydrolyzing of tetraethoxysilane and obtained $BC@SiO_2$ networks that were capable of separation of oil/water emulsions with high separation efficiency. Thus, they developed a promising potential wastewater treatment tool due to renewable and biodegradable nature of BC nanofibers [14]. Jin et al. prepared polyethyleneimine-BC nanofibers for the removal of copper and lead ions from aqueous solutions. Maximum binding capacities of polyethyleneimine-BC for copper and lead ions were obtained as 141 and 148 mg/g, respectively. Adsorption processes of these metals reached equilibrium in 30–60 min and were fitted with the Freundlich model. They demonstrated that polyethyleneimine-BC nanofibers are reusable and good alternative tools as bioadsorbents for heavy metal removal from wastewater [15]. Yang et al. prepared an efficient adsorbent by *in-situ* functionalization of poly(m-phenylenediamine) nanoparticles (NPs) on BC. Optimized poly(m-phenylenediamine) BC showed the Langmuir adsorption model with adsorption capacity of 434.78 mg/g for Cr(VI) ions. Authors observed high stabilization of poly(m-phenylenediamine) on BC and reusability of poly(m-phenylenediamine) BC nanofibers [16]. Shoukat et al. modified BC nanofibers with titanium oxide (TiO_2) and prepared novel TiO_2-BC nanofibers for the removal of lead ions from aqueous solutions. A concentration of 100 mg/L of lead was removed with 90% efficiency by TiO_2-BC nanofibers in 120 min at pH 7.0 at room temperature. TiO_2-BC nanofibers could be used three times without any decrease in lead adsorption capacity. Authors suggested TiO_2-BC nanofibers as an effective, stable, and reusable tool for the removal of lead from wastewater samples [17]. Chen et al. prepared attapulgite/chitosan (ATP/Chi) composite with BC nanofibers exhibiting high binding capacity for metal ions such as lead, copper, chromium, and anionic organic dyes including Congo red. ATP/Chi BC nanofibers were reusable after five adsorption-desorption cycles. ATP/Chi BC nanofibers with their multifunctional properties were effective tools to remove pollutants from environmental wastewater [18]. Hu et al. fabricated an organic–inorganic composite material combining BC and Ca-montmorillonite (Ca-MMT) to remove methylene blue (MB) and tetracycline (TC) from complex wastewater. Adsorption characteristics were well fitted with the Langmuir model and were suitable for a pseudo-second-order model. The removal of both MB and TC was successfully achieved even in the presence of other coexisting molecules. Low-cost Ca-MMT composite BC nanofibers were effectively reusable. The adsorption amount of the binary system containing MB and TC was also investigated and the images of the mixture solutions before and after treatment with the prepared adsorbent demonstrated great separation performance (Fig. 1.2) [19]. Wan et al. prepared bacteria containing BC nanofibers with a novel approach keeping bacteria in the nanofiber network by skipping bacteria removal process. Bacteria containing composite BC nanofibers showed much higher adsorption capacities for lead, copper, nickel, and chromium ions compared to pure BC nanofibers. Authors explain higher adsorption capacity with the presence of functional groups in bacteria such as amide providing coordination interactions. Thus, a simpler and cheaper strategy for producing a novel BC-based adsorbent with high heavy metal binding capacity was applied [20]. Ashour et al. designed flexible BC nanofibers by immobilizing metal organic framework (MOF) containing Fe(III) 1,3,5-benzenetricarboxylate) on BC nanofibers by *in-situ* one-pot green method. MOFs are very popular in recent years and can be modified to be selective adsorbents. The novel MOF BC nanocomposites were used for the removal of arsenic and Rhodamine B from aqueous solutions with binding capacities of 4.81 and 2.77 mg/g, respectively [21].

1.3 Wound dressings

Wound that is a defect in the structure and function of the skin can be caused by trauma, burn, diabetes, etc. Wound healing has four stages of hemostasis, inflammation, proliferation, and tissue remodeling [22]. Hemostasis, which is the first phase of the healing process, starts with the injury and during this phase bleeding is stopped with the clotting system. Inflammation occurs to prepare the wound bed and to remove the bacteria. During proliferation, granulation tissue covers the wound site and keratinocytes and fibroblasts migrate. Finally, the new tissue is remodeled. It is evaluated that 20–60 million people will endure chronic wounds by 2026 in the world [23]. An ideal wound dressing should provide an adequate moist environment for an effective wound healing process as humidity is required to maintain the healing. The properties of water retention capacity of the wound dressings affect substantially the healing to eliminate the excess liquid from the wound area. Also wound dressing with humid environment can act as a barrier against infections, and can be displaced easily on demand. Furthermore wound dressing having moderate water absorption amount can serve as a drug delivery vehicle to facilitate the healing [24].

BC with its unique characteristics such as high mechanical strength, high flexibility, high water-retention capacity, permeability for gas transfer, and high biocompatibility is evaluated as an ideal wound dressing material [25]. BC has been applied as a wound dressing material since 1980s [26]. There exist many commercial BC-based wound dressings namely Biofill that is Bionext, Membracell, and Xcell applied successfully for ulcers and burns. Biofill produced by a Brazilian company Biofill Produtos Bioetecnologicos is the first commercial BC-based wound dressing and is applied to burns, chronic skin ulcers. This company also manufactures Bionext and Membracell

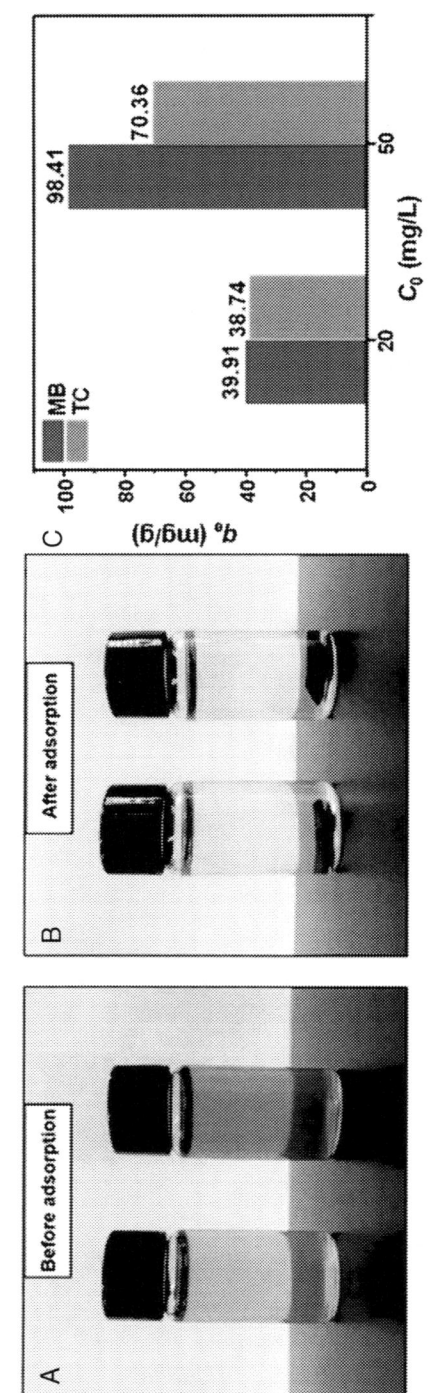

Fig. 1.2 Photographs of the MB and TC mixture solutions before and after the treatment with BC-based adsorbent. (A) A 20 mg/mL MB-TC binary solution (left), 50 mg/mL binary solution (right) before adsorption, (B) after adsorption. (C) The maximum adsorption amounts of MB and TC onto BC-based adsorbent. Adapted with the permission from [19].

for burns and chronic ulcers. Xylos Co. productized Xcell in 1996 as a Food and Drug Administration approved wound dressing having two significant features of absorption of exudate and holding the wound moist [27].

BC ensures efficient wound healing due to high water retention capability. The unbonded water molecules entrapped in the pores of BC function for the hydration [28]. BC with macropores and 3D morphological structure exhibits a suitable platform for the cell proliferation. This nanofibrous network structure also enables gas transfer throughout the membrane. BC can be utilized safely without any toxicity owing to its high biocompatibility. Solway et al. focused on the comparison of BC and commercial wound dressing namely Xeroform in regard to wound healing rate of diabetic foot ulcers. The rate of wound closure of BC was 1.7 times higher than that of Xeroform. BC provided higher water-holding capacity and water removal with high mechanical strength. It was mentioned that the pore size and the nanofibrous network structure of BC enable to start hemostasis and facilitate the formation of granulation tissue and epithelization [29].

To control the humidity of the nanofibers, the incorporation of polymers as a second component throughout BC matrix can be conducted to enhance the water-holding capacity. The decrease of the time period of dehydration decreases the exchange frequency, thus reduces the treatment costs. Sulaeva et al. developed BC/alginate films loaded with poly(hexamethylene biguanide) hydrochloride (PHMB) that is a cationic antimicrobial agent. BC was immersed in the alginate solution of different concentrations with subsequent cross-linking with $CaCl_2$. The feasibility of the large-scale production of BC/alginate membranes loaded with PHMB was also examined and the time required for the preparation of sterilized, and packaged novel BC-based wound dressing was reported as 4 days [30].

Transparency is an important concern for wound dressings as transparent materials enable to inform about the status of the wound and to monitor the healing without detaching of the dressing. Thus, transparent dressings provide patient comfort and efficient healing without interruption and any hospitalization cost. Tabaii et al. prepared transparent antibacterial BC-based wound dressings with silver NPs (AgNPs). The *in-situ* incorporation of AgNPs was conducted using sodium tripolyphosphate (TPP) or sodium hydroxide (NaOH) as reducing agents. The transparent BC-based wound dressings impregnated with AgNPs were formed when utilizing TPP (2%) and NaOH (0.04%). The resultant dressings showed excellent antibacterial activity against *Escherichia coli* and *Staphylococcus aureus* with ease of wound inspection of high transparency [31]. BC was applied as 3D wound dressing via *in-situ* modification with polyvinyl alcohol. It was prepared utilizing oxygen-permeable mold in the shape of a glove [32].

The impregnation of metallic NPs such as silver, zinc oxide, and copper into BC matrix has attracted wide interest approach to obtain antibacterial dressing. These metallic NPs present great antimicrobial activity due to their unique nanoscale characteristics.

Metallic NPs inhibit microbes via disarrangement of cell wall and production of reactive oxygen species inside the organism [33]. AgNPs with broad-spectrum antimicrobial features yield high antibacterial activity due to high surface area and widely utilized to fabricate wound dressings. Silver metal changes to silver ion under physiological conditions and exhibit antibacterial performance by binding the thiol groups of enzymes and proteins thus adjusting the bacterial metabolic pathways. Silver ion displays no toxicity to human cells except high concentrations of Ag^+ as it causes DNA damage to mammalian cells. BC represents an efficient support material by means of unique morphological character for the uniform distribution of AgNPs to avoid the aggregation of the NPs. There exist many reports about the preparation of AgNP-integrated BC using $NaBH_4$, triethanolamine, hydrazine, TEMPO (2,2,6,6-Tetramethylpiperidin-1-yl)oxyl), UV-radiation [34,35]. However, these reducing agents can cause toxicity concerns. Recently, the development of AgNP-incorporated BC membranes via a green route is gaining much attention. The choice of reducing agent, solvent, and stabilizer is the key parameters for green approach. Gupta et al. developed BC-based wound dressing with the integration of AgNP using a curcumin-cyclodextrin complex as a reducing agent. The AgNPs were produced with the curcumin-cyclodextrin complex with subsequent integration into BC. The prepared AgNP-integrated BC dressings exhibited a high survival rate due to MTT (3-(4,5-dimethylthiazol-2-yl)-2,5-diphenyltetrazolium bromide) assay using Panc 1 (human pancreatic ductal adenocarcinoma), U251 (human glioblastoma), and MSTO (human mesothelioma). Also the AgNP-integrated BC hydrogels displayed high any antibacterial activity against *Pseudomonas aeruginosa*, *S. aureus*, and *Candida auris*, while cyclodextrin-integrated BC and BC hydrogels as controls demonstrated antibacterial functionality [36]. The strength of the interactions between BC and AgNPs has great importance as the poor binding occurred between BC and the AgNPs can lead to fast release causing toxic effects. Thus, the design of the AgNPs-incorporated BC dressings should ensure the release of silver ions in a controlled manner. Wu et al. reported the release profile of Ag^+ from AgNP-BC dressings prepared with Tollen's reagent using BC as a nanoreactor (Fig. 1.3). The resultant dressings showed controlled release of silver ions in comparison with the control groups which were neat BC, $AgNO_3$-BC (prepared via immersion of BC into silver nitrate solution), and AgNP-BC (prepared via immersion of BC into the solution of AgNPs). This proves the strong attachment of AgNPs onto BC matrix thus preventing the uncontrolled release of AgNPs [37]. Khalid et al. prepared zinc-oxide NPs and then added them into BC structure with subsequent freeze-drying to fabricate BC/ZnO dressings. BC/ZnO dressings showed 90% bacterial inhibition on average for wound pathogens proving the bactericidal effect of the resultant wound dressings with zinc-oxide NPs. Higher wound healing performance and tissue regeneration was achieved with BC/ZnO dressings rather than that of BC due to participation of zinc during healing process [38]. In addition, copper NPs were incorporated into BC network structure to yield an antibacterial wound dressing via chemical reduction and hydrothermal method [39,40].

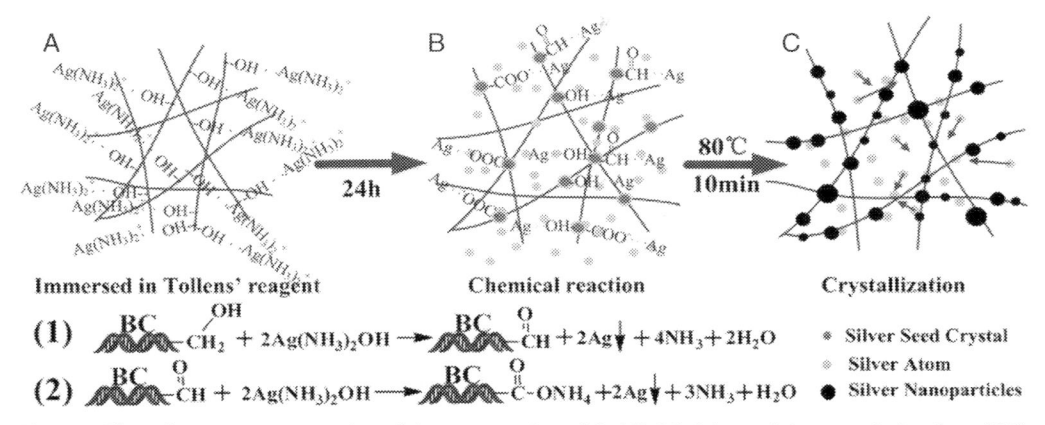

Fig. 1.3 The schematic presentation of the preparation of AgNP-BC. Adapted the permission from [37].

However silver can be integrated in the form of silver sulfadiazine (SSD) that is a common, broad-spectrum topical antibacterial agent utilized widely in the burn treatment. To prevent the cytotoxic side effects of SSD, BC-based wound dressings were formed with sodium alginate. SSD-incorporated BC/alginate hydrogels prepared in a simple route showed good biocompatibility on human embryonic kidney cells and great antibacterial performance againts *E. coli*, *S. aureus,* and *Candida albicans* [41]. In another study, SSD nano- (282 nm) and micro- (2149 nm) particles were synthesized and further integrated with a BC matrix for burn treatment. Owing to antibacterial activity tests, SSD-incorporated BC membranes demonstrated excellent antibacterial performance against Gram-positive and Gram-negative bacteria and high healing rate [42].

Some features of BC can be altered to improve the efficiency of healing performance. For this context, the incorporation of antimicrobial agent into the BC matrix is one of the general approaches to gain antibacterial functionality against the common bacteria that cause infections in the wound region. Amoxicillin is one of the common broad-spectrum antibiotics utilized for the infection. However its usage is restricted owing to the tendency to lose its function easily. Thus, grafting of the antibacterial agent becomes a good alternative. For the grafting of amoxicillin onto BC, first BC was modified with (3-aminopropyl) triethoxysilane and then amoxicillin was covalently bound with EDC/NHS (1-ethyl-3-(3-dimethylaminopropyl)carbodiimide/N-hydroxysuccinimide). Amoxicillin-grafted BC displayed improved antibacterial performance in comparison with neat BC. Also amoxicillin-grafted BC prepared with the drug concentration of 1 mg/mL showed a high wound closure rate of 45% after 3 days of treatment, while bare wound and BC as controls demonstrated 20% [43]. Qiu et al. developed BC-based wound dressings with the incorporation of vaccarin, a major flavonoid, which promotes the endothelial cell proliferation. It was determined that vaccarin-loaded BC membranes showed better healing performance due to rapid epithelization and regeneration [44].

Chi was also utilized to form interpenetrating network within BC network structure to gain antibacterial functionality. Lin et al. prepared BC/Chi membranes by immersion of BC into Chi solution with subsequent lyophilization. After the incorporation of Chi through BC network, BC/Chi membranes demonstrated great antibacterial activity against *E. coli* and *S. aureus*. Also epithelization of the BC/Chi membranes was reported faster than that of BC due to histological evaluations [45]. Doping of Chi onto BC also provides controlled release of antibiotics by altering the dimensions of the pore size of BC network besides antimicrobial activity and healing ability. Cacicedo et al. prepared Chi-impregnated BC hydrogels (BC/Chi) with ciprofloxacin loading for wound management. BC/Chi showed better water vapor transmission rate (WVTR) compared to that of BC. WVTR is one of the significant parameters indicating the capability of the material to keep the wound moist. The delivery of ciprofloxacin from BC/Chi reached an equilibrium at approximately 70% after 6 h offering a fast release for bacterial invasion at the beginning of the treatment. BC/Chi films exhibited high antibacterial activity after ciprofloxacin loading because of the synergistic effect of Chi with the antibiotic [46]. Furthermore, antimicrobial character was gained throughout the BC via covalent attachment of aminoalkyl groups [47].

1.4 Drug delivery vehicle

When the history of drug delivery systems (DDSs) is examined, DDS could be categorized into three stages; in the first generation between 1950 and 1980, researchers tried to understand the releasing mechanisms on DDS and focused on developing the oral and transdermal-based DDS, which were just capable of releasing the drugs in short periods. In the second generation between 1980 and 2010, nanotechnology-based DDS such as smart polymers and hydrogels were developed to use in DDS and the new drug carriers were able to release the drugs by changing the environmental conditions such as pH and temperature. Besides, the protein and peptide releasing studies in long periods were introduced in the second generation by using the other carriers and the first implant; Zoladex Depot was used to deliver goserelin acetate in 1989. The third generation between 2010 and 2040, called modulated delivery systems aims to release drugs to tumors, gene delivery, develop long-term delivery systems, and investigate in the suitable formulations of drugs for clinical trials [48]. DDS aims to transport the drugs needed by organs, cells and tissues at therapeutic concentrations by using suitable carriers. Moreover, this approach brings along various advantages such as increasing the solubility of the drugs, enhancing pharmaceutical activities, maintaining the chemically active agents, and reducing not only the drug aggregations but side effects, so on. However, if aimed to design an effective DDS, some parameters such as the releasing mechanism and releasing kinetic of a selected drug and its carrier selection should be carefully optimized [49].

BC nanofibers having superior properties such as high surface area, high porosity, mechanical properties, and mimicking the extra cellular matrix attracted much attention in drug delivery [50]. Tamahkar et al. aimed to extend the release of gentamicin sulfate by molecular imprinting technique (MIT), allowing designing the tailor-made receptors capable of selectively recognizing the template molecules. For this purpose, gentamicin sulfate–imprinted microparticles were created on BC surface by using methacrylic acid as a functional monomer and methylene bisacrylamide (MBAAm) as a cross-linker. The release of gentamicin sulfate studies were investigated in *in-vitro* drug studies. In the light of the experimental results; the creation of shape-memory effects based on the specific interactions between the functional monomer and drug by using with MIT can not only enhance the drug loading capacity but also extend the releasing of gentamicin sulfate when compared to the non–imprinted BC fiber (NIP) and gentamicin sulfate–imprinted BC fiber (MIP). The loading capacity of gentamicin sulfate on NIP was nearly five fold lower than MIP and gentamicin sulfate–imprinted BC fibers (MIPS) released the drugs within 48 h, whereas NIP released gentamicin sulfate in just 8 h; hence, imprinting cavities could enhance the loading capacity and release of drug of interest by adjusting the experimental conditions [51].

The performance of BC was aimed to hybridize with chitin to yield antibacterial function, and curcumin were formed *in-situ* to obtain the delivery in a controlled manner. The squid pen was chosen as a natural source of chitin, and before the formation of chitin nanofibers, chitin was isolated from its natural source by ultrasonication, after that the supernatant of the isolated solution was centrifuged and lyophilized, respectively. Before the hybridization of the chitin with BC-based nanofibers, a bacteria, *Gluconacetobacter xylinus,* was used for producing of BC pellicles with the suitable growth media, and the alkaline treatment method was applied to fabricate nanofibrillated BC films. After that, to fabricate the hybrid nanofibrillated BC-Chi nanofiber films, the same procedure was used for the formation of the BC, and then the curcumin particles in the ranges of micro- and nano-sizes were created, on which the desired surfaces were formed by the precipitation method. According to the experimental results, the incorporation of chitin could enhance the strength of the fabricated nanofibers, and interestingly the curcumin particles on the hybrid nanofiber surfaces could use an indicator of boric acid that is unsafe food additives, harmful for human health; moreover, the hybrid nanofibers gain the antibacterial properties thanks to the formation of the curcumin particles on their surfaces [52].

In the next study, chlorhexidine (CHX), a widely used model drug, was chosen because of its safety and antiseptic property. In this sense, the researchers investigated the *in-vitro* release profile of CHX and tested its antimicrobial properties against three different microorganisms. For this goal, the BC was produced by growing *Komagataeibacter hansenii* and the fibers were oxidized with the use of different concentrations of $NaIO_4$, following that, CHX was loaded on drugs carrier by using different approaches. In the light of the experimental results, the increase of the oxidation reduced the degree of the crystallinity of the BC, and the thermal stability and the degradation of the BC were dependent on the oxidation process. The addition of the ß-cyclodextrin can extend the releasing performance of a model drug and the oxidation process had no effect on the antimicrobial activity; however, the model drug had an antimicrobial activity toward the three microorganisms [53].

BC having a high adsorption capacity of the water and the biological fluid owing to its network structures has great potential in various purposes including DDS. Hence, in the present study, BC was cross-linked with gelatin because of some good properties of gelatin such as its biocompatibility, high water uptake ratio, nontoxicity, and so on to fabricate the BC-gelatin-based hydrogels for MB as a model drug delivery. Before the preparation of the BC-gelatin hydrogels, glutaraldehyde was selected as a cross-linker, and the decrease of the gelatin can decrease the pore size of the hydrogel and the addition of cellulose can significantly decrease the surface area of the hydrogel. The incorporation of gelatin into cellulose can cause to extend the release of a drug and led to a denser structure of the cellulose [54]. Badshah et al. aimed the release of two drugs: famotidine and tizanidine from BC nanofibers. BC was obtained by growing G. *xylinus*, and the different types of BC fibers were prepared with different conditions such as the freeze-drying method, and the drug loading capacities were related with the preparation methods, and the prepared BC matrices had potential in drug release study when compared with the tablet forms [55]. Beekman et al. reported an efficient approach to scale up the fermentation medium and the drug loading capacity for the production of transparent BC as drug carriers. Polyethylene glycol was used in a pilot scale to enhance the transparency by 41%. The loading amount of diclofenac sodium was increased by 40% and the drug release was also improved by 9% [56].

1.5 Tissue engineering applications

An ideal scaffold is defined as a biomaterial that provides cell attachment and proliferation by mimicking the native tissue in terms of chemical, physical, and mechanical structure. It should present high porosity, proper pore size and interconnectivity, proper chemical structure, high surface area, high mechanical strength, and high biocompatibility [57]. BC with its high surface area, unique nanofibrillar network structure has attracted great attention for the utilization as a scaffold material [58]. The *in-situ* moldability is one of important advantages of BC making it an excellent alternative for tissue engineering applications. Because BC can be formed in a desired shape during synthesis for instance, meniscus and tube form [59]. There exist some commercial BC-based tissue engineering products till today. Among them, Gengiflex was BC-based scaffold incorporated with hydroxyapatite and BASYC was applied as a vascular graft synthesized in the form of a tube.

Due to the similarity with collagen fibers, BC becomes a remarkable alternative as a support material for tissue engineering. However density of nanofibrils along the BC network restricts cell migration inside the matrix. The pore size dimension of BC can be altered with different techniques to obtain efficient cell proliferation. Nge et al. prepared porous scaffolds with BC and Chi and reported the dimensions of pores as 120–280 μm [60]. Also biocompatibility of BC was altered with Chi incorporation. Ul-Islam et al. prepared BC–Chi composite hydrogels via *ex-situ* synthesis with further lyophilization. The obtained porous scaffolds indicated high proliferation of human ovarian cancer cells throughout the matrix [61]. BC in a native form and modified form is successfully applied to various tissue engineering applications such as blood vessels, bone, cartilage, ophthalmic which are explained below in detail.

For the development of artificial blood vessels with diameter smaller than 5 mm, proper tissue proliferation and adequate mechanical strength are the main factors. BC with *in-situ* moldability, high tensile strength, nondegradability, and high biocompatibility draws much attention in recent years. Thus, BC can prevail great alternative as a blood vessel over the common synthetic vascular grafts such as expanded polytetrafluoroethylene, polyurethane, and Dacron since these polymers with high protein adsorption can cause clot formation and blockage. Klemm et al. prepared BC in a tubular shape (patented as BASYC tubes) using glass hollow molds allowing the transfer of the nutrient medium during BC formation at interface between liquid medium and air. BC can be fabricated in the tubular form having demanding length, diameter, and wall thickness [62]. Putra et al. developed a simple method for the synthesis of tubular BC using silicone tubes that allow oxygen transfer by pouring the nutrient medium into the mold with a gas inlet throughout the support [63]. Zang et al. used a polydimethylsiloxane mold for the preparation of BC blood vessels and obtained higher mechanical strength than Putra's work (Fig. 1.4) [64]. Bao investigated and compared the characteristics of tubular BC blood vessels prepared via three ways: using single silicone tube, double silicone tube, and a novel technique of using a central glass rod inside a silicone tube during fermentation [65]. Bodin et al. investigated the fermentation conditions on tubular BC production [66]. Li et al. developed novel artificial small blood vessels based on BC as a multilayered tube constituted of endothelial, smooth muscle and fibroblast cells to mimic the natural blood vessel. BC that was patterned with the cells by microfluidics can roll by self-assembly and it successfully remained in place for 21 days, which was transplanted in the rabbit artery [67].

Ideal bone tissue engineering scaffold should support the formation of new tissue with high porosity enabling the transport of wastes out of the scaffold while transferring the nutrient medium through the matrix. Also high biocompatibility and high mechanical strength are the other major concerns to obtain an efficient bone regeneration. BC with high porosity and biocompatible nature considerably meet the requirements to design the novel scaffolds for bone defects. The elastic modulus of BC was measured as 78 GPa with atomic force microscopy and 114 GPa for individual BC fibers with

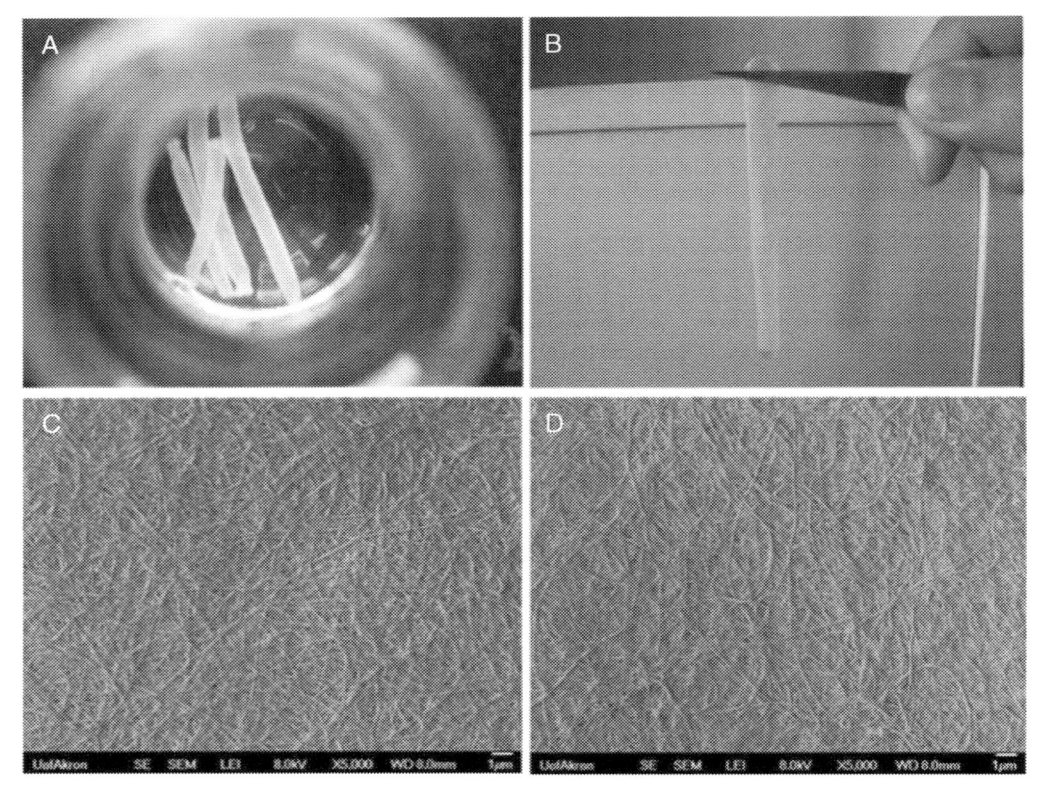

Fig. 1.4 Photographs of BC tubes obtained with polydimethylsiloxane mold (A, B), scanning electron microscope images of BC tubes. Adapted permission from [64].

Raman spectroscopy [68]. In general, the diameter of the pores of the scaffold is considered to be 100 μm to enable the transfer of the metabolites and the nutrient components. To fabricate microporous BC scaffold, several techniques were utilized such as fermentation on patterned agarose and solid particles (iron, aluminum, silica gel, etc.) [69,70]. Zaborowaska et al. prepared BC scaffolds with microscopic pores to enhance osteoblast proliferation for bone tissue engineering. The observed pore diameter of the resultant BC was approximately 100 μm, while using paraffin wax particles the diameter was in range of 300–500 μm. The results showed that microporous BC achieved more mineral deposition rather than nanoporous BC suggesting the great potential of microporous BC membranes for bone regeneration [71]. BC was generally modified with hydroxyapatite (HAp) and collagen (Col) as they are the main components of the bone via *in-situ* synthesis during BC fermentation and *ex-situ* synthesis onto BC nanofibers. Torgbo et al. developed a novel scaffold based on BC. First, BC was modified to gain magnetic property via *in-situ* coprecipitation and then it was incorporated with HAp by biomimetic approach. Magnetic particles gain osteoinduction capability to scaffolds even without any magnetic field. The magnetic BC scaffolds incorporated with HAp NPs demonstrated high biocompatibility [72]. Bayir et al. synthesized BC with HAp incorporation to improve osteoinduction property, and agar powders to enhance the pore diameters by adding the components into growth medium during bacterial incubation. The modified BC prepared with agar powder had a pore size of 275 μm as native BC shows pore diameters of 50–200 nm. Also the mineral deposition of the human osteoblast-like cells onto BC-HAp scaffolds was higher than that of BC [73]. Ran et al. prepared double-network scaffolds with BC and gelatin using hydroxyapatite to obtain high cell proliferation and high mechanical strength. The novel scaffolds were formed by *in-situ* precipitation of HAp particles and then with immersion of BC/HAp into gelatin solution with further lyophilization. The results defined that the resultant composite scaffolds could serve as a bone scaffold with high mechanical strength and high biocompatibility [74].

BC also represents great potential for cartilage tissue engineering because of the ability to mimic collagen matrix. However, to obtain efficient spreading and proliferation of chondrocytes an ideal scaffold should exhibit large pores similar to the bone tissue. Due to the dense structure of nanofibrous BC, it should be modified to improve the morphological features for cell transfer [75]. On the other hand, biocompatibility of BC should be enhanced to completely simulate the extracellular matrix by means of functional groups. Thus, in general BC was fabricated as BC–composites with various biomaterials such as gelatin, collagen, Chi, etc. The fabrication of nanocomposites using hydroxyapatite was also applied and the improvements of chondrocytes growth onto BC/HAp nanocomposite scaffolds were reported. Wu et al. developed BC-based scaffolds for cartilage tissue regeneration via using agarose during incubation to enhance pore size and modification with lotus root starch and hydroxyapatite to increase biocompatibility and mechanical strength. The resultant composite BC scaffolds with pores in the range of 300–500 μm showed enhanced mechanical strength and cell viability over the native BC [76]. Wang et al. prepared double interpenetrating network with BC and silk fibroin (SF) to construct a novel platform for a cartilage tissue with characteristics of high tensile strength and great biocompatibility. SF is one of the remarkable biomaterials owing to its excellent biocompatibility properties and mechanical strength. The BC/SF hydrogels revealed enhanced performance for cartilage regeneration [77]. Ahrem et al. developed a novel strategy to build BC-based scaffolds with large pores. They used spacers and 3D laser to form macropores for uniform cell differentiation [78].

BC can be also applied as artificial cornea due to its unique characteristics; however, poor transparency of BC limits its usage in this field. Bacterial cellulose/polyvinyl alcohol (BC/PVA) scaffolds were produced with a light transmittance of 96%, while light transmittance of BC was 82%. The viability of human corneal stroma tissue was increased with PVA impregnation [79]. Coelho et al. prepared BC-based therapeutic contact lenses with a controlled release of ciprofloxacin and diclofenac sodium using cyclodextrin. BC with Chi NPs showed improved transparency [80].

1.6 Conclusions

BC has drawn considerable interest in several application areas serving as support for separation medium, drug delivery vehicle, scaffold, and wound dressing in the native and modified form. BC presents unique advantages such as high surface area, nanofibrous network structure, high porosity with high interconnectivity, great water-holding capability, high hydrophilicity, high purity, high mechanical strength, high biocompatibility, and *in-situ* moldability. However more research is required in the future regarding the physical, chemical, and mechanical properties to improve the performance of BC. In addition novel approaches are expected to develop scale-up processes and to reduce the production cost for the efficient industrial manufacturing of BC.

References

[1] G.F. Picheth, C.L. Pirich, M.R. Sierakowski, M.A. Woehl, C.N. Sakakibara, C.F. de Souza, A.A. Martin, R. da Silva, R.A. de Freitas, Bacterial cellulose in biomedical applications: a review, Int. J. Biol. Macromol. 104 (2017) 97–106.

[2] C. Sharma, N.K. Bhardwaj, Bacterial nanocellulose: Present status, biomedical applications and future perspectives, Mater. Sci. Eng. C 104 (2019) 109963.

[3] H. Ullah, F. Wahid, H.A. Santos, T. Khan, Advances in biomedical and pharmaceutical applications of functional bacterial cellulose-based nanocomposites, Carbohydr. Polym. 150 (2016) 330–352.

[4] J.M. Rajwade, K.M. Paknikar, J.V. Kumbhar, Applications of bacterial cellulose and its composites in biomedicine, Appl. Microbiol. Biotechnol. 99 (6) (2015) 2491–2511.

[5] Z. Qiu, M. Wang, T. Zhang, D. Yang, F. Qiu, In-situ fabrication of dynamic and recyclable TiO_2 coated bacterial cellulose membranes as an efficient hybrid absorbent for tellurium extraction, Cellulose (2020) 1–18.

[6] X. Xu, X. Chen, L. Yang, Y. Zhao, X. Zhang, R. Shen, D. Sun, J. Qian, Film-like bacterial cellulose based molecularly imprinted materials for highly efficient recognition and adsorption of cresol isomers, Chem. Eng. J. 382 (2020) 123007.

[7] I. Göktürk, E. Tamahkar, F. Yılmaz, A. Denizli, Protein depletion with bacterial cellulose nanofibers, J. Chromatogr. B 1099 (2018) 1–9.

[8] E. Tamahkar, C. Babaç, T. Kutsal, E. Pişkin, A. Denizli, Bacterial cellulose nanofibers for albumin depletion from human serum, Process Biochem. 45 (10) (2010) 1713–1719.

[9] Y. Saylan, E. Tamahkar, A. Denizli, Recognition of lysozyme using surface imprinted bacterial cellulose nanofibers, J. Biomater. Sci. (2017) 1–29 Polymer Edition.

[10] M. Bakhshpour, E. Tamahkar, M. Andaç, A. Denizli, Surface imprinted bacterial cellulose nanofibers for hemoglobin purification, Colloids Surfaces B Biointerfaces 158 (2017) 453–459.

[11] E. Tamahkar, T. Kutsal, A. Denizli, Surface imprinted bacterial cellulose nanofibers for cytochrome c purification, Process Biochem. 50 (12) (2015) 2289–2297.

[12] M. Bakhshpour, E. Tamahkar, M. Andaç, A. Denizli, Affinity binding of proteins to the modified bacterial cellulose nanofibers, J. Chromatogr. B 1052 (2017) 121–127.

[13] A. Derazshamshir, I. Göktürk, E. Tamahkar, F. Yılmaz, N. Sağlam, A. Denizli, Phenol removal from wastewater by surface imprinted bacterial cellulose nanofibres, Environ. Technol. (2019) 1–12.

[14] Y. Hou, C. Duan, G. Zhu, H. Luo, S. Liang, Y. Jin, N. Zhao, J. Xu, Functional bacterial cellulose membranes with 3D porous architectures: conventional drying, tunable wettability and water/oil separation, J. Membrane Sci. 591 (2019) 117312.

[15] X. Jin, Z. Xiang, Q. Liu, Y. Chen, F. Lu, Polyethyleneimine-bacterial cellulose bioadsorbent for effective removal of copper and lead ions from aqueous solution, Bioresour. Technol. 244 (2017) 844–849.

[16] Z. Yang, L. Ren, L. Jin, L. Huang, Y. He, J. Tang, W. Yang, H. Wang, In-situ functionalization of poly(m-phenylenediamine) nanoparticles on bacterial cellulose for chromium removal, Chem. Eng. J. 344 (2018) 441–452.

[17] A. Shoukat, F. Wahid, T. Khan, M. Siddique, S. Nasreen, G. Yang, M.W. Ullah, R. Khan, Titanium oxide-bacterial cellulose bioadsorbent for the removal of lead ions from aqueous solution, Int. J. Biol. Macromol. 129 (2019) 965–971.

[18] X. Chen, J. Cui, X. Xu, B. Sun, L. Zhang, W. Dong, C. Chen, D. Sun, Bacterial cellulose/attapulgite magnetic composites as an efficient adsorbent for heavy metal ions and dye treatment, Carbohydr. Polym. 229 (2020) 115512.

[19] Y. Hu, C. Chen, L. Yang, J. Cui, Q. Hao, D. Sun, Handy purifier based on bacterial cellulose and Ca-montmorillonite composites for efficient removal of dyes and antibiotics, Carbohydr. Polym. 222 (2019) 115017.

[20] Y. Wan, J. Wang, M. Gama, R. Guo, Q. Zhang, P. Zhang, F. Yao, H. Luo, Biofabrication of a novel bacteria/bacterial cellulose composite for improved adsorption capacity, Compos. A Appl. Sci. Manufact. 125 (2019) 105560.

[21] R.M. Ashour, A.F. Abdel-Magied, Q. Wu, R.T. Olsson, K. Forsberg, Green synthesis of metal-organic framework bacterial cellulose nanocomposites for separation applications, Polymers 12 (5) (2020).

[22] J. Ahmed, M. Gultekinoglu, M. Edirisinghe, Bacterial cellulose micro-nano fibres for wound healing applications, Biotechnol. Adv. 41 (2020) 107549.

[23] L.P. da Silva, R.L. Reis, V.M. Correlo, A.P. Marques, Hydrogel-based strategies to advance therapies for chronic skin wounds, Annu. Rev. Biomed. Eng. 21 (1) (2019) 145–169.

[24] E. Liyaskina, V. Revin, E. Paramonova, M. Nazarkina, N. Pestov, N. Revina, S. Kolesnikova, Nanomaterials from bacterial cellulose for antimicrobial wound dressing, J. Phys. Conf. Ser. 784 (2017) 012034.

[25] W. Czaja, A. Krystynowicz, S. Bielecki, R.M. Brown, Microbial cellulose—the natural power to heal wounds, Biomaterials 27 (2) (2006) 145–151.

[26] I. Sulaeva, U. Henniges, T. Rosenau, A. Potthast, Bacterial cellulose as a material for wound treatment: properties and modifications. a review, Biotechnol. Adv. 33 (8) (2015) 1547–1571.

[27] B. Wei, G. Yang, F. Hong, Preparation and evaluation of a kind of bacterial cellulose dry films with antibacterial properties, Carbohydr. Polym. 84 (1) (2011) 533–538.

[28] R. Portela, C.R. Leal, P.L. Almeida, R.G. Sobral, Bacterial cellulose: a versatile biopolymer for wound dressing applications, Microb. Biotechnol. 12 (4) (2019) 586–610.

[29] D.R. Solway, W.A. Clark, D.J. Levinson, A parallel open-label trial to evaluate microbial cellulose wound dressing in the treatment of diabetic foot ulcers, Int. Wound J. 8 (1) (2011) 69–73.

[30] I. Sulaeva, H. Hettegger, A. Bergen, C. Rohrer, M. Kostic, J. Konnerth, T. Rosenau, A. Potthast, Fabrication of bacterial cellulose-based wound dressings with improved performance by impregnation with alginate, Mater. Sci. Eng. C 110 (2020) 110619.

[31] M. Jalili Tabaii, G. Emtiazi, Transparent nontoxic antibacterial wound dressing based on silver nano particle/bacterial cellulose nano composite synthesized in the presence of tripolyphosphate, J. Drug Deliv. Sci. Technol. 44 (2018) 244–253.

[32] M. Osorio, J. Velásquez-Cock, L.M. Restrepo, R. Zuluaga, P. Gañán, O.J. Rojas, I. Ortiz-Trujillo, C. Castro, Bioactive 3D-shaped wound dressings synthesized from bacterial cellulose: effect on cell adhesion of polyvinyl alcohol integrated in situ, Int. J. Polym. Sci. 2017 (2017) 3728485.

[33] M. Ul-Islam, W.A. Khattak, M.W. Ullah, S. Khan, J.K. Park, Synthesis of regenerated bacterial cellulose-zinc oxide nanocomposite films for biomedical applications, Cellulose 21 (1) (2014) 433–447.

[34] N. Eslahi, A. Mahmoodi, N. Mahmoudi, N. Zandi, A. Simchi, Processing and properties of nanofibrous bacterial cellulose-containing polymer composites: a review of recent advances for biomedical applications, Polym. Rev. 60 (1) (2020) 144–170.

[35] S. Pal, R. Nisi, M. Stoppa, A. Licciulli, Silver-functionalized bacterial cellulose as antibacterial membrane for wound-healing applications, ACS Omega 2 (7) (2017) 3632–3639.

[36] A. Gupta, S.M. Briffa, S. Swingler, H. Gibson, V. Kannappan, G. Adamus, M. Kowalczuk, C. Martin, I. Radecka, Synthesis of silver nanoparticles using curcumin-cyclodextrins loaded into bacterial cellulose-based hydrogels for wound dressing applications, Biomacromolecules (2020).

[37] J. Wu, Y. Zheng, W. Song, J. Luan, X. Wen, Z. Wu, X. Chen, Q. Wang, S. Guo, In situ synthesis of silver-nanoparticles/bacterial cellulose composites for slow-released antimicrobial wound dressing, Carbohydr. Polym. 102 (2014) 762–771.

[38] A. Khalid, R. Khan, M. Ul-Islam, T. Khan, F. Wahid, Bacterial cellulose-zinc oxide nanocomposites as a novel dressing system for burn wounds, Carbohydr. Polym. 164 (2017) 214–221.

[39] W. He, X. Huang, Y. Zheng, Y. Sun, Y. Xie, Y. Wang, L. Yue, In situ synthesis of bacterial cellulose/copper nanoparticles composite membranes with long-term antibacterial property, J. Biomater. Sci. 29 (17) (2018) 2137–2153 Polymer Edition.

[40] I.M.S. Araújo, R.R. Silva, G. Pacheco, W.R. Lustri, A. Tercjak, J. Gutierrez, J.R.S. Júnior, F.H.C. Azevedo, G.S. Figuêredo, M.L. Vega, S.J.L. Ribeiro, H.S. Barud, Hydrothermal synthesis of bacterial cellulose–copper oxide nanocomposites and evaluation of their antimicrobial activity, Carbohydr. Polym. 179 (2018) 341–349.

[41] W. Shao, H. Liu, X. Liu, S. Wang, J. Wu, R. Zhang, H. Min, M. Huang, Development of silver sulfadiazine loaded bacterial cellulose/sodium alginate composite films with enhanced antibacterial property, Carbohydr Polym. 132 (2015) 351–358.

[42] X. Wen, Y. Zheng, J. Wu, L. Yue, C. Wang, J. Luan, Z. Wu, K. Wang, In vitro and in vivo investigation of bacterial cellulose dressing containing uniform silver sulfadiazine nanoparticles for burn wound healing, Prog. Nat. Sci. Mater. Int. 25 (3) (2015) 197–203.

[43] S. Ye, L. Jiang, J. Wu, C. Su, C. Huang, X. Liu, W. Shao, Flexible amoxicillin-grafted bacterial cellulose sponges for wound dressing: in vitro and in vivo evaluation, ACS Appl. Mater. Interfaces 10 (6) (2018) 5862–5870.

[44] Y. Qiu, L. Qiu, J. Cui, Q. Wei, Bacterial cellulose and bacterial cellulose-vaccarin membranes for wound healing, Mater. Sci. Eng. C 59 (2016) 303–309.

[45] W.-C. Lin, C.-C. Lien, H.-J. Yeh, C.-M. Yu, S.-h. Hsu, Bacterial cellulose and bacterial cellulose–chitosan membranes for wound dressing applications, Carbohydr. Polym. 94 (1) (2013) 603–611.

[46] M.L. Cacicedo, G. Pacheco, G.A. Islan, V.A. Alvarez, H.S. Barud, G.R. Castro, Chitosan–bacterial cellulose patch of ciprofloxacin for wound dressing: preparation and characterization studies, Int. J. Biol. Macromol. 147 (2020) 1136–1145.

[47] S.C.M. Fernandes, P. Sadocco, A. Alonso-Varona, T. Palomares, A. Eceiza, A.J.D. Silvestre, I. Mondragon, C.S.R. Freire, Bioinspired antimicrobial and biocompatible bacterial cellulose membranes obtained by surface functionalization with aminoalkyl groups, ACS Appl. Mater. Interfaces 5 (8) (2013) 3290–3297.

[48] K. Park, Controlled drug delivery systems: past forward and future back, J. Control. Release 190 (2014) 3–8.

[49] C. Li, J. Wang, Y. Wang, H. Gao, G. Wei, Y. Huang, H. Yu, Y. Gan, Y. Wang, L. Mei, H. Chen, H. Hu, Z. Zhang, Y. Jin, Recent progress in drug delivery, Acta Pharm. Sin. B 9 (6) (2019) 1145–1162.

[50] Z. Li, S. Mei, Y. Dong, F. She, Y. Li, P. Li, L. Kong, Functional nanofibrous biomaterials of tailored structures for drug delivery-a critical review, Pharmaceutics 12 (6) (2020).

[51] E. Tamahkar, M. Bakhshpour, A. Denizli, Molecularly imprinted composite bacterial cellulose nanofibers for antibiotic release, J. Biomater. Sci. 30 (6) (2019) 450–461 Polymer Edition.

[52] Y.-N. Yang, K.-Y. Lu, P. Wang, Y.-C. Ho, M.-L. Tsai, F.-L. Mi, Development of bacterial cellulose/chitin multi-nanofibers based smart films containing natural active microspheres and nanoparticles formed in situ, Carbohydr. Polym. 228 (2020) 115370.

[53] B.S. Inoue, S. Streit, A.L. dos Santos Schneider, M.M. Meier, Bioactive bacterial cellulose membrane with prolonged release of chlorhexidine for dental medical application, Int. J. Biol. Macromol. 148 (2020) 1098–1108.

[54] W. Treesuppharat, P. Rojanapanthu, C. Siangsanoh, H. Manuspiya, S. Ummartyotin, Synthesis and characterization of bacterial cellulose and gelatin-based hydrogel composites for drug-delivery systems, Biotechnol. Rep. 15 (2017) 84–91.

[55] M. Badshah, H. Ullah, S.A. Khan, J.K. Park, T. Khan, Preparation, characterization and in-vitro evaluation of bacterial cellulose matrices for oral drug delivery, Cellulose 24 (11) (2017) 5041–5052.

[56] U. Beekmann, L. Schmölz, S. Lorkowski, O. Werz, J. Thamm, D. Fischer, D. Kralisch, Process control and scale-up of modified bacterial cellulose production for tailor-made anti-inflammatory drug delivery systems, Carbohydr. Polym. 236 (2020) 116062.

[57] W. Liu, H. Du, M. Zhang, K. Liu, H. Liu, H. Xie, X. Zhang, C. Si, Bacterial cellulose-based composite scaffolds for biomedical applications: a review, ACS Sustain. Chem. Eng. 8 (20) (2020) 7536–7562.

[58] H.G. de Oliveira Barud, R.R. da Silva, H. da Silva Barud, A. Tercjak, J. Gutierrez, W.R. Lustri, O.B. de Oliveira, S.J.L. Ribeiro, A multipurpose natural and renewable polymer in medical applications: bacterial cellulose, Carbohydr. Polym. 153 (2016) 406–420.

[59] A. Bodin, H. Bäckdahl, N. Petersen, P. Gatenholm, 2.223 - Bacterial Cellulose as Biomaterial, in: P. Ducheyne (Ed.), Comprehensive Biomaterials, Elsevier, Oxford, 2011, pp. 405–410.

[60] T.T. Nge, M. Nogi, H. Yano, J. Sugiyama, Microstructure and mechanical properties of bacterial cellulose/chitosan porous scaffold, Cellulose 17 (2) (2010) 349–363.

[61] M. Ul-Islam, F. Subhan, S.U. Islam, S. Khan, N. Shah, S. Manan, M.W. Ullah, G. Yang, Development of three-dimensional bacterial cellulose/chitosan scaffolds: analysis of cell-scaffold interaction for potential application in the diagnosis of ovarian cancer, Int. J. Biol. Macromol. 137 (2019) 1050–1059.

[62] D. Klemm, D. Schumann, U. Udhardt, S. Marsch, Bacterial synthesized cellulose—artificial blood vessels for microsurgery, Prog. Polym. Sci. 26 (9) (2001) 1561–1603.

[63] A. Putra, A. Kakugo, H. Furukawa, J.P. Gong, Y. Osada, Tubular bacterial cellulose gel with oriented fibrils on the curved surface, Polymer 49 (7) (2008) 1885–1891.

[64] S. Zang, R. Zhang, H. Chen, Y. Lu, J. Zhou, X. Chang, G. Qiu, Z. Wu, G. Yang, Investigation on artificial blood vessels prepared from bacterial cellulose, Mater. Sci. Eng. C 46 (2015) 111–117.

[65] L. Bao, J. Tang, F.F. Hong, X. Lu, L. Chen, Physicochemical properties and in vitro biocompatibility of three bacterial nanocellulose conduits for blood vessel applications, Carbohydr. Polym. 239 (2020) 116246.

[66] A. Bodin, H. Bäckdahl, H. Fink, L. Gustafsson, B. Risberg, P. Gatenholm, Influence of cultivation conditions on mechanical and morphological properties of bacterial cellulose tubes, Biotechnol. Bioeng. 97 (2) (2007) 425–434.

[67] Y. Li, K. Jiang, J. Feng, J. Liu, R. Huang, Z. Chen, J. Yang, Z. Dai, Y. Chen, N. Wang, W. Zhang, W. Zheng, G. Yang, X. Jiang, Construction of small-diameter vascular graft by shape-memory and self-rolling bacterial cellulose membrane, Adv. Healthc. Mater. 6 (11) (2017) 1601343.

[68] S. Gorgieva, Bacterial cellulose as a versatile platform for research and development of biomedical materials, Processes 8 (5) (2020).

[69] H. Bäckdahl, M. Esguerra, D. Delbro, B. Risberg, P. Gatenholm, Engineering microporosity in bacterial cellulose scaffolds, J. Tissue Eng. Regen. Med. 2 (6) (2008) 320–330.

[70] G. Serafica, R. Mormino, H. Bungay, Inclusion of solid particles in bacterial cellulose, Appl. Microbiol. Biotechnol. 58 (6) (2002) 756–760.

[71] M. Zaborowska, A. Bodin, H. Bäckdahl, J. Popp, A. Goldstein, P. Gatenholm, Microporous bacterial cellulose as a potential scaffold for bone regeneration, Acta Biomater. 6 (7) (2010) 2540–2547.

[72] S. Torgbo, P. Sukyai, Fabrication of microporous bacterial cellulose embedded with magnetite and hydroxyapatite nanocomposite scaffold for bone tissue engineering, Mater. Chem. Phys. 237 (2019) 121868.

[73] E. Bayir, E. Bilgi, E.E. Hames, A. Sendemir, Production of hydroxyapatite–bacterial cellulose composite scaffolds with enhanced pore diameters for bone tissue engineering applications, Cellulose 26 (18) (2019) 9803–9817.

[74] J. Ran, P. Jiang, S. Liu, G. Sun, P. Yan, X. Shen, H. Tong, Constructing multi-component organic/inorganic composite bacterial cellulose-gelatin/hydroxyapatite double-network scaffold platform for stem cell-mediated bone tissue engineering, Mater. Sci. Eng. C 78 (2017) 130–140.

[75] E. Akaraonye, J. Filip, M. Safarikova, V. Salih, T. Keshavarz, J.C. Knowles, I. Roy, Composite scaffolds for cartilage tissue engineering based on natural polymers of bacterial origin, thermoplastic poly(3-hydroxybutyrate) and micro-fibrillated bacterial cellulose, Polym. Int. 65 (7) (2016) 780–791.

[76] J. Wu, N. Yin, S. Chen, D.B. Weibel, H. Wang, Simultaneous 3D cell distribution and bioactivity enhancement of bacterial cellulose (BC) scaffold for articular cartilage tissue engineering, Cellulose 26 (4) (2019) 2513–2528.

[77] K. Wang, Q. Ma, Y.-M. Zhang, G.-T. Han, C.-X. Qu, S.-D. Wang, Preparation of bacterial cellulose/silk fibroin double-network hydrogel with high mechanical strength and biocompatibility for artificial cartilage, Cellulose 27 (4) (2020) 1845–1852.

[78] H. Ahrem, D. Pretzel, M. Endres, D. Conrad, J. Courseau, H. Müller, R. Jaeger, C. Kaps, D.O. Klemm, R.W. Kinne, Laser-structured bacterial nanocellulose hydrogels support ingrowth and differentiation of chondrocytes and show potential as cartilage implants, Acta Biomater. 10 (3) (2014) 1341–1353.

[79] Y. Han, C. Li, Q. Cai, X. Bao, L. Tang, H. Ao, J. Liu, M. Jin, Y. Zhou, Y. Wan, Z. Liu, Studies on bacterial cellulose/poly(vinyl alcohol) hydrogel composites as tissue-engineered corneal stroma, Biomed. Mater. 15 (3) (2020) 035022.

[80] F. Coelho, G.V. do Vale Braido, M. Cavicchioli, L.S. Mendes, S.S. Specian, L.P. Franchi, S.J. Lima Ribeiro, Y. Messaddeq, R.M. Scarel-Caminaga, T.S.O. Capote, Toxicity of therapeutic contact lenses based on bacterial cellulose with coatings to provide transparency, Cont. Lens Anterior Eye 42 (5) (2019) 512–519.

CHAPTER 2

Industrial-scale fabrication and functionalization of nanocellulose

Mani Pujitha Illa, Shivakalyani Adepu, Mudrika Khandelwal
Cellulose and composites group, Department of Materials Science and Metallurgical Engineering, IIT Hyderabad, Telangana, India

2.1 Introduction

Cellulose is a naturally abundant and renewable polymer with 7×10^{10} tons production per year [1,2]. Cellulose is a linear chain polymer comprising repeated β-(1,4) linked D-glucopyranose rings. The polyglucan chains of cellulose assemble through inter- and intramolecular H-bonding to produce a hierarchical fibrous organization (Fig. 2.1). In the process, cellulose chains crystallize, leading to semicrystalline structures comprising crystalline and amorphous regions (Fig. 2.1). The degree of crystallinity and crystallite size varies with the sources and the method of extraction.

Cellulose is produced by prokaryotes, protists, plants, trees, and animals. Cellulose obtained from plants and trees constitutes a major part of commercially available cellulose source. Cellulose can be extracted from plants, including wood, sisal, hemp, palm, coconut husk, flax, and cotton. Agricultural wastes and byproducts are also being explored as a rich source for cellulose. The cellulose content in sources varies significantly, for instance, cotton has >90% cellulose and wood has 40–50% [3]. Extensive chemical and mechanical treatments are needed for the removal of lignin and hemicellulose and other noncellulosic components during the cellulose isolation from plants and trees. Kraft pulping process is mostly used for delignification in which the sodium hydroxide and sodium sulfide are used as delignification agents. Apart from this method,

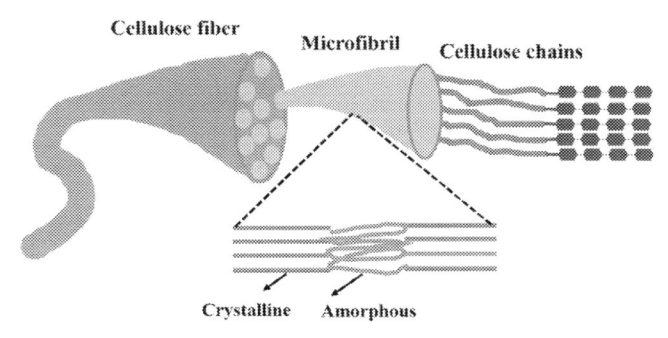

Fig. 2.1 Hierarchical organization of cellulose chains.

delignification can also be done through sulfite pulping processes using sulfur dioxide with different cations, liquor pH, temperature, and bleaching agents [4]. With increased deforestation and pollution from these harsh chemical treatments for isolation from plants and trees, an alternative greener source such as microbial cellulose is actively investigated.

Certain kinds of bacteria produce a pure form of cellulose as a byproduct during their primary metabolic activities. Bacterial cellulose (BC) is produced as a 100% pure hydrogel of a 3D network of highly crystalline nanofibers. Similarly, tunicate-derived cellulose also offers advantages in terms of lower amount of impurities and higher crystallinity [5]. Tunicates or sea squirts are the only animals known to produce cellulose in a microfibrillar form deposited in a multilayered texture with a bundled structure parallel to their epidermis. The tunic contains approximately 60% cellulose and other noncellulosic components are removed by chemical and mechanical treatment to obtain pure cellulose. Cellulose isolated from tunicates has shown high crystallinity and good mechanical properties than plant cellulose [6].

Cellulose, owing to its abundance and ease of usage, is being utilized in industries such as paper and pulp, textiles, reinforcements, cosmetics, food, pharmaceuticals, and so on [7]. However, given the increasing interest in utilizing cellulose in advanced applications such as filtration, specialized membranes, active packaging, and drug delivery, nanodimensional cellulose is actively being explored. Cellulose with at least one of its dimensions in the nanometric scale (1 nm = 10^{-9} m) is defined as nanocellulosic materials. Given the natural fibrous morphology and semicrystalline nature of cellulose, nanocellulose is obtained in the form of nanofibers, nanowhiskers, and nanocrystals. In addition, these can be obtained as dispersions, sheets (aerogels, xerogels, and hydrogels), spheres, and various other morphologies. Nanocellulose possesses unique morphologies, high aspect ratio, high specific surface area, crystallinity, interesting mechanical, optical, thermal, rheological, and barrier properties along with biocompatibility, biodegradability, and nontoxicity [8, 9]. All these characteristics and properties depend on the cellulose origin, isolation, production methods, and pre- or posttreatment conditions [10].

2.2 Nanocellulose production

Given the variety of sources and processing possibilities, nine types of cellulose-based particles are identified. These are wood fiber and plant fiber, microcrystalline cellulose, microfibrillated cellulose, nanofibrillated cellulose (NFC), cellulose nanocrystals (CNC), tunicate cellulose nanocrystals, and BC. Nanocellulose is the nanodimensional and/or nanostructured product or extract from the native cellulose isolated from plants, tunicates, algae, and bacteria [11]. Nanocelluloses include NFC that are cellulose nanofibers obtained from plant and wood fibers, nanofibers obtained from microbes, and cellulose nanocrystals obtained from any cellulose source. There are two important processes in the production of nanocellulose when plant-based celluloses are used—purification (removal of lignin, hemicelluloses, and others) and deaggregation and size reduction (to attain the required dimension) [12].

The most commonly used methods for nanocellulose isolation include mechanical (cryocrushing and/or crushing), chemical (acid or alkaline hydrolysis, organic solvent treatment), and biological (cellulose degrading enzymes) treatments, or a combination of the above. For the purpose of understanding, the nanocellulose production can be categorized into top-down and bottom-up approaches (Fig. 2.2).

The top-down process consists of disintegration/disaggregation of macroscopic materials to nano-sized products. Chemical hydrolysis, enzymatic hydrolysis, and mechanical process are used to breakdown cellulose into nanocrystals and nanofibrils [14]. The CNC are used as rheological modifiers, reinforcement materials, in barrier films, optical films, hybrid composites with other materials, and foams for packaging applications. The NFC is used to produce flexible substrates for electronics, drug delivery, fiberglass, filtration, and so on. Alternatively, in bottom-up approach atoms/molecules/nanoparticles are used as building blocks for nanofiber/nanocrystal fabrication. The examples of the bottom-up process include the electrospinning and production of microbial cellulose. Both the production approaches are elaborated in the following sections.

2.2.1 Bottom-up approaches

2.2.1.1 Bacterial cellulose

BC is an intrinsic nanofibrous cellulose obtained through bacterial fermentation process [15]. BC is obtained as a hydrogel (>90% associated within the fibrous network) at the

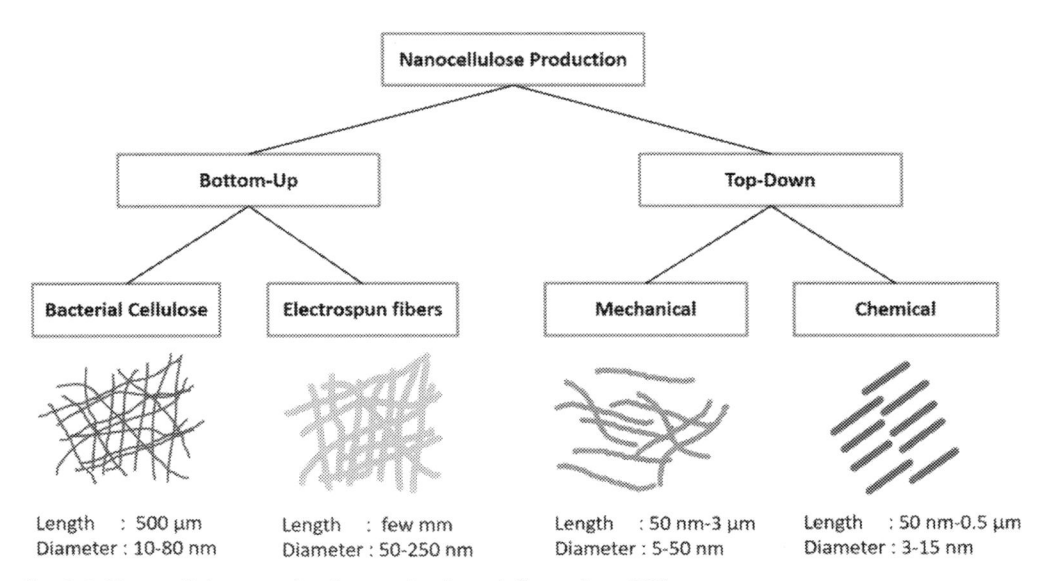

Fig. 2.2 Nanocellulose production methods and dimensions [13].

air-medium interface of the fermentation vessel. BC is highly pure (no treatments are necessary for removing pectin, lignin, and hemicellulose) and possesses high crystallinity (60–90%). The properties of BC such as fiber diameter, arrangement, physicochemical, and mechanical properties depend on the fermentation conditions, type of bacteria, and nutrient media. BC contains high aspect ratio nanofibers, with fiber diameter 10–80 nm and fiber length varying from 200 nm to 500 μm.

Various bacterial strains such as *Acetobacter*, *Rhizobium*, *Achromobacter*, and *Aerobacter* are known to produce cellulose. The fermentation media for bacterial growth consists of sugar and nitrogen sources along with distilled water. In some cases, natural sources such as fruit juices, milk, and wheat weay are used for BC production. BC is produced as a floating pellicle by layer-by-layer formation, where the thickness of cellulose pellicle increases with time and reaches a maximum thickness. Over the past five decades, research is directed toward understanding and optimizing biochemistry and microbiology to maximize cellulose yield and properties [16,17]. BC is commercially available and is being produced by Nata de Coco manufacturers, nanollose, Dermafill, Axcelon, Kasturi food processing, and so on. BC produced by the fermentation of coconut milk is being used Philippines dessert food ingredient for a long time [18].

2.2.1.2 Electrospinning
Electrospinning is a well-known technique to fabricate nanofibers where the polymer solutions are ejected by high voltages to generate nanofibers. Due to the rapid solvent evaporation and substantial stretching and whipping processes, the polymer jet cross-section can be reduced by six orders of the magnitude during jet flight, resulting in nanoscale fibers. Cellulose dissolved in solvents such as N-methyl-morpholine N-oxide, lithium chloride/dimethylacetamide, and ionic liquids are electrospun to produce cellulose nanofibers [19]. The fiber diameters lie in the range of 50–250 nm. The factor such as solution viscosity, type of solution, needle diameter, and rotation speed of a grounded collector will influence the formed fiber diameter significantly [20]. Other nanofiber spinning techniques include centrifugal spinning, solution/melt blow spinning which are being explored for producing cellulose nanofibers [21,22].

2.2.2 Top-down approaches
2.2.2.1 Mechanical processes
Mechanical disintegration is frequently utilized to break cellulose pulp into cellulose nanofibrils, possible owing to the hierarchical structure of cellulose. Mechanical methods such as defibrillation, including refining, homogenization, and grinding, are used to produce cellulose nanofibrils [23]. The high shear gradients in these processes cause transverse cleavage along the cellulose fibers' longitudinal axis and yield nanofibrils.

- *High-pressure homogenization*

In homogenization process, the cellulose suspension is forced to flow through a narrow gap at high pressures. The homogenizer creates rapid pressure drop with subsequent fluid acceleration, cavitation, turbulence, and shear stress, which reduces the fiber size by disintegrating amorphous regions and defibrillation. It is a facile and highly efficient method without involvement of organic solvents. The obtained NFC size can be controlled by altering the number of passages and the homogenization chamber size [24].

- *Microfluidization*

A microfluidizer consists of an intensifier pump that pressurizes the cellulose suspension to very high pressure 4000 bar and then forces to flow through Z-shaped or Y-shaped interactive vessels. The flow creates high shear and impact forces against the chamber walls, resulting in nanoparticle formation [25].

- *Grinding*

In grinders, the cellulose suspension is passed in-between the static and rotating grid stone with 1500 rpm for defibrillating cellulose. The shear forces created by stones' grinding action will rupture the hydrogen bonding in cellulose and liberate NFC [26]. The control over NFC's size can be acquired by adjusting the rotation speed, gap between the grinding stones, grinding time, and the number of passages.

- *Ultrasonication*

Ultrasonication is a simple technique in which the hydrodynamic forces of ultrasound (>20 kHz) result in NFC formation. The transfer of energy to the cellulose chains takes place through cavitation. During the process, the building, growth, and collapse of microbubbles occur inside the water, resulting in an increased temperature and high-pressure levels [27]. The energy generated is sufficient enough to break the hydrogen bonding and promote fibrillation. The power of ultrasound, concentration of pulp, temperature, and distance of the probe from cellulose pulp affect fibrillation.

- *Cryocrushing*

The cryocrushing method is used for cellulose in the frozen state to produce NFC. The aqueous cellulose suspension is frozen using liquid nitrogen and then subsequently crushed. The pressure exerted by ice crystals during mechanical impact will lead to the rupturing of cell walls and liberates the nanofibers.

2.2.2.2 Chemical processes

Mechanical processes are energy-intensive processes and typically result into NFCs. To obtain energy efficiency, high volumetric output, and highly crystalline cellulose, chemical treatments are adopted for CNC production [28]. The idea is to attach the amorphous regions leading to a decrease in dimensions and an increase in crystallinity.

- *Acid hydrolysis*

In acid hydrolysis, the hydrogen ions invade the amorphous cellulose regions and promote the hydrolytic cleavage of the glycosidic bonds. After this, more easily accessible glycosidic bonds in cellulose are hydrolyzed, and the stable crystallites remain intact. The stable crystallites are further isolated as rod-like cellulose nanocrystals. The formed CNCs in acid are further neutralized to remove the free acid. Hydrochloric acid, sulfuric acid, phosphoric acid, hydrobromic acid, and mixed acids are commonly used in hydrolysis process [29]. The geometry of CNCs produced through acid hydrolysis depends on the source of cellulose and acidic agent and conditions. The nanocrystals' width produced through acid hydrolysis is 2–30 nm and could be several hundred nanometers in length. The general conditions for acid hydrolysis involve relatively high acid concentration (50–70 wt%), temperature 40–50 °C, and variable duration of treatment. The variation in these parameters would impact CNC's quality, including length, efficiency, and yield. As an alternative of the acid solution, acid vapor hydrolysis has also been explored. In this method, 35% HCl solution was placed in a desiccator with an open valve to create an HCl environment [30]. Further cotton liners are allowed to undergo hydrolysis for 12 h. The usage of the acidic vapor method significantly reduces the posttreatment steps such as washing.

- *Enzymatic hydrolysis*

Enzymatic hydrolysis is the most economical, environmentally friendly, and significant step for obtaining CNC. For the breakdown of cellulose polymers, three different enzymes act synergistically to separate amorphous and crystalline cellulose segments. These enzymes are endoglucanases, exoglucanases, and cellobiohydrolases [31]. The final products of enzymatic breakdown can be smaller polymer branches or cellobiose and glucose. To obtain CNC, the endoglucanases are highly preferred as they hydrolyze the amorphous regions in cellulose, giving out the cellulose nanocrystals. Combined treatment with endoglucanases and mechanical operations can result in high aspect ratio nanofibers. The enzymatic process is relatively slow and requires more time to produce CNC.

- *Subcritical hydrolysis*

Subcritical hydrolysis has certain advantages such as no requirement of harsh conditions and relatively quick process compared to the enzymatic process [32]. Cellulose pulp in water-filled reactor is heated to temperature of 120–200 °C for 60 min with appropriate pressures, which allows the breakage of amorphous regions. The formed CNC were further filtered and dialyzed to remove any soluble sugars formed during the hydrolysis.

- *Oxidation method*

The hydroxyl groups of cellulose are highly reactive and can be easily oxidized by oxidants such as ketones, aldehydes, and carbonyl groups resulting in depolymerization of cellulose [33–35]. Ammonium persulfate and tetramethyl-piperidin-1-oxyl oxidants are used to produce cellulose nanocrystals. Some researchers have shown the formation of spherical nanocrystals with the usage of the oxidation method. Oxidation method can lead to the formation of NFCs as wells as CNCs.

- *Ionic liquid method*

The ionic liquid is an organic solvent composed of organic cations and anions with a low melting point. The ionic liquid is also known as a green solvent due to its exceptional chemical stability, thermal stability, and low vapor pressure [36]. Cellulose is dissolved in ionic liquid by the interaction of cation with oxygen atoms and anions with hydrogen bonding networks [37]. CNC can be easily regenerated from the ionic liquid through the addition of water, acetone, or ethanol. The ionic liquid is not consumed in the process and can be recovered and reused.

2.2.3 Scale up and industrialization

The industrialization of nanocellulose production is observed where some companies produce tons-per-day. Some of the CNC-producing companies include CelluForce, American Process Inc., Blue Goose Biorefineries, MoRe research, and ICAR-CRICOT and NFC-producing companies include Nippon Paper Group, Oji holdings cooperation, paperlogic, Innventia, SAPPI. However, the production costs are high because of high costs of chemical, maintenance of equipment, and high energy consumption for mechanical treatments. A few hurdles in commercialization are corrosion of equipment, and postprocessing of acid waste. The technological requirements and process duration are high for enzymatic hydrolysis, while chemical consumption and need for wastewater treatment are high in oxidation degradation method. The usage of an enzymatic process and low-cost chemical treatments are being explored to reduce the production costs. However, the commercialization process needs to be accelerated to exploit the advantages of nanocellulose in various applications.

2.3 Functionalization of nanocellulose

Nanocellulosic materials owing to the presence of vast surface hydroxyl groups and high specific surface area find their application in various fields. However, the self-agglomeration and hydrophilicity of nanocellulose pose to be the limiting factors in its extensive application [2]. This demands surface functionalization to increase its widespread usage. Functionalization of nanomaterials can add/enhance their properties such as compatibility with multiple matrices, water absorption capacity, and addition of chemical functionality. Various small molecules of biomedical importance, drug carriers containing nano/microparticles, protective and semipermeable coatings, and hydrophobic polymers can be tethered to nanocellulose for the preparation of functional nanocomposites [38–40].

Various noncovalent and covalent modification methods of nanocellulose have been reported owing to the availability of plentiful active sites [1]. Noncovalent modifications include adsorption of surfactants, oppositely charged entities, and polyelectrolytes. Covalent modifications methods such as acetylation, acylation, oxidation, [41] esterification [42], etherification [43], silylation, polymer grafting [44–46] and fluorescent labeling are used to introduce additional functionality onto nanocellulose or a precursor for further modification to address the adherent issues (Fig. 2.3). In addition, enzymatic and plasma treatment have also been tried to modify the nanocellulose [47,48].

2.3.1 CNC functionalization

• *Noncovalent functionalization*

Noncovalent functionalization is typically achieved by adding surfactants, oppositely charged entities, or polyelectrolytes. In this case, physical adsorption plays a major role, where electrostatic interactions, hydrogen bonding, van der Waals forces, hydrophobic groups association, charge-charge attractions, entropy effects, and solvency exhibit significant influence. However, noncovalent functionalization suffers from the leaching of adsorbed entities and relatively short-time affects. As produced CNC surface contains different groups such as sulfate esters from sulfuric acid hydrolysis, hydroxyl groups resulting from enzymatic or hydrochloric acid–mediated hydrolysis, and carboxyl groups from (2,2,6,6-Tetramethylpiperidin-1-yl)oxyl (TEMPO) oxidation [50]; therefore, the choice of CNC production method determines the surface chemistry and further direction for grafting of chosen molecules.

• *Adsorption of surfactants*

To overcome the challenge of agglomeration in the fluid or solid matrix and for achieving superior properties in advanced systems, surfactants are added to nanocellulose. Multiple surfactants–nanocellulose combinations are possible and the subsequent results are governed by the balance between surfactant factors (structure, electrostatic charges, hydration, and solvency) and their relative affinity. Various surfactants including cetyltrimethylammonium bromide (CTAB) [52], stearic acid [51], and quaternary

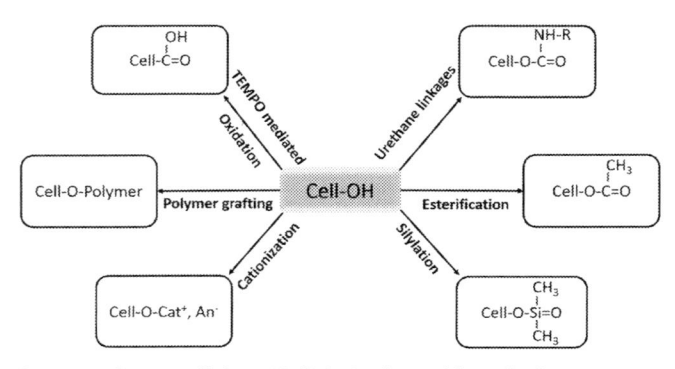

Fig. 2.3 Functionalization of nanocellulose (Cell-OH) adapted from [49].

ammonium salts [53] have been utilized to improve the dispersibility of CNCs in non-polar solvents and polymers.

- *Covalent grafting*

Covalent modifications of nanocellulose are reported using functional groups such as alkoxysilanes, chlorosilanes, acid anhydrides, acid halides, isocyanates, and epoxides. The hydroxyl groups originated from acid and TEMPO hydrolysis make CNC react to covalent grafting easily. The hydroxyl groups on CNC are oxidized to carboxyl groups and are further reacted with primary amines that are utilized as precursors for amidation reactions. The same procedure was reported to synthesize alkyne-based CNC precursors [54–56].

- *Covalent grafting of polymers*

Polymers are grafted to nanocellulose by three approaches. The first approach involves polymerization of monomers to macromolecules and anchoring to nanocellulose, which results in poor grafting density and high polydispersity index. The second approach involves the synthesis of functionalized polymer chains, and then macromolecules are attached to the nanocellulose surface. In the third approach, click chemistry is used to modify the CNCs with different polymers including amidation (CNC-grafted polypeptides) [58], CNC-grafted poly(ethyl ethylene phosphate) [57], or etherification (CNC-grafted (poly(ethylene oxide))) [59]. Polymerization is done using monomer and nanocellulose linked with the initiator in the presence of an activation complex in a suitable solvent, which ensures a controlled grafting process [60]. CNC was grafted and polymerized with styrene using surface-initiated atom transfer free-radical polymerization to increase thermal stability and enhance compatibility with other polymers [61]. By grafting poly(N-vinyl caprolactam) on to CNC, thermoresponsive nanocellulose was obtained [62]. Poly(methyl acrylate) and poly(methyl methacrylate) brushes were grafted on CNC and the grafting improved the mechanical properties of CNCs [63]. In another work, poly(L-lactide) was tagged to CNC and then incorporated into poly (lactic acid) matrix to obtain polymer composites with good mechanical properties [64].

Some of the reported works on covalent and noncovalent functionalization strategies of CNC along with the applications of the respective modifications are enlisted in Tables 2.1 and 2.2.

2.3.2 NFC functionalization

In line with the above discussed motivation for functionalization of CNC and methods reported above, this section discusses functionalization of NFC by polymer grafting, esterification, silylation, TEMPO oxidation, and cationization for various applications [90]. A special focus is laid on altering the wettability and improving dispersion.

Table 2.1 Covalent functionalization of cellulose nanocrystals (CNC) and applications.

Functional groups/active species	Method/materials	Application	Reference
Silver nanoparticles	Tollens' reagent–assisted attachment of silver on CNC	Nonenzymatic glucose detection and antibacterial agent	[65]
Carboxylation	CNC with carbon monolith powder	Dye adsorption (methylene blue)	[66]
Hydroxylation	Irreversible adsorption to oil in water emulsion	Pickering emulsifier	[67]
Silver nanowires	NFC and CNC mixture	Transparent nanopaper for flexible electronics	[68]
L-Leucine and fluorescein	pH-Sensitive fluorescein-grafted CNC with an L-leucine amino acid spacer substitution	pH sensor	[69]
Chitosan and doxorubicin	CNC microcapsules and chitosan layer-by-layer (LBL) assembly	Drug delivery systems	[70]
PVA grafting	CNC and PVA grafting by Thiolene click Michael addition	Fluorescence biosensors	[71]
Polyvinylidene fluoride (PVDF)	Nonsolvent-induced phase separation of PVDF on CNC	Microporous separators for Li-ion batteries	[72]
Methyl esterification of CNC and grafting of Poly(3-hydroxybutyrate-co-3-hydroxyvalerate) (PHBV)	Methyl esterified CNC grafted to poly (3-hydroxybutyrate-co-3-hydroxyvalerate)	Biodegradable food packaging materials	[73]
Graphene oxide LBL assembly	Deposition graphene oxide nanosheets on CNC by spin-assisted LBL assembly technique	Transparent electrically conductive membranes	[74]
Anionic alginate	Cationic CNC electrostatically linked to anionic alginate to get double-membrane structure	Dual drug release	[75]
CNC-PNIPAM grafting	Covalent grafting of [poly (N-isopropylacrylamide) (PNIPAM) to CNC by free-radical polymerization	Thermoresponsive hydrogel for wound dressing with metronidazole release	[76]
Polyacrylamide grafting	Self-assembly of nanocrystalline cellulose with acryl amide hydrogel precursors	Photonic hydrogels with long-range chiral nematic order	[77]
Poly(vinyl alcohol) with pendant viologen	Naphthyl-linked CNC nanorods, PVA with methyl viologen, and cucurbituril cross-linker (macrocyclic supramolecular)	Supra-molecular, self-healing polymer with high modulus	[78]
Graft copolymerization of CNC to PVA-co-PE	Carboxylic acid functionalized CNC grafted to PVA co- PE	Heavy metal adsorption	[79]

Functional groups/active species	Method/materials	Application	Reference
Carboxylation	Carboxyl-functionalized CNC is obtained by reacting CNC with maleic anhydride	Proton exchange membrane	[80]
Poly(vinyl alcohol) (PVA) and sterically hindered polyallylamine	Coupling carboxylic group of CNC with primary amine to yield a stable amide linkage by EDS (1-ethyl-3-(-3-dimethylaminopropyl) carbodiimide hydrochloride)/ NHS (N', N'-dicyclohexyl carbodiimide) chemistry.	Enhanced CO_2 capture	[81]

Table 2.2 Noncovalent functionalization of CNC and applications.

Functional group/active species	Method/materials	Application	Reference
Surfactant	Phosphoric acid mono and diester having alkylphenol end groups	Enhance dispersion in nonpolar solvents	[54]
Anionic surfactant	Phosphate ester of ethoxylated nonylphenol	Improved dispersivity in Polylactic acid (PLA)	[82,83]
Nonionic surfactant	$CaCO_3$ nanoparticles with pluronic P123	Enhance dispersion in polystyrene fibers and polyurethane foams	[84,85]
Quaternary ammonium salts	Grafting of 2-(dimethyl amino) ethyl methacrylate to CNCs and subsequent reaction to alkyl bromides	Hydrophobic antimicrobial materials	[86]
Amphiphilic block copolymers	Xyloglucan oligosaccharide-poly (ethylene glycol)-polystyrene Triblock copolymer	Enhanced dispersion in nonpolar solvents Rheological thickening	[87]
	Hydroxyethylcellulose Polyethylene oxide	High-strength composite with hierarchical biomimetic with tendon	[54]
Polyelectrolytes	Cationic CNC adsorbed with polyacrylamide	Deflocculation of municipal sludge	[88]
	LBL deposition of carboxylated CNC and aminated nanodextran on chemically modified graphene oxide	Chemophotothermal synergistic cancer therapy	[89]

2.3.2.1 Covalent functionalization of NFC

- *Silylation*

Silane functionalization of NFC is done by using silane reagents, namely, 3-amino-propyltriethoxysilane and 3-glycidoxypropyltrimethoxysilane, which convert C–OH bonds of pristine NFC to Si-O-C bonds. As a result, hydrophilic NFC (contact angle ~15°) converts to hydrophobic NFC with a contact angle of ~90° [91]. In another study, NFC's partial silylation was performed using isopropyl dimethyl chlorosilane as a silylating agent, which resulted in the loss of nanostructure [92].

- *Acetylation*

Acetylation involves substituting acetyl groups on to hydrogen of hydroxyl groups of NFC, which changes the wettability behavior and improves chemical affinity with non-polar solvents and polymers. Being more hydrophobic, acetylated nanofibers form stable suspension in solvents such as ethanol or acetone, while sedimentation was noticed with the pristine NFC in the same solvent systems [93]. The degree of acetylation decides the dispersibility of acetylated NFC in nonpolar solvents. Acetylation percentage of 1.5% resulted in a stable suspension of NFC in chloroform, whereas 7.5% acetylation affected the crystal structure as the acetyl groups grafted affected the intermolecular hydrogen bonding in NFC [94]. pH-responsive NFC was obtained using 1,2,3-triazole-4-metha-namine as a functionalizing agent using click chemistry. Aminated NFC was obtained by introducing reactive azide groups and subsequent reaction with propargyl amine [95].

- *Graft copolymerization of polymers*

NFC was grafted with butyl acrylate chains for controlled hydrophobic behavior [96]. Similarly, poly(butyl acrylate) was grafted on NFC as a reinforcement material of biocomposites [97]. Polyacrylic acid was grafted on to epoxide pretreated NFC using potassium persulfate initiator [98]. To improve the interfacial adhesion and dispersibility in nonpolar matrices, Poly(ε-caprolactone) was grafted to NFC by ring-opening polymerization of ε-caprolactone [99]. The acrylic monomer was grafted on NFC increased the surface roughness and overall thickness, retaining the nanofibrillar structure. NFC grafted with 80 wt% poly (glycidyl methacrylate) (PGMA) is visible as a granular structure on the nanofibrils. Grafting of PGMA and poly(methyl methacrylate) to NFC resulted in a stable dispersion of NFC in tetrahydrofuran and acetone [100].

2.3.2.2 Noncovalent modification of NFC

NFC was noncovalently modified with cationic surfactant CTAB by surface adsorption to enhance dispersion behaviour [101]. Alkali (sodium hydroxide at 20 wt%) modified NFC resulted in a change in dimension by 17% as compared to pristine NFC [102]. NFCs are considered as promising candidates as hydrogel biomaterials owing to their large specific surface area, water retention, and sustainability [103]. However, practical application of pristine NFC hydrogel is limited due to poor mechanical properties [104,105]. NFC was chemically cross-linked with ethylenediamine or hexamethylenediamine to improve the mechanical strength; however, they are cytotoxic [106,107]. Biocompatible and eco-friendly cross-linkers such as hemicelluloses have been tried to enhance mechanical strength and hydrogel swelling capacity [108,109].

By presorption method, galactoglucomannan, xyloglucan (XG), and xylan are added to NFC, which improved water absorption capacity, mechanical strength, and cell viability, resulting into a promising scaffold in wound healing application [109]. NFC was used as a reinforcing material to enhance the stiffness and strength of various polymers such as poly(styrene-co-butyl acrylate) and poly(vinyl acetate) latexes [110], hydroxypropyl cellulose [112], polyurethane [111], and poly(lactic acid). For enhancing water sorption and gel strength, NFC was cross-linked with poly (methyl vinyl ether-co-maleic acid) and polyethylene glycol. Moreover, NFC cross-linking with polymers increase mechanical strength, modulus, and thermal stability [113]. Biocomposite hydrogels for the intervertebral disc application are fabricated with Polyvinylpyrrolidone (PVP)-grafted carboxymethylated NFC by UV polymerization of N-vinyl-2- pyrrolidone in the presence of Tween 20 trimethacrylates cross-linker. The resultant composite hydrogels have enhanced mechanical properties compared to pristine hydrogels [113]. Superabsorbent hydrogels were prepared grafting PLA on NFC by UV-mediated surface polymerization of acrylic acid which resulted in a substantial increase in the water absorbency and swelling rate (25% increased) of PLA-grafted NFC [114].

2.3.3 Functionalization of BC

BC, despite having excellent water-holding capacity, nanofibrous microstructure with high porosity, modification specific to application is required. BC is modified by covalent and noncovalent modifications to impart hydrophobicity for compatibility with various nonpolar polymers and organic solvents. Acetylation, silylation, and grafting of hydrophobic polymer brushes are the most common strategies to convert hydrophilic BC to hydrophobic [115]. BC is explored as a suitable material for drug delivery and tissue regeneration due to its biocompatibility and the structure mimics extracellular matrix. However, the practical usage of BC-based composites in drug delivery and tissue regeneration is limited due to its nonbiodegradability inside the human body [116]. Hence, mineralization and oxidization of BC are carried out to enhance the biodegradability. Modifications to impart antibacterial functionality are also of current interest for wound dressing applications. Some of the functionalization strategies of BC are illustrated in Table 2.3

2.4 Conclusions

Cellulose obtained from various sources has different organization, composition, and crystallinity. Nanocellulose is one of the most economical and environmentally friendly renewable material that is easily procured from different resources. The morphology,

Table 2.3 Functionalization strategies of bacterial cellulose and their applications.

Functional group/active species	Method/materials	Application	Reference
ε-poly-L-Lysine (ε-PLL carboxymethyl cellulose grafting	Covalent bonding: carbodiimide chemistry	Wound dressing	[117]
PEI (polyethylenimine)-grafting	Covalent bonding of PEI by EDC (1-ethyl-3-(-3-dimethylaminopropyl) carbodiimide hydrochloride) and NHS (sulfo N-Hydroxysuccinimide) carbodiimide chemistry to TEMPO-oxidized bacterial cellulose (BC)	Gentamicin-controlled delivery	[118]
Polyvinylaniline/polyaniline (PVAN/PANI) bilayer	Surface-initiated ATRP of 4-vinylaniline, with subsequent *in-situ* oxidative polymerization of aniline	Electrochemical biosensors	[119]
Silver alkoxysilane	Alkoxysilane polycondensation using (3-aminopropyl) triethoxysilane	Antibacterial membranes	[120]
Phosphoric acid	Covalent phosphorylation	Adsorption of heavy metals (Uranium VI)	[121]
Poly(fluorophenol)	*In-situ* polymerization grafting of fluorophenol with laccase	Improved hydrophobicity and strength	[122]
Poly(m-phenylenediamine) nanoparticles	*In-situ* pre-adsorption oxidative polymerization grafting	Bioadsorbent with high chromium adsorption capacity	[123]
Hydroxyapatite and anti-BMP-2 antibody	Adsorption	Bone tissue regeneration	[124]
Polihexanide and povidone-iodine functionalization	Adsorption	Antibacterial activity	[125]
Collagen	Oxidized bacterial cellulose tagged to collagen with EDS/NHS chemistry	Anticancer application	[126]
Gelatin	Covalent grafting of gelatin to BC by consecutive periodate oxidation and carbodiimide cross-linking with freeze-thawing	Guided tissue regeneration	[127]
2,3 dialdehyde cellulose and hydroxyapatite	Oxidized BC tagged with hydroxyapatite	Enhanced degradation and bone regeneration	[128]
Trimethyichlorosilane	Trimethylsilylation reaction of BC with trimethyichlorosilane in liquid phase and freeze-drying	Oil/water separation	[129]
Acetic anhydride	Acetylation by using acetic anhydride with iodine as catalyst	Hydrophobic surface and good mechanical properties	[130]
Zein	Green solution impregnation with subsequent self-assembly of adsorbed zein protein	Controlled hydrophobicity and surface roughness	[131]

size, and physicochemical properties of nanocellulose rely on the preparation and the cellulose source. The CNC and NFC are produced through chemical treatments and mechanical disintegration methods. The high aspect ratio cellulose nanofibers are obtained through electrospinning and bacterial fermentation process. Nanocellulose finds its applications in healthcare, electronics, paper, filtration, cosmetics, pharmaceuticals, sensors, and food packaging industries. Further work is needed to scale up nanocellulose production through greener methods involving minimal chemical treatments and energy input.

To further enhance the surface functionalities, reduce self-agglomeration, and improve structural properties, both covalent and noncovalent functionalization strategies are adopted. Given the large variety of sources and functionalization possibilities, nanocellulose is bound to play a vital role in several advanced applications.

References

[1] T. Abitbol, A. Rivkin, Y. Cao, Y. Nevo, E. Abraham, T. Ben-Shalom, S. Lapidot, O. Shoseyov, Nanocellulose, a tiny fiber with huge applications, Curr. Opin. Biotechnol. 39 (2016) 76–88.

[2] P. Phanthong, P. Reubroycharoen, X. Hao, G. Xu, A. Abudula, G. Guan, Nanocellulose: extraction and application, Carbon Resour. Convers. 1 (2018) 32–43.

[3] M. Jonoobi, R. Oladi, Y. Davoudpour, K. Oksman, A. Dufresne, Y. Hamzeh, R. Davoodi, Different preparation methods and properties of nanostructured cellulose from various natural resources and residues: a review, Cellulose 22 (2015) 935–969.

[4] E.O. Fernandez, R.A. Young, Properties of cellulose pulps from acidic and basic processes, Cellulose 3 (1996) 21–44.

[5] M. Khandelwal, A.H. Windle, N. Hessler, In situ tunability of bacteria produced cellulose by additives in the culture media, J. Mater. Sci. 51 (2016) i1–i6.

[6] M. Khandelwal, A.H. Windle, Small angle X-ray study of cellulose macromolecules produced by tunicates and bacteria, Int. J. Biol. Macromol. 68 (2014) 215–217.

[7] S. Khattak, F. Wahid, L.P. Liu, S.R. Jia, L.Q. Chu, Y.Y. Xie, Z.X. Li, C. Zhong, Applications of cellulose and chitin/chitosan derivatives and composites as antibacterial materials: current state and perspectives, Appl. Microbiol. Biotechnol. 103 (2019) 1989–2006.

[8] A. Dufresne, Nanocellulose: a new ageless bionanomaterial, Mater. Today. 16 (2013) 220–227.

[9] D. Lasrado, S. Ahankari, K. Kar, Nanocellulose-based polymer composites for energy applications—a review, J. Appl. Polym. Sci. 137 (2020) 1–14.

[10] D. Trache, A.F. Tarchoun, M. Derradji, T.S. Hamidon, N. Masruchin, N. Brosse, M.H. Hussin, Nanocellulose: from fundamentals to advanced applications, Front. Chem. 8 (2020) 392.

[11] H. Kargarzadeh, M. Ioelovich, I. Ahmad, S. Thomas, A. Dufresne, Methods for Extraction of Nanocellulose from Various Sources, In: H. Kargarzadeh, I. Ahmad, S. Thomas, A. Dufresne (Eds.), Handbook of Nanocellulose and Cellulose Nanocomposites. Wiley-VCH Verlag GmbH & Co. KGaA, 2017, pp. 1–49.

[12] D. Klemm, F. Kramer, S. Moritz, T. Lindström, M. Ankerfors, D. Gray, A. Dorris, Nanocelluloses: a new family of nature-based materials, Angew. Chemie Int Ed. 50 (2011) 5438–5466

[13] K. Zhang, A. Barhoum, C. Xiaoqing, H. Li, Cellulose nanofibers: fabrication and surface functionalization techniques, In: A. Barhoum, M. Bechelany, A.H. Makhlouf (Eds.), Handbook of Nanofibers. Springer, Cham, 2019, pp. 409–449.

[14] H. Xie, H. Du, X. Yang, C. Si, Recent strategies in preparation of cellulose nanocrystals and cellulose nanofibrils derived from raw cellulose materials, Int. J. Polym. Sci. 2018 (2018).

[15] S.M. Choi, E.J. Shin, The nanofication and functionalization of bacterial cellulose and its applications, Nanomaterials 10 (2020) 406.

[16] A. Krystynowicz, W. Czaja, A. Wiktorowska-Jezierska, M. Gonçalves-Miśkiewicz, M. Turkiewicz, S. Bielecki, Factors affecting the yield and properties of bacterial cellulose, J. Ind. Microbiol. Biotechnol. 29 (2002) 189–195.

[17] D. Mikkelsen, B.M. Flanagan, G.A. Dykes, M.J. Gidley, Influence of different carbon sources on bacterial cellulose production by *Gluconacetobacter xylinus* strain ATCC 53524, J. Appl. Microbiol. 107 (2009) 576–583.

[18] M. Gama, P. Gatenholm, D. Klemm, Bacterial nanocellulose: a sophisticated multifunctional material, CRC Press, FL, 2012.

[19] R. Prasanth, S. Nageswaran, V.K. Thakur, J.H. Ahn, Electrospinning of Cellulose: Process and Applications, In: V.K. Thakur (Ed.), Nanocellulose Polymer Nanocomposites: Fundamentals and Applications, John Wiley and sons, NJ, 2014, pp. 311–340.

[20] C.-W. Kim, D.-S. Kim, S.-Y. Kang, M. Marquez, Y.L. Joo, Structural studies of electrospun cellulose nanofibers, Polymer (Guildf) 47 (2006) 5097–5107.

[21] S. Iwamoto, A. Isogai, T. Iwata, Structure and Mechanical properties of wet-spun fibers made from natural cellulose nanofibers, Biomacromolecules 12 (2011) 831–836.

[22] X. Zhuang, X. Yang, L. Shi, B. Cheng, K. Guan, W. Kang, Solution blowing of submicron-scale cellulose fibers, Carbohydr. Polym. 90 (2012) 982–987.

[23] R.K. Mishra, A. Sabu, S.K. Tiwari, Materials chemistry and the futurist eco-friendly applications of nanocellulose: status and prospect, J. Saudi Chem. Soc. 22 (2018) 949–978.

[24] J. Li, X. Wei, Q. Wang, J. Chen, G. Chang, L. Kong, J. Su, Y. Liu, Homogeneous isolation of nanocellulose from sugarcane bagasse by high pressure homogenization, Carbohydr. Polym. 90 (2012) 1609–1613.

[25] Q. Wang, W. Wei, F. Chang, J. Sun, S. Xie, Q. Zhu, Controlling the size and film strength of individualized cellulose nanofibrils prepared by combined enzymatic pretreatment and high pressure microfluidization, BioResources 11 (2016) 2536–2547.

[26] V. da C. Correia, V. dos Santos, M. Sain, S.F. Santos, A.L. Leão, H. Savastano Junior, Grinding process for the production of nanofibrillated cellulose based on unbleached and bleached bamboo organosolv pulp, Cellulose 23 (2016) 2971–2987.

[27] H. Wang, X. Zhang, Z. Jiang, Z. Yu, Y. Yu, Isolating nanocellulose fibrills from bamboo parenchymal cells with high intensity ultrasonication, Holzforschung 70 (2015) 401–409.

[28] A.K. Bharimalla, S.P. Deshmukh, P.G. Patil, N. Vigneshwaran, Energy efficient manufacturing of nanocellulose by chemo- and bio-mechanical processes: a review, World J. Nano Sci. Eng. 05 (2015) 204–212.

[29] Y. Yang, Z. Chen, J. Zhang, G. Wang, R. Zhang, D. Suo, Preparation and applications of the cellulose nanocrystal, Int. J. Polym. Sci. (2019) 2019.

[30] W.Y. Hamad, Cellulose nanocrystals: properties, production and applications, 2017.

[31] R.S.A. Ribeiro, B.C. Pohlmann, V. Calado, N. Bojorge, N. Pereira, Production of nanocellulose by enzymatic hydrolysis: trends and challenges, Eng. Life Sci. 19 (2019) 279–291.

[32] L.P. Novo, J. Bras, A. García, N. Belgacem, A.A.S. Curvelo, Subcritical water: a method for green production of cellulose nanocrystals, ACS Sustain. Chem. Eng. 3 (2015) 2839–2846.

[33] I. Filipova, F. Serra, Q. Tarrés, P. Mutjé, M. Delgado-Aguilar, Oxidative treatments for cellulose nanofibers production: a comparative study between TEMPO-mediated and ammonium persulfate oxidation, Cellulose (2020).

[34] L. Rozenberga, L. Andze, L. Vecbiskena, I. Filipova, M. Laka, U. Grinfelds, Preparation of nanocellulose using ammonium persulfate and method's comparison with other techniques, Key Eng. Mater. 674 (2016) 21–25.

[35] R. Salminen, M. Reza, T. Pääkkönen, J. Peyre, E. Kontturi, TEMPO-mediated oxidation of microcrystalline cellulose: limiting factors for cellulose nanocrystal yield, Cellulose 24 (2017) 1657–1667.

[36] F. Hermanutz, M.P. Vocht, N. Panzier, M.R. Buchmeiser, Processing of cellulose using ionic liquids, Macromol. Mater. Eng. 304 (2019) 1800450.

[37] M. Babicka, M. Woźniak, K. Dwiecki, S. Borysiak, I. Ratajczak, Preparation of nanocellulose using ionic liquids: 1-propyl-3-methylimidazolium chloride and 1-ethyl-3-methylimidazolium chloride, Molecules 25 (2020) 1–13.

[38] B. Thomas, M.C. Raj, J. Joy, A. Moores, G.L. Drisko, C. Sanchez, Nanocellulose, a versatile green platform: from biosources to materials and their applications, Chem. Rev. 118 (2018) 11575–11625.

[39] S. Adepu, M. Khandelwal, Bacterial cellulose with microencapsulated antifungal essential oils: a novel double barrier release system, Materialia 9 (2020) 100585.

[40] S. Adepu, M. Khandelwal, Broad-spectrum antimicrobial activity of bacterial cellulose silver nanocomposites with sustained release, J. Mater. Sci. 53 (2018) 1596–1609.

[41] B. Sun, Q. Hou, Z. Liu, Y. Ni, Sodium periodate oxidation of cellulose nanocrystal and its application as a paper wet strength additive, Cellulose 22 (2015) 1135–1146.

[42] A. Salam, L.A. Lucia, H. Jameel, Fluorine-based surface decorated cellulose nanocrystals as potential hydrophobic and oleophobic materials, Cellulose 22 (2015) 397–406.

[43] B. Sun, Q. Hou, Z. Liu, Z. He, Y. Ni, Stability and efficiency improvement of ASA in internal sizing of cellulosic paper by using cationically modified cellulose nanocrystals, Cellulose 21 (2014) 2879–2887.

[44] M.P. Illa, A.D. Pathak, C.S. Sharma, M. Khandelwal, Bacterial cellulose–polyaniline composite derived hierarchical nitrogen-doped porous carbon nanofibers as anode for high-rate lithium-ion batteries, ACS Appl. Energy Mater. 3 (2020) 8676–8687.

[45] U.D. Hemraz, K.A. Campbell, J.S. Burdick, K. Ckless, Y. Boluk, R. Sunasee, Cationic poly (2-aminoethylmethacrylate) and poly (N-(2-aminoethylmethacrylamide) modified cellulose nanocrystals: synthesis, characterization, and cytotoxicity, Biomacromolecules 16 (2015) 319–325.

[46] U.D. Hemraz, A. Lu, R. Sunasee, Y. Boluk, Structure of poly (N-isopropylacrylamide) brushes and steric stability of their grafted cellulose nanocrystal dispersions, J. Colloid Interface Sci. 430 (2014) 157–165.

[47] D.M. Panaitescu, E.R. Ionita, C.-A. Nicolae, A.R. Gabor, M.D. Ionita, R. Trusca, B.-E. Lixandru, I. Codita, G. Dinescu, Poly (3-hydroxybutyrate) modified by nanocellulose and plasma treatment for packaging applications, Polymers (Basel) 10 (2018) 1249.

[48] S. Afrin, Z. Karim, Isolation and surface modification of nanocellulose: necessity of enzymes over chemicals, Chem. Bio. Eng. Rev. 4 (2017) 289–303.

[49] S. Rebouillat, F. Pla, State of the art manufacturing and engineering of nanocellulose: a review of available data and industrial applications, J. Biomater. Nanobiotechnol. 04 (2013) 165–188.

[50] C. Fraschini, G. Chauve, J. Bouchard, TEMPO-mediated surface oxidation of cellulose nanocrystals (CNCs), Cellulose 24 (2017) 2775–2790.

[51] S. Spoljaric, A. Genovese, R.A. Shanks, Polypropylene–microcrystalline cellulose composites with enhanced compatibility and properties, Compos. Part A Appl. Sci. Manuf. 40 (2009) 791–799.

[52] S. Padalkar, J.R. Capadona, S.J. Rowan, C. Weder, Y.-H. Won, L.A. Stanciu, R.J. Moon, Natural biopolymers: novel templates for the synthesis of nanostructures, Langmuir 26 (2010) 8497–8502.

[53] M. Salajková, L.A. Berglund, Q. Zhou, Hydrophobic cellulose nanocrystals modified with quaternary ammonium salts, J. Mater. Chem. 22 (2012) 19798–19805.

[54] Y. Habibi, Key advances in the chemical modification of nanocelluloses, Chem. Soc. Rev. 43 (2014) 1519–1542.

[55] K. Chin, S. Sung Ting, H.L. Ong, M. Omar, Surface functionalized nanocellulose as a veritable inclusionary material in contemporary bioinspired applications: a review, J. Appl. Polym. Sci. 135 (2018) 46065.

[56] J. Lewandowska-Łańcucka, A. Karewicz, K. Wolski, S. Zapotoczny, Surface functionalization of nanocellulose-based hydrogels, in: M. Mondal (Ed.), Surface functionalization of nanocellulose-based hydrogels, Cellulose-Based Superabsorbent Hydrogels. Polymers and Polymeric Composites: A Reference Series (2018) 1–29.

[57] H. Wang, J. He, M. Zhang, K.C. Tam, P. Ni, A new pathway towards polymer modified cellulose nanocrystals via a "grafting onto" process for drug delivery, Polym. Chem. 6 (2015) 4206–4209.

[58] F. Azzam, E. Siqueira, S. Fort, R. Hassaini, F. Pignon, C. Travelet, J.-L. Putaux, B. Jean, Tunable aggregation and gelation of thermoresponsive suspensions of polymer-grafted cellulose nanocrystals, Biomacromolecules 17 (2016) 2112–2119.

[59] E. Kloser, D.G. Gray, Surface grafting of cellulose nanocrystals with poly (ethylene oxide) in aqueous media, Langmuir 26 (2010) 13450–13456.

[60] J.O. Zoppe, N.C. Ataman, P. Mocny, J. Wang, J. Moraes, H.-A. Klok, Surface-initiated controlled radical polymerization: state-of-the-art, opportunities, and challenges in surface and interface engineering with polymer brushes, Chem. Rev. 117 (2017) 1105–1318.

[61] Y. Yin, X. Tian, X. Jiang, H. Wang, W. Gao, Modification of cellulose nanocrystal via SI-ATRP of styrene and the mechanism of its reinforcement of polymethylmethacrylate, Carbohydr. Polym. 142 (2016) 206–212.

[62] R.D. Roeder, O. Garcia-Valdez, R.A. Whitney, P. Champagne, M.F. Cunningham, Graft modification of cellulose nanocrystals via nitroxide-mediated polymerisation, Polym. Chem. 7 (2016) 6383–6390.

[63] A. Boujemaoui, S. Mazières, E. Malmström, M. Destarac, A. Carlmark, SI-RAFT/MADIX polymerization of vinyl acetate on cellulose nanocrystals for nanocomposite applications, Polymer (Guildf) 99 (2016) 240–249.

[64] E. Lizundia, E. Fortunati, F. Dominici, J.L. Vilas, L.M. León, I. Armentano, L. Torre, J.M. Kenny, PLLA-grafted cellulose nanocrystals: role of the CNC content and grafting on the PLA bionanocomposite film properties, Carbohydr. Polym. 142 (2016) 105–113.

[65] S. Wang, J. Sun, Y. Jia, L. Yang, N. Wang, Y. Xianyu, W. Chen, X. Li, R. Cha, X. Jiang, Nanocrystalline cellulose-assisted generation of silver nanoparticles for nonenzymatic glucose detection and antibacterial agent, Biomacromolecules 17 (2016) 2472–2478.

[66] X. He, K.B. Male, P.N. Nesterenko, D. Brabazon, B. Paull, J.H.T. Luong, Adsorption and desorption of methylene blue on porous carbon monoliths and nanocrystalline cellulose, ACS Appl. Mater. Interfaces. 5 (2013) 8796–8804.

[67] I. Kalashnikova, H. Bizot, P. Bertoncini, B. Cathala, I. Capron, Cellulosic nanorods of various aspect ratios for oil in water pickering emulsions, Soft Matter 9 (2013) 952–959.

[68] X. Xu, J. Zhou, L. Jiang, G. Lubineau, T. Ng, B.S. Ooi, H.-Y. Liao, C. Shen, L. Chen, J.Y. Zhu, Highly transparent, low-haze, hybrid cellulose nanopaper as electrodes for flexible electronics, Nanoscale 8 (2016) 12294–12306.

[69] L. Tang, T. Li, S. Zhuang, Q. Lu, P. Li, B. Huang, Synthesis of pH-sensitive fluorescein grafted cellulose nanocrystals with an amino acid spacer, ACS Sustain. Chem. Eng. 4 (2016) 4842–4849.

[70] V. Mohanta, G. Madras, S. Patil, Layer-by-layer assembled thin films and microcapsules of nanocrystalline cellulose for hydrophobic drug delivery, ACS Appl. Mater. Interfaces. 6 (2014) 20093–20101.

[71] B. Schyrr, S. Pasche, G. Voirin, C. Weder, Y.C. Simon, E.J. Foster, Biosensors based on porous cellulose nanocrystal–poly (vinyl alcohol) scaffolds, ACS Appl. Mater. Interfaces. 6 (2014) 12674–12683.

[72] M. Bolloli, C. Antonelli, Y. Molméret, F. Alloin, C. Iojoiu, J.-Y. Sanchez, Nanocomposite poly (vynilidene fluoride)/nanocrystalline cellulose porous membranes as separators for lithium-ion batteries, Electrochim. Acta. 214 (2016) 38–48.

[73] H. Yu, C. Yan, J. Yao, Fully biodegradable food packaging materials based on functionalized cellulose nanocrystals/poly (3-hydroxybutyrate-co-3-hydroxyvalerate) nanocomposites, RSC Adv 4 (2014) 59792–59802.

[74] R. Xiong, K. Hu, A.M. Grant, R. Ma, W. Xu, C. Lu, X. Zhang, V.V Tsukruk, Ultrarobust transparent cellulose nanocrystal-graphene membranes with high electrical conductivity, Adv. Mater. 28 (2016) 1501–1509.

[75] N. Lin, A. Gèze, D. Wouessidjewe, J. Huang, A. Dufresne, Biocompatible double-membrane hydrogels from cationic cellulose nanocrystals and anionic alginate as complexing drugs codelivery, ACS Appl. Mater. Interfaces. 8 (2016) 6880–6889.

[76] K. Zubik, P. Singhsa, Y. Wang, H. Manuspiya, R. Narain, Thermo-responsive poly (N-isopropylacrylamide)-cellulose nanocrystals hybrid hydrogels for wound dressing, Polymers (Basel) 9 (2017) 119.

[77] J.A. Kelly, A.M. Shukaliak, C.C.Y. Cheung, K.E. Shopsowitz, W.Y. Hamad, M.J. MacLachlan, Responsive photonic hydrogels based on nanocrystalline cellulose, Angew. Chemie Int. Ed. 52 (2013) 8912–8916.

[78] J.R. McKee, E.A. Appel, J. Seitsonen, E. Kontturi, O.A. Scherman, O. Ikkala, Healable, stable and stiff hydrogels: combining conflicting properties using dynamic and selective three-component recognition with reinforcing cellulose nanorods, Adv. Funct. Mater. 24 (2014) 2706–2713.

[79] Q. Zhu, Y. Wang, M. Li, K. Liu, C. Hu, K. Yan, G. Sun, D. Wang, Activable carboxylic acid functionalized crystalline nanocellulose/PVA-co-PE composite nanofibrous membrane with enhanced adsorption for heavy metal ions, Sep. Purif. Technol. 186 (2017) 70–77.

[80] X. Yang, Y. Wei, X. Chu, Q. Zhao, W. Yan, C. Dong, B. Liu, Z. Sun, W. Hu, N. Zhang, Carboxyl-functionalized nanocellulose reinforced nanocomposite proton exchange membrane, Chem. Res. Chinese Univ. 35 (2019) 735–741.

[81] S. Janakiram, X. Yu, L. Ansaloni, Z. Dai, L. Deng, Manipulation of fibril surfaces in nanocellulose-based facilitated transport membranes for enhanced CO_2 capture, ACS Appl. Mater. Interfaces. 11 (2019) 33302–33313.

[82] F.V. Ferreira, I.F. Pinheiro, R.F. Gouveia, G.P. Thim, L.M.F. Lona, Functionalized cellulose nanocrystals as reinforcement in biodegradable polymer nanocomposites, Polym. Compos. 39 (2018) E9–E29.

[83] K. Chi, J.M. Catchmark, Enhanced dispersion and interface compatibilization of crystalline nanocellulose in polylactide by surfactant adsorption, Cellulose 24 (2017) 4845–4860.

[84] B.L. Tardy, S. Yokota, M. Ago, W. Xiang, T. Kondo, R. Bordes, O.J. Rojas, Nanocellulose–surfactant interactions, Curr. Opin. Colloid Interface Sci. 29 (2017) 57–67.

[85] K.S. Gordeyeva, A.B. Fall, S. Hall, B. Wicklein, L. Bergström, Stabilizing nanocellulose-nonionic surfactant composite foams by delayed Ca-induced gelation, J. Colloid Interface Sci. 472 (2016) 44–51.

[86] M. Li, X. Liu, N. Liu, Z. Guo, P.K. Singh, S. Fu, Effect of surface wettability on the antibacterial activity of nanocellulose-based material with quaternary ammonium groups, Colloids Surf. A Physicochem. Eng. Asp. 554 (2018) 122–128.

[87] T. Xia, Y. Huang, P. Lan, L. Lan, N. Lin, Physical modification of cellulose nanocrystals with a synthesized triblock copolymer and rheological thickening in silicone oil/grease, Biomacromolecules 20 (2019) 4457–4465.

[88] T. Suopajärvi, J.A. Sirviö, H. Liimatainen, Cationic nanocelluloses in dewatering of municipal activated sludge, J. Environ. Chem. Eng. 5 (2017) 86–92.

[89] T.S. Anirudhan, F. Shainy, J.P. Thomas, Effect of dual stimuli responsive dextran/nanocellulose polyelectrolyte complexes for chemophotothermal synergistic cancer therapy, Int. J. Biol. Macromol. 135 (2019) 776–789.

[90] S. Kalia, S. Boufi, A. Celli, S. Kango, Nanofibrillated cellulose: surface modification and potential applications, Colloid Polym. Sci. 292 (2014) 5–31.

[91] J. Lu, P. Askeland, L.T. Drzal, Surface modification of microfibrillated cellulose for epoxy composite applications, Polymer (Guildf) 49 (2008) 1285–1296.

[92] M. Andresen, L.-S. Johansson, B.S. Tanem, P. Stenius, Properties and characterization of hydrophobized microfibrillated cellulose, Cellulose 13 (2006) 665–677.

[93] M. Jonoobi, J. Harun, A.P. Mathew, M.Z.B. Hussein, K. Oksman, Preparation of cellulose nanofibers with hydrophobic surface characteristics, Cellulose 17 (2010) 299–307.

[94] P. Tingaut, T. Zimmermann, F. Lopez-Suevos, Synthesis and characterization of bionanocomposites with tunable properties from poly (lactic acid) and acetylated microfibrillated cellulose, Biomacromolecules 11 (2010) 454–464.

[95] N. Pahimanolis, U. Hippi, L.-S. Johansson, T. Saarinen, N. Houbenov, J. Ruokolainen, J. Seppälä, Surface functionalization of nanofibrillated cellulose using click-chemistry approach in aqueous media, Cellulose 18 (2011) 1201.

[96] M. Xiao, S. Li, W. Chanklin, A. Zheng, H. Xiao, Surface-initiated atom transfer radical polymerization of butyl acrylate on cellulose microfibrils, Carbohydr. Polym. 83 (2011) 512–519.

[97] S. Li, M. Xiao, A. Zheng, H. Xiao, Cellulose microfibrils grafted with PBA via surface-initiated atom transfer radical polymerization for biocomposite reinforcement, Biomacromolecules 12 (2011) 3305–3312.

[98] T. Toledano-Thompson, M.I. Loria-Bastarrachea, M.J. Aguilar-Vega, Characterization of henequen cellulose microfibers treated with an epoxide and grafted with poly (acrylic acid), Carbohydr. Polym. 62 (2005) 67–73.

[99] H. Lönnberg, K. Larsson, T. Lindstrom, A. Hult, E. Malmström, Synthesis of polycaprolactone-grafted microfibrillated cellulose for use in novel bionanocomposites–influence of the graft length on the mechanical properties, ACS Appl. Mater. Interfaces. 3 (2011) 1426–1433.

[100] K. Littunen, U. Hippi, L.-S. Johansson, M. Österberg, T. Tammelin, J. Laine, J. Seppälä, Free radical graft copolymerization of nanofibrillated cellulose with acrylic monomers, Carbohydr. Polym. 84 (2011) 1039–1047.

[101] K. Syverud, K. Xhanari, G. Chinga-Carrasco, Y. Yu, P. Stenius, Films made of cellulose nanofibrils: surface modification by adsorption of a cationic surfactant and characterization by computer-assisted electron microscopy, J. Nanopart. Res. 13 (2011) 773–782.

[102] A.N. Nakagaito, H. Yano, Toughness enhancement of cellulose nanocomposites by alkali treatment of the reinforcing cellulose nanofibers, Cellulose 15 (2008) 323–331.

[103] E. Abraham, B. Deepa, L.A. Pothan, M. Jacob, S. Thomas, U. Cvelbar, R. Anandjiwala, Extraction of nanocellulose fibrils from lignocellulosic fibres: a novel approach, Carbohydr. Polym. 86 (2011) 1468–1475.

[104] J. Kopecek, Hydrogels: from soft contact lenses and implants to self-assembled nanomaterials, J. Polym. Sci. A Polym. Chem. 47 (2009) 5929–5946.

[105] H. Mertaniemi, C. Escobedo-Lucea, A. Sanz-Garcia, C. Gandía, A. Mäkitie, J. Partanen, O. Ikkala, M. Yliperttula, Human stem cell decorated nanocellulose threads for biomedical applications, Biomaterials 82 (2016) 208–220.

[106] L. Alexandrescu, K. Syverud, A. Gatti, G. Chinga-Carrasco, Cytotoxicity tests of cellulose nanofibril-based structures, Cellulose 20 (2013) 1765–1775.

[107] K. Syverud, G. Chinga-Carrasco, J. Toledo, P.G. Toledo, A comparative study of Eucalyptus and *Pinus radiata* pulp fibres as raw materials for production of cellulose nanofibrils, Carbohydr. Polym. 84 (2011) 1033–1038.

[108] J. Liu, R. Korpinen, K.S. Mikkonen, S. Willför, C. Xu, Nanofibrillated cellulose originated from birch sawdust after sequential extractions: a promising polymeric material from waste to films, Cellulose 21 (2014) 2587–2598.

[109] J. Liu, G. Chinga-Carrasco, F. Cheng, W. Xu, S. Willför, K. Syverud, C. Xu, Hemicellulose-reinforced nanocellulose hydrogels for wound healing application, Cellulose 23 (2016) 3129–3143.

[110] F. López-Suevos, C. Eyholzer, N. Bordeanu, K. Richter, DMA analysis and wood bonding of PVAc latex reinforced with cellulose nanofibrils, Cellulose 17 (2010) 387–398.

[111] M.Ö. Seydibeyoğlu, K. Oksman, Novel nanocomposites based on polyurethane and micro fibrillated cellulose, Compos. Sci. Technol. 68 (2008) 908–914.

[112] C. Eyholzer, F. Lopez-Suevos, P. Tingaut, T. Zimmermann, K. Oksman, Reinforcing effect of carboxymethylated nanofibrillated cellulose powder on hydroxypropyl cellulose, Cellulose 17 (2010) 793–802.

[113] C. Eyholzer, A. Borges de Couraça, F. Duc, P.E. Bourban, P. Tingaut, T. Zimmermann, J.A.E. Manson, K. Oksman, Biocomposite hydrogels with carboxymethylated, nanofibrillated cellulose powder for replacement of the nucleus pulposus, Biomacromolecules 12 (2011) 1419–1427.

[114] Y. Wen, X. Zhu, D.E. Gauthier, X. An, D. Cheng, Y. Ni, L. Yin, Development of poly (acrylic acid)/nanofibrillated cellulose superabsorbent composites by ultraviolet light induced polymerization, Cellulose 22 (2015) 2499–2506.

[115] Y. Huang, C. Zhu, J. Yang, Y. Nie, C. Chen, D. Sun, Recent advances in bacterial cellulose, Cellulose 21 (2014) 1–30.

[116] S. Torgbo, P. Sukyai, Biodegradation and thermal stability of bacterial cellulose as biomaterial: the relevance in biomedical applications, Polym. Degrad. Stabil. 179 (2020) 109232.

[117] M. Fürsatz, M. Skog, P. Sivlér, E. Palm, C. Aronsson, A. Skallberg, G. Greczynski, H. Khalaf, T. Bengtsson, D. Aili, Functionalization of bacterial cellulose wound dressings with the antimicrobial peptide ε-poly-L-Lysine, Biomed. Mater. 13 (2018) 25014.

[118] X. Chen, X. Xu, W. Li, B. Sun, J. Yan, C. Chen, J. Liu, J. Qian, D. Sun, Effective drug carrier based on polyethylenimine-functionalized bacterial cellulose with controllable release properties, ACS Appl. Bio Mater. 1 (2018) 42–50.

[119] A.R. Rebelo, C. Liu, K.-H. Schäfer, M. Saumer, G. Yang, Y. Liu, Poly (4-vinylaniline)/polyaniline bilayer-functionalized bacterial cellulose for flexible electrochemical biosensors, Langmuir 35 (2019) 10354–10366.

[120] W. Shao, J. Wu, H. Liu, S. Ye, L. Jiang, X. Liu, Novel bioactive surface functionalization of bacterial cellulose membrane, Carbohydr. Polym. 178 (2017) 270–276.

[121] S. Zhuang, J. Wang, Removal of U (VI) from aqueous solution using phosphate functionalized bacterial cellulose as efficient adsorbent, Radiochim. Acta. 107 (2019) 459–467.

[122] J.E. Song, C. Silva, A.M. Cavaco-Paulo, H.R. Kim, Functionalization of bacterial cellulose nonwoven by poly (fluorophenol) to improve its hydrophobicity and durability, Front. Bioeng. Biotechnol. 7 (2019) 332.

[123] Z. Yang, L. Ren, L. Jin, L. Huang, Y. He, J. Tang, W. Yang, H. Wang, In-situ functionalization of poly (m-phenylenediamine) nanoparticles on bacterial cellulose for chromium removal, Chem. Eng. J. 344 (2018) 441–452.

[124] F. Coelho, M. Cavicchioli, S.S. Specian, R.M. Scarel-Caminaga, L. de A. Penteado, A.I. de Medeiros, S.J. de L. Ribeiro, T.S. de O. Capote, Bacterial cellulose membrane functionalized with hydroxiapatite and anti-bone morphogenetic protein 2: a promising material for bone regeneration, PLoS One 14 (2019) e0221286.

[125] C. Wiegand, S. Moritz, N. Hessler, D. Kralisch, F. Wesarg, F.A. Müller, D. Fischer, U.-C. Hipler, Antimicrobial functionalization of bacterial nanocellulose by loading with polihexanide and povidone-iodine, J. Mater. Sci. Mater. Med. 26 (2015) 245.

[126] M.D.A. Peixoto, E.M. Dos Reis, K. Cesca, L.M. Porto, Study of melanoma cell behavior in vitro in collagen functionalized bacterial nanocellulose hydrogels, Matrix 20 (2020) 21.

[127] S. Gorgieva, S. Hribernik, Microstructured and degradable bacterial cellulose–gelatin composite membranes: mineralization aspects and biomedical relevance, Nanomaterials 9 (2019) 303.

[128] E.P.C.G. Luz, P.H.S. Chaves, L. de A.P. Vieira, S.F. Ribeiro, M. de Fátima Borges, F.K. Andrade, C.R. Muniz, A. Infantes-Molina, E. Rodríguez-Castellón, M. de Freitas Rosa, In vitro degradability and bioactivity of oxidized bacterial cellulose-hydroxyapatite composites, Carbohydr. Polym (2020) 116174.

[129] H. Sai, R. Fu, L. Xing, J. Xiang, Z. Li, F. Li, T. Zhang, Surface modification of bacterial cellulose aerogels' web-like skeleton for oil/water separation, ACS Appl. Mater. Interfaces. 7 (2015) 7373–7381.

[130] W. Hu, S. Chen, Q. Xu, H. Wang, Solvent-free acetylation of bacterial cellulose under moderate conditions, Carbohydr. Polym. 83 (2011) 1575–1581.

[131] Z. Wan, L. Wang, L. Ma, Y. Sun, X. Yang, Controlled hydrophobic biosurface of bacterial cellulose nanofibers through self-assembly of natural zein protein, ACS Biomater. Sci. Eng. 3 (2017) 1595–1604.

CHAPTER 3

Nanocellulose in packaging industry

Riddhi Trivedi, Prajesh Prajapati
School of Pharmacy, National Forensic Sciences University, Gandhinagar, Gujarat, India

3.1 Introduction

Nanocellulose is defined as the nano-organized cellulose that can either be in the form of cellulose nanocrystals (CNC or NCC), cellulose nanofibers that is CNF (nano-fibrillated cellulose) or in the form of bacterial nanocellulose (BNC), which are mainly derived from microbes. CNF is a material made with a great percentage nanocellulose fibrils. Fibrils are 5–20 nm in their thickness and their length for the formation of the regular cellulosic fibers is within the micrometers range. The pseudoplastic and thixotropic properties of these materials express gels and fluids under common conditions. Be that as it may, it turns out to be less gooey when shaken or disrupted. Furthermore, when shear powers are given to the solution its gels get converted into its one fine state. The fibers from the cellulose get isolated and get incorporated with the plant base through various techniques such as extreme pressure, an ample amount of the temperature, and an immediate influence of the homogenization or may be microfluidization [1]. Furthermost the nanocellulosic particles are stabilized with the process of destructive hydrolysis; the fibers obtained offered to rise exceptionally glass–like and are inflexible nanoparticles in its structure, which are almost diminished in size such as 100s to 1000s nm compared to the cellulosic nanofibers (CNF) obtained through the same process such as homogenization, microfluidization, and pounding courses. These materials are called cellulosic nanocrystals (CNC) [2]. The utilization of bundling materials can be legitimized for anticipating the crumbling of food and drink, beauty care products, medicines or healthcare, and other shopper merchandise because of physical, biochemical, and microbiological factors. What is more, they should likewise be supportive and provide adequate hindrance against oxygen, water, gases, oil, and microorganisms, among others [3]. The packaging business by and by uses materials subject to glass, aluminum, tin, and fossil–induced made plastics, which offers rise to stresses from both functional and ecological points of view [4]. These facts confirm that the materials referred above show high quality and boundary properties; be that as it may, they additionally have a few disadvantages, for example, impracticality, delicacy, and in some cases, they are extremely overwhelming, and that builds vitality costs for transportation [5–7]. The proceeds with utilization of oil-based items will mean in the long run a declined inaccessibility and in this way an expansion in the cost of the crude materials. Also, in

43

light of their absence of biodegradability, oil-based items can create significant garbage removal issues in certain zones. The turn of events furthermore, the mission of creating the new methodologies, new materials have become a tremendous need to initiate on the green sciences, eco-efficiency and norms of legitimacy with an extension to the biological concerns on the top of end-of-life evacuation challenges and viability, and obtaining the materials from highly maintainable resources inflexible to possess some replacements. The cellulose structure is been portrayed in Fig. 3.1.

The structure represents around 40% of lignocellulose biomass [8]. The constitution of cellulosic material with polymer is the abundant reasonable characteristic material in biosphere and it is assessed per year in excess of 75 billion tons [9–11]. Yearly cellulosic creation for the mechanical change has been consistence around 2 billion for human usage including various potential uses [9]. The cellulosic-based materials and its tremendous portion are utilized in the packaging industry with different purposes such as wrapping of materials and holders, rigid packaging, helper groups, and versatility [10]. Surely, there are limitless points of interest when cellulose is used for paper-based pack-developing, such as sophistication in weight, insignificant exertion, and by and large noteworthy, reasonability. Amazingly, there are a couple of flaws inherent in the utilization of the ordinary paper superimposed from lignocellulosic fibers. These papers with very low limit toward water, smoke, oil, and oxygen insufficiencies should be tended to deliver top-notch bundles that meet different details [12–14]. The bundling business by and by utilizes fundamentally impractical plastics, wax coatings, or aluminum sheets and innumerable different materials that are subject to paper for making genuine packs. Cellophane is highly used in packaging business, which is the fundamental material as of now utilized as a film for bundling, as it gives a successful gas boundary chiefly in dry conditions. Notwithstanding, and regardless of the obvious advantages of utilizing an item dependent on photosynthetically sustainable cellulose, the creation of cellophane is destructive to the earth; the gooey course to cellophane creation generates unsafe results

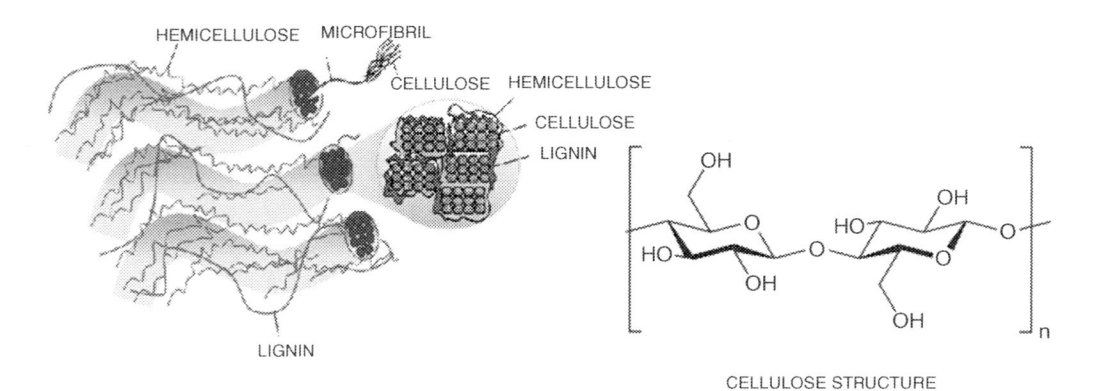

Fig. 3.1 Cellulose structure.

and uses sulfur–based synthetic compounds [15]. Considering the issues, nano–fibrillated cellulose (NFC), cellulose nanocrystals (CNCs), and BNC have been developed as key parts that ought to be considered for preparation of the cellulosic-based materials for packaging purposes [16]. For the mechanical and compound activities nanocellulose can be obtained by various plant resources [17]. Cellulosic nanomaterials have huge explicit within the ability to outline hydrogen bonds and surface domains. The holding capacity of the hydrogen will allow the material to be strong and thick, and will allow to undergo framework for various instances of particle experience. Hence it can be considered remarkable for limited applications and therefore packaging business should undergo cellulosic materials that have potential applications over a few mechanical areas furthermore, grants inventive material improvement and overall customary material properties nanocellulose are utilized, in composites, fillers, and the coverings by achieving extraordinary charming and promising properties along with essential perspectives, for example, its inexhaustible nature, nonfood agrarian based sources, biodegradability as well as biocompatibility, ease and low vitality utilization have pulled in a great deal of consideration, which is motivated additionally by the overall enthusiasm toward a practical economy ready to defeat the present reliance on fossil sources [18].

This chapter reviews the utilization of nanocellulose for bundling purposes just as to sum up the ongoing treatments in different boundary films dependent on nanocellulose with uncommon spotlight on oxygen and water fume obstruction properties.

3.2 Preparation and types of Nano cellulose

Plant cell walls are unpredictable structures involved in differing designs of between locking polysaccharides [11] as shown in Fig. 3.2. The figure shows rearranged structure of the cell wall and the cellulose course of action in the plant cell [19]. The cell wall has three unique layers: basic wall, helper wall, and middle lamella [20].

The middle lamina possesses very great proportion of the active constituent that is lignin, which is the foremost risk for the neighboring cells [21]. Hence, the fundamental divider that is around 30–1000 nm in width contains important constituents such as cellulose, gelatin, and hemicellulose, which generate cellulose microfibrils that are arranged across [22]. The helper cell wall is partitioned into three layers: S1 which is external, S2 central portion, and S3 inward layer that differentiates microfibrils with the fibers that remain intact in the center [8]. The S2, the external most layer, is having the vital portion inside the cellulose which leads to the most significant layer among all the layers of the cell wall. The mass of the wood strands of the cell is shaped by rehashed glass-like structures coming about because of the accumulation of cellulose chains, otherwise called microfibrils [19]. The secondary divider contains the majority of cellulosic microfibrils, stuffed thick level helix [8]. These are considered to be the humblest morphological units of the fiber [23] Cellulose polymers possess chain that is

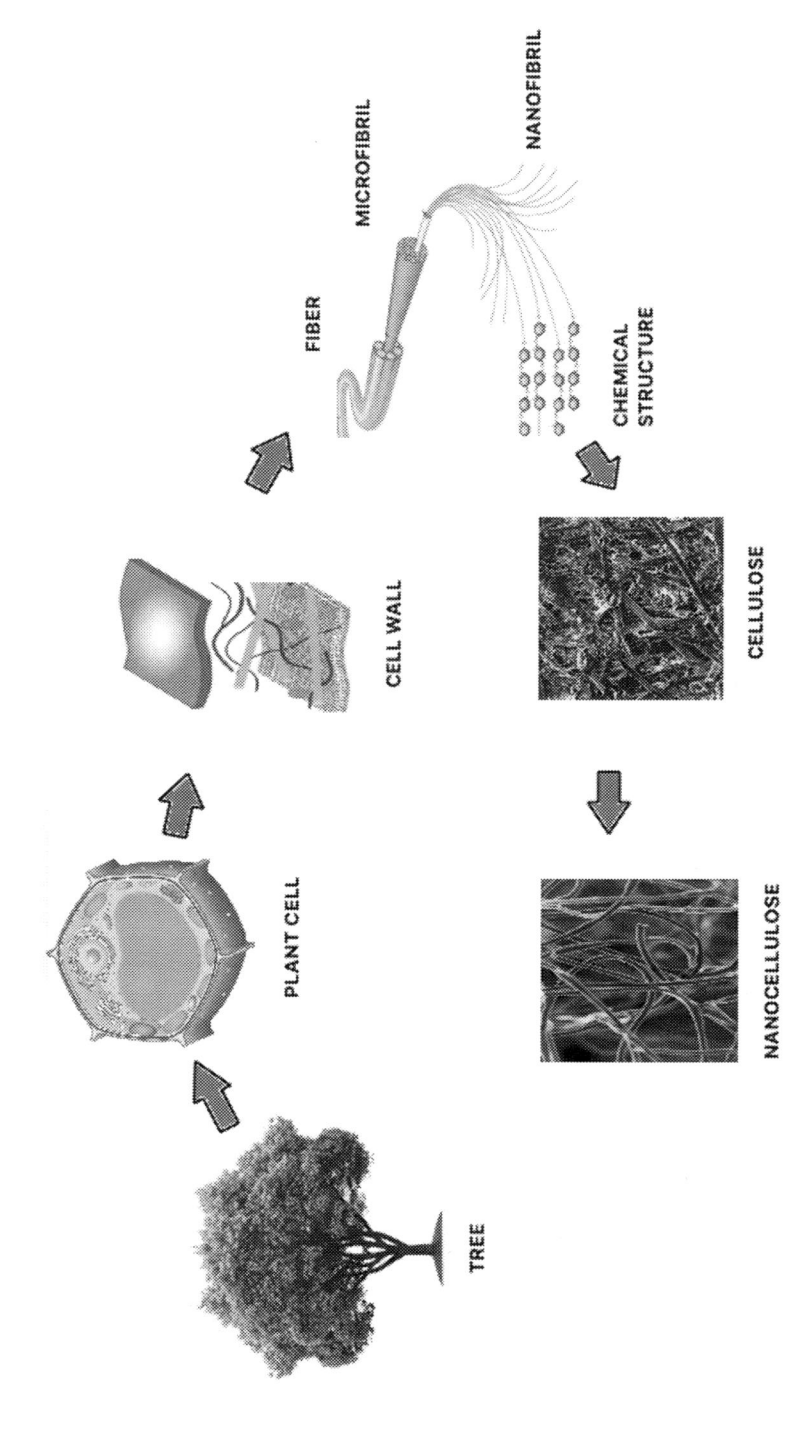

Fig. 3.2 Preparation of the nanocellulose.

expelled during CNF creation [24]. And furthermost it forms simple fibrils and microfibrils that have uniform width distribution of 2–20 nm; in any case, this is inconsistent in the microfibrils [25,26]. Depending upon the method and union states of nanocellulose, which possess their preparation and their applications, essential classes of CNCs, in any case called cellulose stubbles, cellulose nanofibrils (CNFs), in any case called NFC the crystals which are smaller in scale are known microfibrillated cellulose or nanofibers which are equivalent known as bacterial cellulose (BC) and CNFs. Its main classification underlies two main components: CNC and CNF that are made by breaking down of cellulose filaments into nanoscale particles, similar to BC and ECNF that possess low nuclear weight sugars or are separated cellulose created by microscopic organisms or electrospinning, individually [25]. In this way, huge scope creation of BC and ECNF (Electronic cellulose nanofibrils) is troublesome and commercialization stays sketchy. The business creation of nanocellulose has just started and is currently the focal point of exploration is increasingly more on modern applications. Prior CNF seclusion was viewed as a more costly procedure because of the high vitality requests required in mechanical deterioration. Be that as it may, having divulgence with various pretreatment strategies for 2,2,6,6-tetramethylpiperidine-N-oxyl (TEMPO), enzymatic hydrolysis [27–31] with furthermost mechanical disintegrations by CNF becomes an all the more engaging material for business applications. Examiners are presently centered on the enhancement of the current strategies to grow earth-agreeable techniques that have beneficial impact, which leads to the creation of nanocellulose with their properties.

3.2.1 Classification of Nano cellulose

3.2.1.1 Nanocrystalline cellulose

Nanocrystalline celluloses (CNCs) are formed by destructive hydrolysis of the celluloses dissolved in water. With everything taken into account, concentrated sulfuric destructive is used, which separates the vague regions of cellulose and the glass-like regions are dismissed [32]. Despite the fact that this procedure makes rigid cellulosic nanocrystals with 90% of the sulfate packs that are annexed at the outer strands, it leads to contaminating impacts [33]. The length that possesses separation across CNC by and large moves from a length of 200–500 nm to an estimated length of 3–35 nm.

3.2.1.2 Nanofibrils cellulose

Nanofibrils possess high pressure, which has capabilities to overcome cellulosic pound suspension and hence the trapped fibrils are estimates to be within nanometers range [25]. Dissimilar to CNCs, which have close perfect crystallinity, it is similar to glass-like cellulose spaces inside the single strands [34], which have the size of 5–50 nm and a length of two or three micrometers [33]. Nanofibrils can be extracted by three sorts of systems: mechanical medicines; synthetic medicines; and furthermore, a blend of mixture and mechanical prescriptions [22].

3.2.1.3 Bacterial cellulose

BC is generally called microbial cellulose. It is regularly supportive of produced from microorganisms, (e.g., *Acetobacter xylinum*) as an alternate molecule and does not require extra getting ready to oust contaminants such as lignin, gelatin, and hemicellulose. Besides, nanofibrils and nanocrystals biosynthesis of the BC are in the Armstrong units. The glucose chains are bacterial body in its inner layer, which have capability to expelled out through the micropores present on cell wall [35]. They are shaped as ribbon-like web-shaped structure 20–100 nm nanofibers framework.

3.2.2 Preparation of Nano cellulose

3.2.2.1 Pretreatment techniques

Pretreatment of the cellulose obtained from plants is essential to utilize the process of mechanical fibrillation that leads to improved degree of fibrillation [36]. As essentialness use is the guideline drawback for the formation of nanofibers by mechanical isolation shapes, the pretreatment technique has become a significant advance. Besides, the progress of nanofibers is increased to a great extent as it improves the fibrillation methodology [37].

3.2.2.1.1 Hydrolysis by enzyme

Limitation of explicit hydrolysis: Proteins, for instance laccase, have the properties that can easily degrade or alter the lignin and hemicellulose substance without any cellulose content [38,39]. As cellulosic fibers contain a wide range of natural mixes as a composite structure, a solitary explicit protein cannot corrupt the fiber. The accompanying arrangement of chemicals is required to rot additional cellulose compound [40]:

Cellobiohydrolases: A and B type cellulases—assault extraordinarily on translucent cellulose.

Endoglucanases: C and D type cellulases—assault confused structure of cellulose.

3.2.2.1.2 Alkaline acid

Alkali acid pretreatment is the most widely recognized strategy utilized for lignin, hemicellulose, and gelatin solubilization before a mechanical separation of nanofibrils [41–43]. This method is primarily uses [44] sodium hydroxide (NaOH). Retaining fibers of about 12–17.5 wt% results after 2 h, which gives rise to the surface region of cellulosic strands and encourages the hydrolysis, hydrochloric destructive. Engrossing fibers are produced at 60–80°C, which furthermost get themselves solubilize when treated with 2% of the sodium hydroxide solution for 2 h at 60–80°C temperature, which causes disturbances in the lignin structure and breaks the linkages among sugar and lignin which is also an essential pretreatment and a powerful strategy that can improve cellulose yield from 43% to 84% [45]. It is like manner that serves to remove lignin and hemicelluloses for the most part from soy structure strands and wheat straw.

3.2.2.1.3 Ionic fluids

Ionic liquids (ILs) are common salts having uncommon properties, for instance, non-flammability, warm, and compound reliability, with low pressure [46,47]. The experts focus on the cellulosic material [48] that is treated with the sugarcane and 1 butyl - 3-methylimidazolium chloride combine as high ionics followed again by high-pressure homogenization (HPH) to form a design lattice of the nanofiberic cellulose. These strands get crushed in the homogenizer ceaselessly, which leads to the rush in the water and recuperated by freeze-drying.

3.2.2.2 Mechanical process

During this process cellulosic materials have all the mechanical treatments for the defibrillation. Pretreatment of this will encourage all the fibrillation strategies and are mixed [49]. These compound medications can help in growing vast space in between the hydroxyl groups at the internal surface which extends the crystallinity and the breakage of the cellulosic hydrogen bonds, in this way upgrading surface locales, which helps bolster the reactivity of the strands [50]. There are various mechanical methods for changing cellulosic fiber to nanocellulose, for instance, homogenizing [51–53], micro fluidization [54,81], beating [56], cryo pulverizing [57], and high-power ultrasonication (HIUS) [58–60].

3.2.2.2.1 High-pressure homogenization

For refining of cellulosic fibers, homogenization is an effective strategy. The utilization of nanofibril cellulose from wood squash first took place in the year 1983 [61,62]. This technique is extremely basic and does not include the expansion of any natural solvents [63]. In this procedure, the cellulosic mash has gone through an exceptionally little spout at high weight. There are numerous sorts of powers that can be applied on cellulose mash, for example, at high weight. There are various technologies applied in the preparation of cellulosic pounds that possess properties such as light weight similar to impact and shear powers which have the abilities to deliver some of the shear rates within the stream to decrease the size of fibers to the nanoscale [58]. Various researchers have used HPH for some other rough materials, for example, dyed sugar beet by extraction of thorny pear [64,65], there are a few downsides in HPH, for example, fiber obstructing. Cellulosic filaments must be cleaved into small pieces and pass through HPH to resolve an issue of fiber obstructing [66].

3.2.2.2.2 Micro fluidization

Microfluidizers works on the basic principle of homogeneity in the formation of the nanocellulosic fiber. Microfluidizers uses intensifiers that have direct pressures, while the coordinated effort chamber is used for shear and impact powers against affecting streams to defibrillate the strands [55]. The large surface area of cellulose fiber and nanocrystals fibers could be gained using the microfluidizer. The cellulose fibrils exhibited a higher proportion of hydroxyl (OH) bundles similar to agglomeration conduct because of more surface zone. The size of NFC and its surface area depends on the homogenizer used [67].

3.2.2.2.3 Grinding
Pulverizing is another framework to break cellulose into nanosized fibers. In this procedure, the mash goes through two or three stones, where one stone is fixed while the other stones turn. This instrument gives shear forces to isolate the hydrogen security and cell divider structure of fibers and convert pound into nanoscale strands. Amaral Labat et al. [51] used a business stone processor to make NFC from the dyed eucalyptus mash, which assists with contemplating the association between essentialness usage and fibrillation time as a component of crystallinity [56]. Grinding of stones produces heat due to the fibrillation, which helps to evaporate water and remains the solid substance, which takes 11 h to increase fibrillation process. Various different strategies and their impacts are described with number of HPH cycles hence crushed fibers are beaten for 10 times to make the uniform size of NFC [68].

3.2.2.2.4 Cryocrushing
The term mechanical methodology is utilized to break the cell wall into nano-sized fibers that are kept in water and the cellulose gets acclimatized into the water gap which is later immersed in liquid nitrogen and later on crushed in a mortar and pestle [69]. This leads to high influence on cellulosic filaments prompts break because of applied weight through ice gems bringing about change to nanocellulose [51]. By using the HPH and cryo crushing methodology nanofibers from soybean can be produced with size up to 50–100 nm which can be confirmed by microscopy such as transmission electron microscopy.

3.2.2.2.5 High-force ultrasonication
Hydrodynamic forces induced by ultrasound have ability to separate cellulose fibrils [59]. Gaps in cells of cellulose readily get transformed into a mechanical state. This readily happens as the cellulosic particles absorb this high–power hydrodynamic force, which leads to implosion, augmentation, and advancement of scaled-down minuscule gas bubbles [70]. Various examinations have been accounted for managing the amalgamation of nanocellulose fiber from cellulose through HIUS and wavering force. Chen et al. [71] considered the impact of temperature, fixation, power, size, time, and good ways from test tip on the level of fibrillation utilizing HIUS for cellulosic materials. It is in the like manner declared that a mix of HPH and HIUS extended the fibrillation and gave consistency in nanofibers obtained from HIUS alone.

3.2.2.2.6 Ball-processing process
It is a mechanical methodology in which round and hollow holder is utilized for the making of nanofibers.

Cylinders having different compartments possess balls that are made up of ceramic zirconia and other metals, the holder rotates and allows the breakage of the cell wall with high shear between the balls resulting into various nanofibers. Number of balls to cellulose content, time of ball milling and carboxylic acid are critical parameter in this method [72].

3.2.2.3 Chemical hydrolysis

For the separation of the lengthwise longitudinal microfibrils, controlled destructive hydrolysis treatment is performed that is categorized under strong treatments. The procured nanocellulose undergoes destructive hydrolysis process; later on the hydronium molecule enters the indistinct regions of cellulose chains and advances the hydrolytic cleavage of the glyosidic bonds. A mechanical treatment for nanocellulose scattering, for example, sonication, is required to forestall agglomeration. Different solid acids have been concentrated effectively to debase cellulose fiber however the most widely recognized are hydrochloric and sulfuric acids. Be that as it may, cellulosic nanoparticles possess phosphoric destructive [73], similar to hydrobromic and nitric acids [74,75]. The usage of sulfuric acid leads to destructive and a hydrolyzing administrator as it began advancement to other joining of anionic sulfate ester and undergoes esterification on the surface of the cellulosic material. They are negatively charged and due to that they are dissolved in water. Be that as it may, it diminishes the agglomeration and decreases the thermostability of the nanoparticles [76]. Destructive hydrolysis is the most straightforward and most established technique for CNC readiness. As of late, some different strategies have been thought about various vague spaces from the cellulosic strands, which are degenerative in the nature and undergoes hydrolysis treatment, TEMPO-mediated oxidation, and the treatment with the ILs [77].

3.3 Types of packaging

Packaging can be characterized as prudent methods for giving an introduction, security, ID data, regulation, accommodation, and consistence for an item during capacity, carriage, show, and until the item is expended. Bundling must give assurance against climatic conditions organic, physical, and compound perils, and should be prudent. The bundle must guarantee the satisfactory dependability of the item all through the timeframe of realistic usability [78].

- The outside picture of the package must provide details of item; however, information package should contain are as follows:
- Package ought to give satisfactory data identified with the substance including legitimate necessities, course of organization, stockpiling conditions, clump number, expiry date, fabricates name and address, and item permit number.
- Package should aid tolerant consistence.
- Package ought to ideally have a tastefully worthy structure.

The essential bundling comprises those bundling parts that have immediate contact with the item. The principle elements of the essential bundle are to contain and to confine any concoction, climatic, or organic or every so often mechanical danger that may cause or lead item to disintegrate. Bundling must likewise work as a method for tranquilizing organizations [79].

Nanocellulose is used with different polymers to increase the self-life of the packaging, which can restrain the oxygen entrapment. Various packaging materials used are summarized in Table 3.1.

3.3.1 Innovations for nano cellulose-based materials production

The lengths are up to a couple of micrometers, dependent upon the material and method used for their extraction. This material has phenomenal optical and mechanical properties, which allows the age of a monstrous combination of unrivaled materials [80–82]. For the difference in cellulosic fibers into nanofibrils, it is required to utilize serious mechanical medicines [83]; in any case, a chemical pretreatment, for example, enzymatic treatment [84] might be applied to lessen the vitality request, contingent upon the crude materials and procedure. In the writing, one can discover a few strategies utilized for the creation of CNFs, for example, refining and high–pressure homogenizers [27,85], microfluidizers, and processors. On the other hand, sulfuric corrosive or hydrochloric corrosive is required to get ready CNCs [54,86,87]. BNC, which is sans lignin and does not require compound medicines or escalated processing to accomplish nano–scale measurements, so far has gotten considerably less examination consideration as an approach to make hindrance films for bundling [88,89] therefore expanding their timeframe of realistic usability. Due to such encouraging outcomes, notwithstanding the "for the most part viewed as shielded" there is strong motivation to consider BNC for food-contact packaging applications. In late writing, one can find different methodology hypothetically appropriate for the making of nanocellulose-based materials.

3.3.1.1 Layer-by-layer (LbL) assembly

Layer-by-layer (LbL) get together is a nonexclusive technique for a humble film covering of functional materials onto surfaces. This system on a very basic level allows

Table 3.1 Types of packaging.

Nanocellulose	Types of packaging
Cellulosic Nanocrystals Bacterial Cellulose Cellulosic Nanofibers	Nanocellulose based Production Layer-by-layer Assembly Electrospinning Composite expulsion Casting from evaporation and solution Coating

the development of multipart films on solid sponsorships by controlling adsorption from game plans or scatterings [90]. Thusly it possible to ensure the properties that are critical for bundling applications, as gas obstruction and wet-quality [91–93]. To accomplish the impacts referred to over, the layers of nanocellulose were traded with poly-(ethylene imine), cationic starch, or polyamide-amine wet-quality pitch. Various researchers have masterminded LbL motion pictures of nanocellulose pivoting with chitosan [94,95] or different cationic polyelectrolytes, for example, poly-(ethylene imine), poly(allylamine hydrochloride), and poly-(diallyldimethylammonium chloride). On a fundamental level, the self-gathering of nanomaterials, because of electrostatic interest, can be envisioned as a component possibly provoking a thick, ultra-small nanocellulose layer during the trading steps of the LbL building process [96,97]. This methodology can in like manner make different layer structures of extraordinarily small trading films subject to electrostatic correspondences and hydrogen protections between a polyanion and a polycation [93]. This strategy shows different inclinations such as for the test, ease, and versatility; thickness control at the nanoscale; and the capacity of covering successfully on tridimensional objects, for example, little jugs, cups, and plate. Taking into account the numerous means required to execute the sorts of LbL structures in the above-mentioned articles, there likewise has been solid enthusiasm for self-gathered structures, for instance, arranged nanocomposites with montmorillonite soil nanoplatelets and NFC. This blend, energized by the codevelopment of chitosan, yielded a promising blend of high durability, imperviousness to fire, and insurance from oxygen invasion. Layering including the normal and inorganic nanoparts was evident in the composite [98].

3.3.1.2 Electrospinning (ES)

Electrospinning (ES) is a broadly used advancement for electro-static fiber improvement that utilizes electrical forces to convey polymer strands with estimations going from 2 nm to a couple of micrometers using polymer courses of action of the two attributes and manufactured polymers [99]. The gooey stream is responsible for the course of action of the individual polymer chains. In the composition, there are a huge amount of works showing that ES can be used to get different polymers together with nanocellulose, conquering issues of similitude, and protecting the most entrancing properties of the biopolymer. All the more, unequivocally mulls over have been based on the making of nanocellulose composites with both bio-based and built polymers. Furthermore, the warm and mechanical properties ought to be redesigned, and basically this could be cultivated by the alteration in crystallinity and the game plan of nanocellulose along the fiber length [100]. For example, cellulose microfibrils were brought into poly(ethylene oxide) fibers by ES, and in 2010, electrospun poly(vinyl liquor) with CNCs were obtained from fortified nanofibers [101,102].

3.3.1.3 Composite expulsion

The ejection methodology changes materials from solid to liquid and back again with the perfect thickness. This methodology is by all accounts more flexible and more straightforward than LbL or ES; anyway, the helpless similarity of nanocellulose with made polymers can cause process issues. The crucial target of this technique is to join nanocellulose in different polymeric systems and misusing all their unprecedented properties such as high viewpoint proportion, explicit surface region, and sustainability, among others [103]. Since 1990, one can discover a wide range of works identified with this theme clarify in detail how the nanocellulose can brace the polymeric grid [79].

3.3.1.4 Casting from solution and evaporation

This technique relies on the usage of moderate temperature and a dissolvable to control the nanocellulose focus [104–106]. Great dispersibility of the parts should be accomplished, and this is fundamentally the primary disadvantage that can be discovered. Models where scientists had the option to scatter CNCs from a fluid solution containing innately water–cherishing framework materials, for example, chitosan [107–109]. Even though projecting from the arrangement is not a method commonly utilized in the bundling business, there are a few works in the composing that report how the nanocellulose has viably been projected. In this sense, nanocellulose got together with a 50/50 amylopectin-glycerol blend [110,111].

3.3.1.5 Coating

This is a strategy is commonly used to improve various properties of packaging materials [112,113]. When covering is used in the packaging industry, the central point that ought to be looked for after is including wobbly layers, which may be either external or sandwiched between two substrates. These layers ordinarily show thicknesses that run from tenths nanometers to a couple of micrometers. The phenomenal dispersibility of nanocellulose in water makes it alluring as a watery covering that can be applied as an unadulterated nanocellulose meager layer or as a composite with other conventional covering materials. Additionally, by and by there is an example to endeavor to diminish the thickness of oil-based plastic motion pictures by using a thin layer of down to the earth and high performing bio-based material. In this sense, nanocellulose seems, by all accounts, to be a perfect material because of its high crystallinity, among different properties. Indeed, there are a few reports in the composing that have adequately attempted the use of nanocellulose for covering. Among all the models that can be found in the composition, partners secured carboxymethylated miniaturized scale fibrillar cellulose on papers, and they choose a low oxygen permeability of the films at low relative dampness regards. The characteristics were like those that appeared by conventionally manufactured movies, for example, ethylene vinyl liquor. Be that as it may, a similar pattern was not seen at high relative dampness esteems, as the microfibers were expanded and plasticized [114].

3.4 Applications of Nano cellulose in packaging

3.4.1 Nanocellulose in the paper industry

Around 100 million tons for each annum of monetarily gathered cellulose is utilized for the creation of paper and paperboard [104]. The paper making procedure includes steps including setting up the paper segments, wet refining, pressing, calendaring, shaping of wet sheet and wrapping up. These refining of cellulose strands in water medium leads to a tremendous development of the preformation of the solid paper. Recent Developments [42] have demonstrated the probability to build the quality of paper with the added substance of nanocellulose particles to paper synthesis. Such sheets show commendable mechanical properties. These properties, as indicated [115] are in any event 2–5 times higher than those of customary papers surrounded from normal refining structures.

3.4.2 Nanocellulose in the composite industry

Composite industry possesses their own internal clarification and interest toward the nano-sized cellulose. These industries redesign their mechanical properties for the materials that are having higher consistency and less disfigurements and are able to set the conditions for the cultivation of the various nano-sized cellulosic fibers that are also known as nanocellulosic fibers [116]. These cellulosic fibers are used by this industry to prepare various polymer composites that are readily able to change the mechanical properties to make the polymer water dissolvable by reinforcing fillers, which leads to thickness and augmentation of dry composites. Significance for the development of nanocellulose within biodegradable polymers possesses licenses to improve mechanical properties to increase the rate of biodegradation [117].

3.4.3 Nanocellulose in the biomedical industry

Nanocellulose is a trademark biodegradable material, significantly sensible for the biomedical business. Unadulterated nanocellulose is nontoxic for people and is biocompatible. Consequently, it tends to be used for human services applications, for example, individual cleanliness items, cosmetics, and other various different types of biomedicine. The strongest utilization of the nanocellulose is influenced in the modification of clinical suspensions, which is the stage completely opposing the separation and sedimentation which leads to overpowering fixations. Distorted changes in the flow of the cellulose can be a promising transporter for the immobilization of proteins and different medications. In like manner, it very well may be utilized as a delicate yet dynamic stripping operator in beautifying agents [118].

3.4.4 Nanocellulose in nanoparticulate drug delivery

Medication conveyance research throughout the years has gotten exceptionally interdisciplinary. Analysts from different fields, for example, biomedical designing, pharmaceutical sciences, and life sciences, examine plenty of examination questions relating to their experience. One of the fascinating discoveries, subsequently, is the impact of nanoparticles' calculation on the viability of the conveyance framework. At the point when polymeric micelles of adaptable fiber types were contrasted and the circular sorts, the fiber types display multiple times longer dissemination time and are likewise taken up more promptly by cells because of their all-inclusive stream [119]. The anticancer medication paclitaxel was successfully conveyed, which brought about the contracting of the human-derived tumors in the mouse model. Other stretched novel transporters, for example, prolonged liposomes, carbon nanotubes, and others, are additionally answered to display any longer leeway time when contrasted and the round frameworks [120].

3.4.5 Nanocellulose in tablet formulation

Cellulose and its subordinates in various structures have been essential components in the arrangement of tablets for quite a while. Cellulose subsidiaries, for example, micro-crystalline cellulose (MCC), hydroxypropyl methylcellulose, ethylcellulose, carboxy-methylcellulose, and others are broadly utilized in ordinary just as controlled-release tablet plans. With the reasonable edge provided by nanocellulose in various practical properties being acknowledged, a couple of examinations have investigated its potential as utilitarian excipients in tablet plans. The potential of spray-dried CNFs as novel tablet excipients was assessed and looked at against two business MCC: Avicel PH-101 and Avicel PH-102, which are the two most ordinarily utilized direct pressure excipients [120]. CNFs were found to have fantastic compressibility and were amendable to both wet granulation and direct pressure techniques for tablet arrangements. CNFs arranged through direct pressure strategy indicated quicker crumbling and medication discharge demonstrating its potential as immediate pressure excipients. Freeze-dried CNC arranged from water sugarcane bagasse was likewise appeared to improve the disintegration of diltiazem hydrochloride tablets arranged with the nanocellulose [121].

3.4.6 Nanocellulose in aerogels

Aerogels are lightweight materials with the exceptional surface territory and open porosity, appropriate for high stacking of dynamic mixes [122]. They are nanoporous frameworks acquired from the wet gels or hydrogels through an appropriate drying innovation that keeps the permeable surface of the wet material unblemished. Because of their web-like structure, high porosity, and high surface reactivity, aerogels arranged from nanocellulose have high mechanical adaptability and malleability with capacity for water uptake, which makes them a brilliant possibility for the evacuation of color toxins, warm protection materials, and medication conveyance framework [123]. Therefore, various sorts of nanocellulose, because of their magnificent and appropriate properties, have become the subject of distinct fascination for the planning of aerogels for sedate conveyance. The freeze-drying strategy was applied to make profoundly permeable aerogels from nanofibrillar cellulose acquired from four unique sources, and contrasted and MCC as nanoparticulate oral medication conveyance frameworks [124]. The arrival of the beclomethasone dipropionate sedate nanoparticle incorporated into the aerogel framework was seen as brisk and prompt for red pepper–based aerogel and MCC, while BC, quince seed, and TEMPO-oxidized birch cellulose-based aerogels show supported medication discharge.

3.4.7 Nanocellulose in food industry

Nanocellulose as a characteristic emulsifying and balancing out fixing is of the incredible enthusiasm for food items, for example, a plate of mixed greens dressings, whipped garnishes, sauces, froths, soups, puddings, plunges, and numerous others [125–127]. In 2013, the report "nanocellulose as an added substance in staple" was distributed in participation between the Swedish Institute for Food and Biotechnology (SIK) and Innventia AB [128,129]. In this report, the creators inferred that nanocellulose has fascinating potential as a stabilizing operator for emulsions Unctuousness [130,131]. Nanocellulose additionally replaces stabilizers and emulsifiers like hydrophilic polysaccharides extricated from kelp, vegetable seeds, microorganisms, or something like that; insoluble polysaccharides, for example, microcrystalline cellulose; and engineered stabilizers, for example, carboxymethylcellulose [132]. A few focal points for utilizing nanocellulose as a stabilizer have been explored. Turbak and colleagues distinguished cost reserve funds in setting up the nanocellulose and the suspension in a solitary stage activity. They shaped the nanocellulose *in situ* in the suspension by blending stringy cellulose in with the elements of the suspension and later blending it through a homogenizer [133,134]. In a few cases, suspensions might be made that were not beforehand conceivable. For instance, modest quantities of nanocellulose (0.94 wt.%) can balance out emulsions with soybean oil up to 71.5 wt.%. The patent utilized nanocellulose to expand the timeframe in which a solidified sweet can hold its shape. The improvement in shape maintenance with stabilizers can be accomplished by expanding the measure of stabilizers. Sometimes, use of high concentration of stabilizers make the surfaces pale and make the nanocellulose like solidified pastry. Including somewhere in the range of 0.10 and 0.30 wt.% of vegetable nanocellulose expanded the shape maintenance time without antagonistically influencing different properties. In all cases, nanocellulose expanded the time before the item dissolved and fell. A significant improvement was accomplished in a hard–frozen yogurt with 0.30 wt.% of nanocellulose The time before softening and falling was expanded from 23 min 52 s to 35 min 25 s". As a result of the little amount of nanocellulose, it was pointless to change the organization of the treat blend, so great flavor and surface could be accomplished, and the properties of the solidified pastry were not unfavorably influenced when the shape maintenance was improved [135].

Applications are summarized in Table 3.2.

Table 3.2 Applications of nanocellulose.

Nanocellulose	Applications of nanocellulose
Cellulosic nanocrystals	Composites reinforcing Coatings Paints Packaging Rheology Films
Bacterial cellulose	Rheology modifiers Paintings Pharmaceutical industry (packaging) Food packaging industry Biomedicinal materials
Cellulosic nanofibers	Flexible soft electronics Electronic packaging Piezoelectric films Sensors Actuators Paper Industry Food packaging industry

3.5 Conclusion

Nanocellulose obtained from the plants possesses very great properties such as high crystallinity, perspective extent and other unbending characteristics which are obtained from the natural source. Various topographical modifications lead to the formation of the nanocellulose which possesses properties such as regular mechanical strategies convention partner used to make CNFs, for example, granulating or then again homogenization. Other promising methods that are discussed in our chapter need extra research and industrialization of these materials will lead to the good quality packaging for various industries. Pretreatment techniques are a significant advance for monetarily effective creation cellulosic materials as they can unequivocally decrease the essentialness use required during the mechanical weakening procedure. New advancement in these leads to greater impact on packaging and naturally cordial pretreatments stays a significant goal. Subtleties and similar investigation of three extraordinary kinds of nanocellulose will choose their different applications. It is accessible, allowing use of all its unprecedented properties. Many developed similarly as proposed methods have risen as far as the enormous scope creation of nanocellulose. Nanocellulose-based materials are carbon-unbiased, nontoxic, supportable, and recyclable.

References

[1] H. Zhu, W. Luo, Ciesielski, N. Peter, Z. Fang, J.Y. Zhu, Gunnar. Henriksson, Michael E. Himmel, Liangbing Hu, Wood-derived materials for green electronics, biological devices, and energy applications, Chem. Rev. 116 (16) (2016) 9305–9374, doi:10.1021/acs.chemrev.6b00225.

[2] B. Peng, N. Dhar, H. Liu, K. Tam, "Chemistry and applications of nanocrystalline cellulose and its derivatives: a nanotechnology perspective" (PDF), Can. J. Chem. Eng 89 (5) (2011) 1191–1206, doi:10.1002/cjce.20554 Archived from the original (PDF) on 2016-10-24. Retrieved 2012-08-28.

[3] S. Nair, J. Zhu, Y. Deng, A. Ragauskas, High performance green barriers based on nanocellulose, Sustain. Chem. Process. 2 (2014) 23. https://doi.org/10.1186/s40508-014-0023-0.

[4] C. Johansson, J. Bras, I. Mondragon, P. Nechita, D. Plackett, P. Simon, D.G. Svetec, S. Virtanen, M.G. Baschetti, C. Breen, F. Clegg, S. Aucejo, Renewable fibers and bio-based materials for packaging applications −A review of recent developments, BioResources 7 (2) (2012) 2506–2552, doi:10.15376/biores.7.2.2506-2552.

[5] I. Bayer, D. Fragouli, A. Attanasio, B. Sorce, G. Bertoni, R. Brescia, C.R. Di, T. Pellegrino, M. Kalyva, S. Sabella, P.P. Pompa, R. Cingolani, A. Athanassiou, ACS Appl. Mater. Interf. 3 (2011) 4024–4031. https://doi.org/10.1021/acsapm.8b00106.

[6] A. Reis, C. Yoshida, A. Reis, T.T. Franco, Application of chitosan emulsion as a coating on Kraft paper, Polym. Int. 60 (2011) 963–969. https://doi.org/10.1002/pi.3023.

[7] G. Rodionova, M. Lenes, O. Eriksen, O. Gregersen, Surface chemical modification of microfibrillated cellulose: improvement of barrier properties for packaging applications, Cellulose 18 (2011) 127–134. https://doi.org/10.1007/s10570-010-9474-y.

[8] H. Abdul, A. Bhat, A. Ireana, Green composites from sustainable cellulose nanofibrils: a review, Carbohydr. Polym. 87 (2012) 963–979. https://doi.org/10.1016/j.carbpol.2011.08.078.

[9] A. French, N. Bertoniere, R. Brown, H. Chanzy, D. Gray, K. Hattori, W. Glasser, Kirk-Othmer Encyclopedia of Chemical Technology, Vol. 5 Seidel, A. (Ed.), 5th edn, John Wiley & Sons, Inc., New York, 2004, doi:10.1039/c3cs60204d.

[10] Y. Habibi, Key advances in the chemical modification of nanocelluloses, Chem. Soc. Rev 43 (5) (2014) 1519–1542, doi:10.1039/C3CS60204D PMID 24316693.

[11] F. Li, E. Mascheroni, L. Piergiovanni, The potential of nanocellulose in the packaging field: a review, Pack. Technol. Sci. 28 (2015) 475–508. https://doi.org/10.1002/pts.2121.

[12] E.R.P. Keijsers, G. Yilmaz, J.E.G. van Dam, The cellulose resource matrix, Carbohydr. Polym. 93 (1) (2013) 9–21. https://doi.org/10.1016/j.carbpol.2012.08.110.

[13] D. Lee, K. Yam, L. Piergiovanni, Food Packaging Science and Technology, CRC Press, Boca Raton: New York, 2008, pp. 243–274. https://doi.org/10.1533/9780857095664.3.274.

[14] S. Nair, S. Wang, D. Hurley, Nanoscale characterization of natural fibers and their composites using contact-resonance force microscopy, Compos. Part A 41 (2010) 624–631. https://doi.org/10.1016/j.compositesa.2010.01.009.

[15] W. Hyden, Manufacture and properties of regenerated cellulose films, Ind. Eng. Chem. 21 (1929) 405–410. https://doi.org/10.1021/ie50233a003.

[16] I. Hoeger, S. Nair, A. Ragauskas, Y. Deng, O. Rojas, J. Zhu, Mechanical deconstruction of lignocellulose cell walls and their enzymatic saccharification, Cellulose 20 (2013) 807–818. https://doi.org/10.1007/s10570-013-9867-9.

[17] W. Stelte, A.R. Sanadi, Preparation and characterization of cellulose nanofibers from two commercial hardwood and softwood pulps, Ind. Eng. Chem. Res. 48 (2009) 11211–11219. https://doi.org/10.1021/ie9011672.

[18] D. Klemm, F. Kramer, S. Mortiz, T. Lindstrom, M. Ankerfors, D. Gray, A. Dorris, Nanocelluloses: a new family of nature-based materials, Angew. Chem. Int. Ed. 50 (24) (2011) 5438–5466. https://doi.org/10.1002/anie.201001273.

[19] H. Gilbert, J. Knox, A. Boraston, Advances in understanding the molecular basis of plant cell wall polysaccharide recognition by carbohydrate-binding modules, Curr. Opin. Struct. Biol. 23 (2013) 669–677. https://doi.org/10.1016/j.sbi.2013.05.005.

[20] A. Dufresne, Nanocellulose: a new ageless bionanomaterial, Mater. Today 16 (2013) 220–227. https://doi.org/10.1016/j.mattod.2013.06.004.

[21] R. Moon, A. Martini, J. Nairn, J. Simonsen, J. Youngblood, Cellulose nanomaterials review: structure, properties and nanocomposites, Chem. Soc. Rev. 40 (2011) 3941–3994. https://doi.org/10.1039/C0CS00108B.

[22] W. Boerjan, J. Ralph, M. Baucher, Lignin biosynthesis, Annu. Rev. Plant. Biol. 54 (2003) 519–546. https://doi.org/10.1146/annurev.arplant.54.031902.134938.

[23] N. Lin, A. Dufresne, Nanocellulose in biomedicine: current status and future prospect, Eur. Polym. J. 59 (2014) 302–325. https://doi.org/10.1016/j.eurpolymj.2014.07.025.

[24] A. Frey-Wyssling, K. Muhlethaler, Ultrastructural Plant Cytology with an Introduction to Molecular Biology, Elsevier Pub. Co, Amsterdam and New York, 1965, pp. 377.

[25] D. Klemm, B. Heublein, H. Fink, A. Bohn, Cellulose: fascinating biopolymer and sustainable raw material, Angew. Chem. Int. Ed. 44 (2005) 3358–3393. https://doi.org/10.1002/anie.200460587.

[26] I. Usov, G. Nyström, J. Adamcik, et al., Understanding nanocellulose chirality and structure-properties relationship at the single fibril level, Nat. Commun. 6 (2015) 7564. https://doi.org/10.1038/ncomms8564.

[27] M. Henriksson, G. Henriksson, L.A. Berglund, T. Lindström, An environmentally friendly method for enzyme-assisted preparation of microfibrillated cellulose (MFC) nanofibers, Eur. Polym. J. 43 (2007) 3434–3441. https://doi.org/10.1016/j.eurpolymj.2007.05.038.

[28] A. Isogai, T. Saito, H. Fukuzumi, TEMPO-oxidized cellulose nanofibers, Nanoscale 3 (2011) 71–85, doi:10.1039/C0NR00583E.

[29] S. Janardhnan, M. Sain, Isolation of cellulose microfibrils–an enzymatic approach, Bio. Resources 1 (2007) 176–188 2007 ISSN: 1930-2126.

[30] O. Nechyporchuk, M.N. Belgacem, J. Bras, Production of cellulose nanofibrils: a review of recent advances, Ind. Crops. Prod. 93 (2016) 2–25. https://doi.org/10.1016/j.indcrop.2016.02.016.

[31] T. Saito, Y. Nishiyama, J. Putaux, M. Vignon, A. Isogai, Homogeneous suspensions of individualized microfibrils from TEMPO-catalyzed oxidation of native cellulose, Biomacromolecules 7 (2006) 1687–1691. https://doi.org/10.1021/bm060154s.

[32] M. Paakko, M. Ankerfors, H. Kosonen, et al., Enzymatic hydrolysis combined with mechanical shearing and high-pressure homogenization for nanoscale cellulose fibrils and strong gels, Biomacromolecules 8 (2007) 1934–1941. https://doi.org/10.1021/bm061215p.

[33] P. Lu, Y. Hsieh, Preparation and characterization of cellulose nanocrystals from rice straw, Carbohydr. Polym. 87 (2012) 564–573. https://doi.org/10.1016/j.carbpol.2011.08.022.

[34] G. Tonoli, E. Teixeira, A.C. Corrêa, et al., Cellulose micro/nanofibres from Eucalyptus kraft pulp: preparation and properties, Carbohydr. Polym. 89 (2012) 80–88. https://doi.org/10.1016/j.carbpol.2012.02.052.

[35] F Jiang, Y. Hsieh, Chemically and mechanically isolated nano cellulose and their self- assembled structures, Carbohydr. Polym. 95 (2013) 32–40. https://doi.org/10.1016/j.carbpol.02.022.

[36] S. Lin, I. Calvar, J. Catchmark, J. Liu, A. Demirci, K. Cheng, Biosynthesis, production and applications of bacterial cellulose, Cellulose 20 (2013) 2191–2219. https://doi.org/10.1007/s10570-013-9994-3.

[37] H. Khalil, Y. Davoudpour, M. Islam, et al., Production and modification of nanofibrillated cellulose using various mechanical processes: a review, Carbohydr. Polym. 99 (2014) 649–665. https://doi.org/10.1016/j.carbpol.2013.08.069.

[38] G. Chinga-Carrasco, Cellulose fibres, nanofibrils and microfibrils: the morphological sequence of MFC components from a plant physiology and fibre technology point of view, Nanoscale Res. Lett 6 (2011) 1–7. https://doi.org/10.1186/1556-276X-6-417.

[39] M. Nasir, R. Hashim, O. Sulaiman, N.A. Nordin, J. Lamaming, M. Asim, Laccase, an emerging tool to fabricate green composites: a review, BioResources 10 (2015) 6262–6284 2015 ISSN: 1930–2126.

[40] M. Nasir, A. Gupta, M.D.H. Beg, G.K. Chua, M. Asim, Laccase application in medium density fibre board to prepare a bio-composite, RSC Adv. 4 (2014) 11520–11527. https://doi.org/10.1039/C3RA40593A.

[41] M. Henriksson, L.A. Berglund, Structure and properties of cellulose nanocomposite films containing melamine formaldehyde, J. Appl. Polym. Sci. 106 (2007) 2817–2824. https://doi.org/10.1002/app.26946.

[42] S. Osong, S. Norgren, P. Engstrand, Processing of wood-based microfibrillated cellulose and nanofibrillated cellulose, and applications relating to papermaking: a review, Cellulose 23 (2015) 93–123.

[43] A. Oun, J. Rhim, Characterization of nanocelluloses isolated from Ushar (*Calotropis procera*) seed fiber: effect of isolation method, Mater. Lett. 168 (2016) 146–150. https://doi.org/10.1016/j.matlet.2016.01.052.

[44] B. Wang, M. Sain, Dispersion of soybean stock-based nanofiber in a plastic matrix, Polym. Int. 56 (2007) 538–546. https://doi.org/10.1002/pi.2167.

[45] A Bhatnagar, M. Sain, Processing of cellulose nanofiber reinforced composites, J. Reinf. Plast. Compos. 24 (2005) 1259–1268. https://doi.org/10.1177/0731684405049864.

[46] A. Alemdar, M. Sain, Isolation and characterization of nanofibers from agricultural residues—wheat straw and soy hulls, Bioresour. Technol. 99 (2008) 1664–1671. https://doi.org/10.1016/j.biortech.2007.04.029.

[47] A. Pinkert, K. Marsh, S. Pang, M.P. Staiger, Ionic liquids and their interaction with cellulose, Chem. Rev. 109 (2009) 6712–6728. https://doi.org/10.1021/cr9001947.

[48] S. Kuzina, I. Shilova, A. Mikhailov, Chemical and radiation-chemical radical reactions in lignocellulose materials, Radiat. Phys. Chem. 80 (2011) 937–946. https://doi.org/10.1016/j.radphyschem 2011.04.005.

[49] J. Li, X. Wei, Q. Wang, et al., Homogeneous isolation of nanocellulose from sugarcane bagasse by high pressure homogenization, Carbohydr. Polym. 90 (2012) 1609–1613. https://doi.org/10.1016/j.carbpol.2012.07.038.

[50] V. Chauhan, S. Chakrabarti, Use of nanotechnology for high performance cellulosic and papermaking products, Cellul. Chem. Technol. 46 (2012) 389–400.

[51] G. Amaral Labat, A. Szczurek, V. Fierro, A. Pizzi, E. Masson, A. Celzard, "Blue glue": a new precursor of carbon aerogels, Microporous Mesoporous Mater. 158 (2012) 272–280. https://doi.org/10.1016/j.micromeso.2012.03.051.

[52] I. Siró, D. Plackett, Microfibrillated cellulose and new nanocomposite materials: a review, Cellulose 17 (2010) 459–494. https://doi.org/10.1007/s10570-010-9405-y.

[53] R. Zuluaga, J.-L. Putaux, A. Restrepo, I. Mondragon, P. Ganán, Cellulose microfibrils from banana farming residues: isolation and characterization, Cellulose 14 (2007) 585–592. https://doi.org/10.1007/s10570-007-9118-z.

[54] A. Ferrer, I. Filpponen, A. Rodríguez, J. Laine, O. Rojas, Valorization of residual empty palm fruit bunch fibers (EPFBF) by microfluidization: production of nanofibrillated cellulose and EPFBF nanopaper, Bioresour. Technol 125 (2012) 249–255. https://doi.org/10.1016/j.biortech.2012.08.108.

[55] M. Malainine, M. Mahrouz, A. Dufresne, Thermoplastic nanocomposites based on cellulose microfibrils from Opuntia *ficus-indica* parenchyma cell, Compos. Sci. Technol 65 (2005) 1520–1526. 2005. https://doi.org/10.1016/j.compscitech.2005.01.003.

[56] S. Lee, S. Chun, G. Doh, I. Kang, S. Lee, K. Paik, Influence of chemical modification and filler loading on fundamental properties of bamboo fibers reinforced polypropylene composites, J. Compos. Mater. 43 (2009) 1639–1657. https://doi.org/10.1177/0021998309339352.

[57] S. Panthapulakkal, M. Sain, Preparation and characterization of cellulose nanofibril films from wood fibre and their thermoplastic polycarbonate composites, Int. J. Polym. Sci. (2005) 6. https://doi.org/10.1155/2012/381342.

[58] A. Chakraborty, M. Sain, M. Kortschot, Cellulose microfibrils: a novel method of preparation using high shear refining and cryocrushing, Holzforschung 59 (2005) 102–107. https://doi.org/10.1515/HF.2005.016.

[59] A. Frone, D. Panaitescu, D. Donescu, et al., Preparation and characterization of PVA composites with cellulose nanofibers obtained by ultrasonication, BioResources 6 (2011) 487–512 2011 ISSN: 1930–2126.

[60] R. Johnson, A. Zink-Sharp, S. Renneckar, W. Glasser, A new bio-based nanocomposite: fibrillated TEMPO-oxidized celluloses in hydroxypropylcellulose matrix, Cellulose 16 (2009) 227–238. https://doi.org/10.1007/s10570-008-9269-6.

[61] F. Herrick, R. Casebier, J. Hamilton, K. Sandberg, Microfibrillated cellulose: morphology and accessibility, J. Appl. Polym. Sci. Appl. Polym. Symp. 37 (1983) 797–813. https://doi.org/10.1002/app.30116.

[62] E. Qua, P. Hornsby, H. Sharma, G. Lyons, R. McCall, Preparation and characterization of poly(vinyl alcohol) nanocomposites made from cellulose nanofibers, J. Appl. Polym. Sci. 113 (2009) 2238–2247. https://doi.org/10.1002/app.30116.

[63] A. Turbak, F. Snyder, K. Sandberg, Microfibrillated cellulose, a new cellulose product: properties, uses, and commercial potential, J. Appl. Polym. Sci. Appl. Polym. Symp. 37 (1983) 815–827. 1983CONF-8205234-Vol.2.

[64] M. Keerati-U-Rai, M. Corredig, Effect of dynamic high pressure homogenization on the aggregation state of soy protein, J. Agric. Food Chem. 57 (2009) 3556–3562. https://doi.org/10.1021/jf803562q.

[65] J. Leitner, B. Hinterstoisser, M. Wastyn, J. Keckes, W. Gindl, Sugar beet cellulose nanofibril- reinforced composites, Cellulose 14 (2007) 419–425. https://doi.org/10.1007/s10570-007-9131-2.

[66] Y. Habibi, M. Mahrouz, M. Vignon, Microfibrillated cellulose from the peel of prickly pear fruits, Food. Chem. 115 (2009) 423–429. 2009. https://doi.org/10.1016/j.foodchem.2008.12.034.

[67] M. Jonoobi, J. Harun, A. Mathew, M. Hussein, K. Oksman, Preparation of cellulose nanofibers with hydrophobic surface characteristics, Cellulose 17 (2010) 299–307. https://doi.org/10.1007/s10570-009-9387-9.

[68] T. Wang, L. Drzal, Cellulose-nanofiber-reinforced poly(lactic acid) composites prepared by a water-based approach, ACS Appl. Mater. Interf. 4 (2012) 5079–5085. https://doi.org/10.1021/am301438g.

[69] S. Iwamoto, W. Kai, A. Isogai, T. Iwata, Elastic modulus of single cellulose microfibrils fromtunicate measured by atomic force microscopy, Biomacromolecules 10 (2009) 2571–2576. https://doi.org/10.1021/bm900520n.

[70] Q. Cheng, S. Wang, T. Rials, Poly(vinyl alcohol) nanocomposites reinforced with cellulose fibrils isolated by high intensity ultrasonication, Compos. Part A 40 (2009) 218–224. https://doi.org/10.1016/j.compositesa.2008.11.009.

[71] P. Chen, H. Yu, Y. Liu, W. Chen, X. Wang, M. Ouyang, Concentration effects on the isolation and dynamic rheological behavior of cellulose nanofibers via ultrasonic processing, Cellulose 20 (2013) 149–157. https://doi.org/10.1007/s10570-012-9829-7.

[72] S. Wang, Q. Cheng, A novel process to isolate fibrils from cellulose fibers by high-intensity ultrasonication, Part 1: process optimization, J. Appl. Polym. Sci. 113 (2009) 1270–1275. https://doi.org/10.1002/app.30072.

[73] L. Zhang, T. Tsuzuki, X. Wang, Preparation of cellulose nanofiber from softwood pulp byball milling, Cellulose 22 (2015) 1729–1741. 2015. https://doi.org/10.1007/s10570-015-0582-6.

[74] J. Araki, M. Wada, S. Kuga, T. Okano, Flow properties of microcrystalline cellulose suspension prepared by acid treatment of native cellulose, Coll. Surf. A 142 (1998) 75–82. https://doi.org/10.1016/S0927-7757(98)00404-X.

[75] I. Filpponen, D. Argyropoulos, Regular linking of cellulose nanocrystals via click chemistry: synthesis and formation of cellulose nanoplatelet gels, Biomacromolecules 11 (2010) 1060–1066. https://doi.org/10.1016/S0927-7757(98)00404-X.

[76] D. Liu, T. Zhong, P.R. Chang, K. Li, Q. Wu, Starch composites reinforced by bamboo cellulosic crystals, Bioresour. Technol. 101 (2010) 2529–2536. https://doi.org/10.1016/j.biortech.2009.11.058.

[77] M. Roman, W. Winter, Effect of sulfate groups from sulfuric acid hydrolysis on the thermal degradation behavior of bacterial cellulose, Biomacromolecules 5 (2004) 1671–16777. https://doi.org/10.1021/bm034519+.

[78] G. Siqueira, S. Tapin-Lingua, J. Bras, D. da Silva Perez, A. Dufresne, Morphological investigation of nanoparticles obtained from combined mechanical shearing, and enzymatic and acid hydrolysis of sisal fibers, Cellulose 17 (2010) 1147–1158. https://doi.org/10.1007/s10570-010-9449-z.

[79] E. Fortunati, F. Luzi, D. Puglia, F. Dominici, C. Santulli, J.M. Kenny, L. Torre, Investigation of thermo-mechanical, chemical and degradative properties of PLA-limonene films reinforced with cellulose nanocrystals extracted from Phormium tenax leaves, Eur. Polym. J. 56 (2014) 77–91. https://doi.org/10.1016/j.eurpolymj.2014.03.030.

[80] E. Abraham, B. Deepa, L.A. Pothan, M. Jacob, S. Thomas, U. Cvelbar, R. Anandjiwala, Extraction of nanocellulose fibrils from lignocellulosic fibres: a novel approach, Carbohydr. Polym. 86 (2011) 1468–1475. https://doi.org/10.1016/j.carbpol.2011.06.034.

[81] E. Harry, Automating Management Information Systems, Barcode Engineering and Implementation - Harry E. Burke, Thomson Learning, 2011. ISBN 0-442-20712-3.

[82] L. He, X. Li, W. Li, J. Yuan, H. Zhou, A method for determining reactive hydroxyl groups in natural fibers: application to ramie fiber and its modification, Carbohydr. Res. 348 (2012) 95–98. https://doi.org/10.1016/j.carres.2011.10.035.

[83] A. Kaushik, M. Singh, Isolation and characterization of cellulose nanofibrils from wheat straw using steam explosion coupled with high shear homogenization, Carbohydr. Res. 346 (2011) 76–85. https://doi.org/10.1016/j.carres.2010.10.020.

[84] K. Uetani, H. Yano, Nanofibrillation of wood pulp using a high-speed blender, Biomacromolecules 12 (2011) 348–353. https://doi.org/10.1021/bm101103p.

[85] G. Chinga-Carrasco, K. Syverud, On the structure and oxygen transmission rate of biodegradable cellulose nanobarriers, Nanoscale Res. Lett. 7 (2012) 192. https://doi.org/10.1186/1556-276X-7-192.

[86] V. Karande, S. Mhaske, A. Bharimalla, G. Hadge, N. Vigneshwaran, Evaluation of two-stage process (refining and homogenization) for nanofibrillation of cotton fibers, Polym. Eng. Sci. 5 (8) (2012). https://doi.org/10.1002/pen.23413.

[87] Q. Wang, J. Zhu, R. Gleisner, T. Kuster, U. Baxa, S. McNeil, Morphological development of cellulose fibrils of a bleached eucalyptus pulp by mechanical fibrillation, Cellulose 19 (2012) 1631–1643. https://doi.org/10.1007/s10570-012-9745-x.

[88] L.M. Dobre, A. Stoica-Guzun, M. Stroescu, I.M. Jipa, T. Dobre, M. Ferdes, S. Ciumpiliac, Modelling of sorbic acid diffusion through bacterial cellulose-based antimicrobial films, Chem. Pap. 66 (2) (2012) 144–151. https://doi.org/10.2478/s11696-011-0086-2.

[89] A. Saxena, T.J. Elder, J. Kenvin, A.J. Ragauskas, High oxygen nanocomposite barrier films based on xylan and nanocrystalline cellulose, Nano-Micro. Lett 2 (2010) 235–241, doi:10.3786/nml.v2i4.p235-241.

[90] J. Padrao, S. Goncalves, J.P. Silva, V. Sencadas, S. Lanceros-Mendez, A. Pinheiro, A. Vicente, L. Rodrigues, F. Dourado, Bacterial cellulose-lactoferrin as an antimicrobial edible packaging, Food. Hydrocoll. 58 (2016) 126–140. https://doi.org/10.1016/j.foodhyd.2016.02.019.

[91] C. Aulin, E. Karabulut, A. Tran, L. Wågberg, T. Lindström, Transparent nanocellulosic multilayer thin films on polylactic acid with tunable gas barrier properties, ACS Appl. Mat. Interf. 5 (15) (2013) 7352–7359. https://doi.org/10.1021/am401700n.

[92] G. Decher, Fuzzy nanoassemblies: toward layered polymeric multicomposites, Science 277 (1997) 1232–1237, doi:10.1126/science.277.5330.1232.

[93] A. Marais, S. Utsel, E. Gustafsson, L. Wågberg, Towards a super-strainable paper using the Layer-by-Layer technique, Carbohydr. Polym. 100 (2014) 218–224. https://doi.org/10.1016/j.carbpol.2013.03.049.

[94] M. Ankerfors, T. Lindström, G. Glad Nordmark, Multilayer assembly onto pulp fibres using oppositely charged microfibrillated celluloses, starches, and wet-strength resins – effect on mechanical properties of CTMP-sheets, Nordic Pulp Pap. Res. J. 31 (1) (2016) 135–141. https://doi.org/10.3183/npprj-2016-31-01-p135-141.

[95] J. DeMesquita, C. Donnici, F. Pereira, Biobased nanocomposites from layer-by-layer assembly of cellulose nanowhiskers with chitosan, Biomacromolecules 11 (2) (2010) 473–480. https://doi.org/10.1021/bm9011985.

[96] F. Li, P. Biagioni, M. Finazzi, S. Tavazzi, L. Piergiovanni, Tunable green oxygen barrier through layer-by-layer self-assembly of chitosan and cellulose, Carbohydr. Polym. 92 (2) (2013) 2128–2134. https://doi.org/10.1016/j.carbpol.2012.11.091.

[97] L. Wågberg, G. Decher, M. Norgren, T. Lindstrom, M. Ankerfors, K. Axnas, The build-up of polyelectrolyte multilayers of microfibrillated cellulose and cationic polyelectrolytes, Langmuir 24 (3) (2008) 784–795. https://doi.org/10.1021/la702481v.

[98] B. Jean, L. Heux, F. Dubreuil, G. Chambat, F. Cousin, Non-electrostatic building of biomimetic cellulose-xyloglucan multilayers, Langmuir 25 (7) (2009) 3920–3923. https://doi.org/10.1021/la802801q.

[99] A.D. Liu, L.A. Berglund, Clay nanopaper composites of nacre-like structure based on montmorilonite and cellulose nanofibers-Improvements due to chitosan addition, Carbohydr. Polym. 87 (2012) 53–60. https://doi.org/10.1016/j.carbpol.2011.07.019.

[100] N. Bhardwaj, S.C. Kundu, Electrospinning: a fascinating fiber fabrication technique, Biotechnol. Adv. 28 (3) (2010) 325–347. https://doi.org/10.1016/j.biotechadv.2010.01.004.

[101] G. Fortunato, T. Zimmermann, J. Lubben, N. Bordeanu, R. Hufenus, Reinforcement of polymeric submicrometer-sized fibers by microfibrillated cellulose, Macromol. Mater. Eng. 297 (6) (2012) 576–584. https://doi.org/10.1002/mame.201100408.

[102] W. Park, M. Kang, H. Kim, H. Jin, Electro-spinning of poly(ethylene oxide) with bacterial cellulose whiskers, Macromol. Symp. 249–250 (1) (2007) 289–294. https://doi.org/10.1002/masy.200750347.

[103] M. Peresin, Y. Habibi, A. Vesterinen, O. Rojas, J. Pawlak, J. Seppala, Effect of moisture on electrospun nanofiber composites of poly(vinyl alcohol) and cellulose nanocrystals, Biomacromolecules 11 (9) (2010) 2471–2477. https://doi.org/10.1021/bm1006689.

[104] C. Aulin, M. Gallstedt, T. Lindstrom, Oxygen and oil barrier properties of microfibrillated cellulose films and coatings, Cellulose 17 (2010) 559–574. https://doi.org/10.1007/s10570-009-9393-y.

[105] A. Khan, R.A. Khan, S. Salmieri, C. Le Tien, B. Riedl, J. Bouchard, G. Chauve, V. Tan, M.R. Kamal, M. Lacroix, Mechanical and barrier properties of nanocrystalline cellulose reinforced chitosan based nanocomposite films, Carbohydr. Polym. 90 (4) (2012) 1601–1608. https://doi.org/10.1016/j.carbpol.2012.07.037.

[106] C. Miao, W. Hamad, Cellulose reinforced polymer composites and nanocomposites: a critical review, Cellulose 20 (5) (2013) 2221–2262. https://doi.org/10.1007/s10570-013-0007-3.

[107] P. Lu, H.N. Xiao, W.W. Zhang, G. Gong, Reactive coating of soybean oil-based polymer on nanofibrillated cellulose film for water vapor barrier packaging, Carbohydr. Polym. 111 (2014) 524–529. https://doi.org/10.1016/j.carbpol.2014.04.071.

[108] M. Pereda, A. Dufresne, M.I. Aranguren, N.E. Marcovich, Polyelectrolyte films based on chitosan/olive oil and reinforced with cellulose nanocrystals, Carbohydr. Polym. 101 (2014) 1018–1026. https://doi.org/10.1016/j.carbpol.2013.10.046.

[109] G. Rodionova, S. Roudot, O. Eriksen, F. Mannle, O. Gregersen, The formation and characterization of sustainable layered films incorporating microfibrillated cellulose (MFC), BioResources 7 (3) (2012) 3690–3700 ISSN: 1930-2126.

[110] F. Rafieian, J. Simonsen, Fabrication and characterization of carboxylated cellulose nanocrystals reinforced glutenin nanocomposite, Cellulose 21 (6) (2014). 4167–4180. https://doi.org/10.1007/s10570-014-0305-4.

[111] A. Svagan, M Azizi Samir, L Berglund, Biomimetic polysaccharide nanocomposites of high cellulose content and high toughness, Biomacromolecules 8 (8) (2007) 2556–2563. https://doi.org/10.1021/bm0703160.

[112] European Pharmacopoeia, 4th Ed., Council of Europe, Strasbourg S. Farris, L. Piergiovanni, Emerging coating technologies for food and beverage packaging materials, in: K.L. Yam, D.S. Lee (Eds.), Emerging Food Packaging Technologies: Principles and Practice, Woodhead Publishing Ltd, Oxford, UK, 2012, pp. 274–302. https://doi.org/10.1533/9780857095664.3.274.

[113] A.J. Svagan, M.S. Hedenqvist, L.A. Berglund, Reduced water vapour sorption in cellulose nanocomposites with starch matrix, Compos. Sci. Technol. 69 (3–4) (2009) 500–506. https://doi.org/10.1016/j.compscitech.2008.11.016.

[114] L. Pal, M.K. Joyce, P.D. Fleming, S.A. Cretté, C. Ruffner, High barrier sustainable co-polymerized coatings, JCT Research, 4, Springer, Boston, 2008, pp. 279–489. https://doi.org/10.1007/s11998-008-9101-0.

[115] M. Ioelovich, A. Leykin, Nano-cellulose and its application, J. SITA 6 (2004) 17–24 2004.

[116] M. Henriksson, L.A. Berglund, P. Isaksson, T. Lindstrom, T. Nishino, Cellulose nanopaper structures of high toughness, Biomacromolecules 9 (2008) 1579–1585. https://doi.org/10.1021/bm800038n.

[117] K. Spence, Y. Habibi, A. Dufresne, Nanocellulose-based Composites Cellulose Fibers: Bio-and Nanopolymer Composites, Springer, Berlin, Heidelberg and New York, 2011, pp. 179–213. https://doi.org/10.1007/978-3-642-17370-7_7.

[118] N. Saba, T.M. Paridah, K. Abdan, N.A. Ibrahim, Preparation and characterization of fire retardant nano-filler from oil palm empty fruit bunch fibers, BioResources 10 (2015) 4530–4543. ISSN: 1930-2126.

[119] Y. Geng, P. Dalhaimer, S. Cai, et al., Shape effects of filaments versus spherical particles in flow and drug delivery, Nat. Nanotechnol. 2 (2007) 249–255. https://doi.org/10.1038/nnano.2007.70.

[120] E.A. Simone, T.D. Dzubiula, V.R. Muzykantov, Polymeric carriers: role of geometry in drug delivery, Expert Opin. Drug. Deliv. 5 (2008) 1283–1300. https://doi.org/10.1517/17425240802567846.

[121] R. Kolakovic, L. Peltonen, T. Laaksonen, K. Putkisto, A. Laukkanen, J. Hirvonen, Spray-dried cellulose nanofibers as novel tablet excipients, AAPS PharmSciTech. 12 (2011) 1366–1373. https://doi.org/10.1208/s12249-011-9705-z.

[122] L. Emara, A. El-Ashmawy, N. Taha, K. El-Shaffei, S. Mahdey, H. El-Kholly, Freeze-dried nanocrystalline cellulose derived from water sugar-cane bagasse as a novel tablet excipients, The 41st Annual Meeting & Exposition of the Controlled Release Society, Chicago, IL, USA, July 2014, 13–16.

[123] C.A. Garcia-Gonzalez, M.A.I Smirnova, Polysaccharide-based aerogels-promising biodegradable carriers for drug delivery systems, Carbohydr. Polym. 86 (2011) 1425–1438. https://doi.org/10.1016/j.carbpol.2011.06.066.

[124] W. Chen, Q. Li, Y. Wang, et al., Comparative study of aerogels obtained from differently prepared nanocellulose fibers, ChemSusChem 7 (2014) 154–161. https://doi.org/10.1002/cssc.201300950.

[125] K. Mizuguchi, I. Fujioka, & H. Kobayashi, (1983). JP58190352. Japan: Patent and Trademark Office. https://doi.org/10.1016/j.foodhyd.2016.01.023.

[126] K. Mizuguchi, I. Fujioka, & H. Kobayashi, (1983). JP58190369. Japan: Patent and Trademark Office. https://doi.org/10.1016/j.foodhyd.2016.01.023.

[127] H. Valo, S. Arola, P. Laaksonen, et al., Drug release from nanoparticles embedded in four different nanofibrillar cellulose aerogels, Eur. J. Pharm. Sci. 50 (2013) 69–77. https://doi.org/10.1016/j.ejps.2013.02.023.

[128] K. Mizuguchi, I. Fujioka, & H. Kobayashi, (1983). JP58190382. Japan: Patent and Trademark Office. https://doi.org/10.1016/j.foodhyd.2016.01.023.

[129] A.F. Turbak, F.W. Snyder, K.R. Sandberg, US 4.487.634, US: Patent and Trademark Office, Washington, DC, 1984.

[130] G. Ström, C. Öhgren, M. Ankerfors, Nanocellulose as an additive in foodstuff, Stockholm (2013). Retrieved from http://www.innventia.com/Documents/Rapporter/Innventiareport403.pdf.

[131] T. Winuprasith, M. Suphantharika, Properties and stability of oil-in-water emulsions stabilized by microfibrillated cellulose from mangosteen rind, Food Hydrocoll. 43 (2015) 690–699. https://doi.org/10.1016/j.foodhyd.2014.07.027.

[132] D. C. Kleinschmidt, B. A. Roberts, D. L. Fuqua, & J. R. Melchion, (1988). US 4.774.095. Washington D.C: U.S: Patent and Trademark Office.

[133] A.F. Turbak, F.W. Snyder, K.R. Sandberg, US 4.341.807, US: Patent and Trademark Office, Washington, DC, 1982.

[134] H. Yano, K. Abe, Y. Kase, S. Kikkawa, Y. Onishi, US 20140342075 A1, 344, U.S. Patent Application No, Washington DC, 2012, p. 158.

[135] A.F. Turbak, F.W. Snyder, K.R. Sandberg, US 4.378.381, US: Patent and Trademark Office, Washington DC, 1983.

CHAPTER 4

Nanocellulose hybrid systems: carriers of active compounds and aerogel/cryogel applications

Nadia Obrownick Okamoto-Schalch[a,b], Natalia Cristina da Silva[a,b], Rafael Belasque Canedo da Silva[a,b], Milena Martelli Tosi[a,b,c]

[a]Departamento de Engenharia de Alimentos, Faculdade de Zootecnia e Engenharia de Alimentos, Universidade de São Paulo, Rua Duque de Caxias Norte 225, Pirassununga, SP, Brazil
[b]Postgraduate Programme in Materials Science and Engineering, University of São Paulo, USP/FZEA, Av. Duque de Caxias Norte, 225, Pirassununga, Brazil
[c]Departamento de Química, Faculdade de Filosofia, Ciências e Letras, Universidade de São Paulo, Av. Bandeirantes, 3900, Ribeirão Preto, SP, Brazil

4.1 Introduction

Cellulose is the most abundant polysaccharide on the planet, being the main constituent of plant cell walls [1]. In the cell wall of vegetables, the cellulose microfibrils are cross-linked to hemicelluloses and lignin by physical interactions and covalent bonds. From these interactions, successive structures are formed, giving rise to the cell wall of the fibers. Cellulose molecules aggregate in the form of microfibrils in which highly ordered (crystalline) regions alternate with less ordered (amorphous) regions [2]. As a consequence of this fibrous structure, cellulose has high tensile strength and is insoluble in most solvents.

Nanocellulose suspensions have been obtained from several lignocellulosic sources [3], including mercerized soybean straw [4,5]. The production of soybeans reached 349 million tons in 2018 [6], with straw corresponding to 1.2 −1.5 times the amount of grains [7], and it is usually discarded in the field during harvest. Soybean straw is an important source of a lignocellulosic material (biomass) with 35% cellulose, 21% acid insoluble lignin, 17% hemicelluloses, 11% ash, 1% soluble lignin, and other constituents such as proteins, pectin, and substitutes glucuronic acid [4,8,9], which vary according to the plant's genotype [10]. Cellulose nanocrystals (CNCs) or nanowhiskers are normally obtained by hydrolysis using sulfuric acid and have OSO^{3-} groups that characterize the anionic character of these particles. Other processes can be used (mechanical or enzymatic) to defibrillate cellulose in nanofibrils (CNFs) such as enzimatic (coquetels of xylanases and celluloses) and mechanical treatments (ultrahomogeneization and sonication) [5].

CNCs have been used as drug delivery systems [11–15]. The hydroxyl groups (OH) on the CNC surface are able to bind to the hydrophobic or nonionized groups of other compounds. Gels or nanocomposite films were also used as carrier matrices for water-insoluble drugs through the absorption of these substances in the fibrous matrix [14].

Another important carrier of active compounds is chitosan particles. Chitosan is a highly organized cationic polysaccharide, composed of units of glucosamine (β-(1-4)-2-amino-2-deoxy-D-glucose and β-(1-4)-2-acetamide-2-deoxy- D-glucose) and obtained through the deacetylation of chitin in an alkaline medium [16]. The main characteristics of this polymer are the insolubility in aqueous medium and organic solvents, antibacterial properties, biodegradability, and low toxicity [17]. Chitosan nanoparticles (ChNPs) can be obtained by ionic gelation using sodium tripolyphosphate (TPP) as a cross-linking agent [18]. They can also be produced through polymerization in mold by the addition of methacrylic acid, where protonation of the chitosan groups (NH_2) occurs [19].

The hybrid system associating ChNPs and CNCs has the potential for a new compound with the capacity to increase stability and enable controlled release of the asset [20]. The main objective was an exploratory study to verify the association of these polymers and their action on the stability of folic acid (FA). The association of CNCs and CNFs with other polymers leads to physical and chemical cross-linking, improving their properties. Therefore, this chapter focuses mainly on the production of CNCs, CNFs, and their hybrid systems. The properties of other matrices associated to nanocelluloses are summarized in comprehensible tables. Finally, we present a case study of the production of chitosan cryogels containing the hybrid system ChNP-CNC.

4.2 Production of CNCs or CNFs and application in hybrid systems

For nanocellulose productions, cellulose microfibrils are cleaved across the amorphous regions (2–10 nm thicker and 100% 1000 nm in length) by enzymatic or acid hydrolysis, followed by mechanical processes. When the size of a particle is reduced, new intrinsic phenomena are observed and become predominant at the nanoscale [21]. In general, nanocelluloses are normally produced in three stages: (1) chemical treatment to prepare the material (hydrolysis of hemicelluloses and delignification), (2) acid hydrolysis [10,22–24] or enzymatic [4,5] to act on amorphous structures, and (3) mechanical fragmentation. Fig. 4.1 shows the general system for producing cellulose nanofibrils by enzymatic hydrolysis and CNCs by chemical hydrolysis.

In general, acid hydrolyses are able to produce CNCs with lower lengths (L) and aspect ratio (L/D, D: diameter), whereas enzymatic and mechanical treatments produced CNFs with higher lengths and aspect ratios. This is because, as shown in Fig. 4.1, acids are more effective at breaking down and separating lignocellulosic material. However the enzymes still result in fibers with low levels of lignin and hemicellulose associated with the crystalline portion. However, the fiber diameters can change according to the lignocellulosic source and production process (types of reagents and/or number of processing steps) [4,25,26].

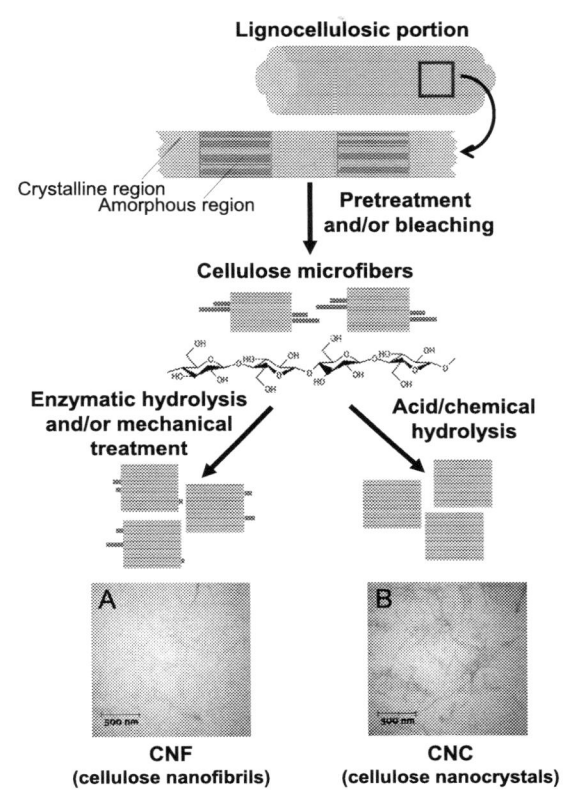

Fig. 4.1 Production mechanisms of cellulose nanofibrils (CNFs) by enzymatic hydrolysis/chemical treatments and cellulose nanocrystals (CNCs) by acid/chemical hydrolysis. Microscopy images were adapted from Martelli-Tosi et al. [5] and Martelli-Tosi (2020).

Residual products with high levels of fibers are excellent sources of cellulose, as they are a sustainable alternative for obtaining this polymer. For pretreatment and/or bleaching of fibers, basic and less aggressive reagents are usually used, for example: sodium hydroxide (NaOH), sodium chlorite ($NaClO_2$), hydrogen peroxide (H_2O_2), potassium hydroxide (KOH), and sodium nitrite ($NaNO_2$) [27–30]. In the hydrolysis stage, sulfuric acid (H_2SO_4) is normally used to obtain the CNCs. This guarantees the total breakdown of the matrix; nevertheless, recent works studied its substitution by acids less aggressive to the environment (such as the acetic acid (CH_3COOH)) [5,27]. For the production of CNF, enzymatic hydrolysis occurs from the addition of enzymes or enzymatic cocktails [5,29]. Finally, a mechanical process (microfluidizer, sonication, or steam–explosion) can be used to guarantee the separation of the fibers/crystals [31].

Table 4.1 shows some studies on CNFs and CNCs suspensions that were obtained from different sources using different processes and the interference of these factors on the final dimensions of the fibers.

Table 4.1 Production steps and dimensions (L = length and D = diameter) of nanocelluloses (CNFs and CNCs) from different sources.

Source	Production steps	Dimensions	References
Brewed green tea	Chemical pretreatment (NaOH), bleaching ($NaClO_2$), and CH_3COOH hydrolysis	L: 200 nm D: 20 nm	[27]
Lemongrass	Chemical pretreatment (NaOH) and enzymatic hydrolysis (Viscozyme L)	L: 105 nm D: –	[32]
Pineapple crown leaf	Chemical pretreatment (NaOH), bleaching (H_2O_2), and H_2SO_4 hydrolysis	L: 100–400 nm D: 20–60 nm	[28]
Banana peel	Chemical pretreatment (NaOH and KOH) and enzymatic hydrolysis (xylanase)	L: 1490–1544 nm D: 3.7–8.8 nm	[29]
Soybean	Chemical pretreatment (NaOH), bleaching (H_2O_2), and H_2SO_4 hydrolysis	L: 100–600 nm D: 9 nm	[5]
Soybean	Chemical pretreatment (NaOH), bleaching (H_2O_2), and enzymatic hydrolysis (Optimash VR)	L: > 1 μm D: 9 nm	[5]
Sugar beet pulp	Bleaching ($NaClO_2$) and enzymatic hydrolysis (Viscozyme L, Pectinex Ultra Clear, Pulpzyme HC, Fibercare R, and Aquazym 240 L)	L: 5 nm D: 1 μm	[31]
Pistachio shells	Chemical pretreatment (NaOH), bleaching (ethanol/toluene), and H_2SO_4 hydrolysis	L: 187 nm D: 12 nm	[33]
Banana peel	Chemical pretreatment (NaOH), bleaching ($NaClO_2$), and H_2SO_4 hydrolysis	L: 335–454 nm D: 10–22 nm	[34]
Cotton	Chemical pretreatment (NaOH), bleaching (H_2O_2), and H_2SO_4 hydrolysis	L: 17–230 nm D: 2–25 nm	[35]
Cassava peel	Chemical pretreatment (HNO_3 and $NaNO_2$), bleaching (H_2O_2), and H_2SO_4 hydrolysis	L: 100–300 nm D: 3–8 nm	[32]
Sugarcane bagasse	Liquid hot water pretreatment, enzymatic hydrolysis (Cellic), and H_2SO_4 hydrolysis	L: 195–250 nm D: 16–17 nm	[36]
Barley straw	Enzymatic pretreatment (*Aspergillus niger*), bleaching (H_2O_2), and H_2SO_4 hydrolysis	L: 270 nm D: 15 nm	[37]
Husk	Chemical pretreatment (NaOH), bleaching (H_2O_2), and H_2SO_4 hydrolysis	L: 280 nm D: 30 nm	[37]

The use of strong acids, which is normal in the hydrolysis stage, has important disadvantages, such as cellulose degradation, corrosivity, in addition to the generation of waste. It is for this reason that enzymatic hydrolysis has been exploited for being less aggressive and more environmentally friendly [38–41].

CNFs and CNCs have been widely studied due to their excellent mechanical properties and biodegradability. Most studies in the literature describe that, according their dimensions, CNFs are useful in composite materials and CNCs (where there is greater degradation of the amorphous portion) usually result in colloidal suspensions [31].

4.3 Production of CNC or CNF hybrid systems

Nanocellulose-based functional hybrid materials can be prepared in an integrative approach from aqueous dispersions of CFNs or CNCs associated to other matrices. The objective of integration with nanocellulose is to improve the devastated characteristics of the pure matrix. In this way, it is possible to associate the individual properties of each component and create a new functionalization possibility [42,43].

In general, the preparation of new hybrid systems based on nanocelluloses consists of the interaction of this material with a second material that serves as a dispensing medium [44]. The advantage of nanocelluloses as incorporation material is that they are of biological origin, so they can offer advantages such as bioactivity, biocompatibility, and biodegradability [42,45]. In recent years, nanocellulose-based hybrid systems are the focus of studies in a wide variety of applications fields, such as biomedicine [46], films and coatings [5,29], or metal carrier particles or other nutrients [47,48].

For the production of films and blends, nanocelluloses are generally associated with polymers such as starch [29], protein [5], polyester [27], among others. The association of CNFs or CNCs in these matrices results in materials with greater Young's modulus (MY), tensile strength, fracture strength, and break strain. For gels, such as aerogels or hydrogels, the integration of polymers with CNFs or CNCs results in materials with larger pores. This is feasible for several applications, as it positively affects the effects of liquid absorption [49], controlled release [50], and electrical conduction capacity [51]. Other associations are also possible between nanocelluloses and different matrices, resulting in different materials. In Table 4.2, some of the possibilities of associating CNF or CNC and other matrices for the formation of hybrid systems are presented. The properties positively affected by the addition of the nanocelluloses are also highlighted.

The properties of the final product depend on the good dissociation of nanocellulose in the association matrix, in addition to the processing method. Our research group recently used CNCs to produce nanocomposites associated to chitosan ChNPs [20]. Chitosan are protonated in acid medium (NH_3^+) and CNCs have hydroxyl (OH)

Table 4.2 Association of nanocelluloses (CNFs and CNCs) with other matrices for the production of new hybrid materials.

Association matrix		Production	Final product	Properties/effects	References
CNF/	polyacryl-amide	Free-radical polymerization	Hydrogel	Tensile strength, fracture strength, and break strain	[51]
	starch	Casting	Film	Young's modulus (MY) and tensile strength	[29]
	essential oil	Freeze drying	Emulsion	Controlled release of oils	[52]
	chitosan	Homogenization	Hydrogel	Viscosity and shear rate	[46]
	protein	Casting	Film	Water vapor permeability, MY, and tensile strength	[5]
	xyloglucan/pectin	Homogenization	Hydrogel	Controlled release of drugs	[50]
	alginate	metal cation cross-linking,	Hydrogel	Controlled release of drugs	[52]
	sepiolite	Fusion by high-shear disperser	Nanopaper	Hydrophilic character and thermal degradation	[42]
	quercetin	Freeze drying	Nanoparticle	Quercetin encapsulation	[48]
CNC/	xyloglucan	Freeze drying	Aerogel	Pore size and elastic modulus	[53]
	chitosan	Ionic gelation	Nanoparticle	Hyaluronic acid encapsulation	[47]
	polyester	*in-situ* polymerization	Film	Tensile strength, elongation, and toughness	[27]
	polyvinyl acetate	Freeze drying	Aerogel	Pore size and absorption of nonpolar solvents	[54]
	polyvinyl acetate	Freeze drying	Aerogel	Pore size, water absorption capacity, and stability	[49]
	protein	Casting	Film	MY and tensile strength	[5]
	alginate	Extrusion	Hydrogel	Controlled release of drugs	[55]
	starch	Casting	Film	MY, tensile strength, and elongation	[34]
	gelatin	Free-radical polymerization	Hydrogel	Controlled release of drugs	[56]
	polylactic acid/chitosan	Casting	Film	MY, tensile strength and elongation	[37]

and OSO^{3-} groups when they are obtained via hydrolysis with sulfuric acid [14]. In our study, ChNPs were produced by ionic gelation using sodium TPP [57]. ChNPs were already studied as carriers of vitamins C, B9, B12, and D3 and the results indicate a slow vitamin degradation in aqueous medium when they were encapsulated in ChNPs [58–60]; however, few studies have studied the effect of ChNP-CNC hybrid system for vitamin carriers. The objective of this study was to produce ChNP-CNC hybrid systems for using as a carrier of FA (pteroyl–L-glutamic acid) r(FA). The following steps were carried out to produce these particles:

1. Washed and dried pods (50 °C for 24 h) were treated with 17.5% NaOH (w/v) solution for 15 h [4]. The suspension was washed with distilled water approximately four times until pH 6.0–7.0. The solution with 4% H_2O_2 (w/v) 2% NaOH (w/v), and 0.3% $MgSO_4.7H_2O$ (w/v) was added to bleach the fibers for 3 h at 90 °C. Then, the fibers were washed until neutral pH, dried, and triturated.

2. CNCs were produced by acid hydrolysis of pretreated pods: sulfuric acid (H_2SO_4) 64% (v/v) at 70 °C was used for the acid hydrolysis process for 40 min. Then, the material was dialyzed to neutral pH. Mechanical fragmentation was carried out using Ultra turrax (24,000 rpm for 5 min) and tip sonication (3 min, 550 W) obtained colloidal suspension concentration of 1.3 g/L of CNCs.

3. After the CNC production, Okamoto-Schalch et al. [20] produced ChNPs by an ionic gelation method [59]. The TPP solution (0.6 g/L) was drop added to chitosan solution (2 g/L in 1% acetic acid solution) for cross-linking and nanoparticles formation (Fig. 4.1). To produce the hybrid system, they added CNC colloidal dispersion in the chitosan solution, and TPP was then dropped into the dispersion under continuous stirring (Fig. 4.2).

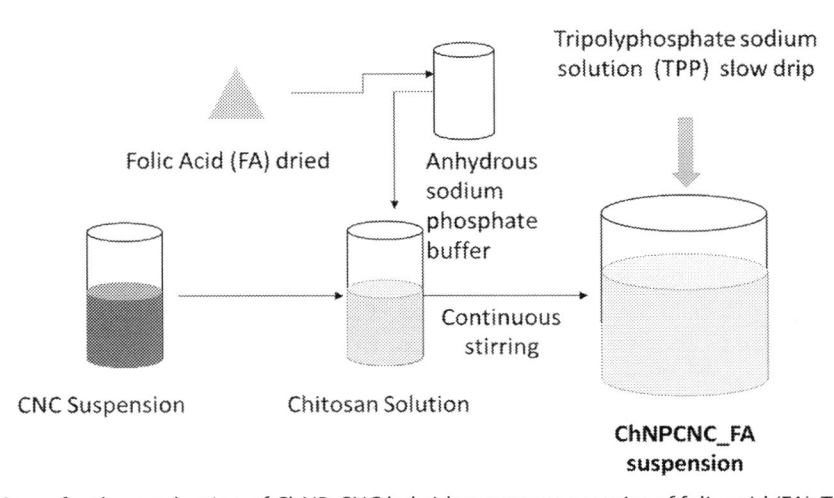

Fig. 4.2 Steps for the production of ChNP-CNC hybrid systems as a carrier of folic acid (FA). The figure was adapted from the work of Okamoto-Schalch (2019).

The encapsulation efficiency (EE) was 62 ± 4% and 72 ± 8% for ChNP_FA (without CNCs) and ChNP-CNC_FA, respectively. The final concentration of the hybrid system was 0.1 g/L of FA solution (pH 6.38), 0.6 g/L of TPP solution (pH 9.28), 2.0 g/L of Ch solution (pH 3.59), and 0.3 g/L of CNC dispersion (pH 5.89).

To evaluate the morphology, Okamoto-Schalch et al. [20] used field emission scanning electron microscopy and observed that unreacted fibers formed a three-dimensional network in suspension ChNP-CNC_FA. This network increased the protection of FA-loaded particles against ultraviolet light degradation. The protection may have occurred due to the ability of CNCs to scatter light. The FA release kinetics was determined in acidic media (pH 3.0) and neutral media (pH 7.4). All the suspensions were capable of retaining the vitamin in a neutral pH in determined times. In acid media, CNC–chitosan–based nanoparticles present good protective action as a barrier according to Fick's second law of diffusion.

Akhlaghi et al. [61] used CNCs and chitosan oligosaccharide to encapsulate vitamin C and obtained 91% of EE at pH 5 and 72% at pH 3. Xu et al. [62] produced a nanocomposite hydrogel based on CNC and Ch through a Schiff base reaction between aldehyde groups and free amino groups as a carrier of theophylline, and the EE ranged from 85% to 92% [63] also produced a hybrid nanoparticle using alginate and CNC loaded with rifampicin and their EE ranged from 43% to 70%.

Some works also studied the release of active compounds in hybrid systems. Akhlaghi et al. [61] used CNCs and chitosan oligosaccharide to encapsulated vitamin C and observed at pH 5 that a more controlled and a higher concentration of vitamin was released than at pH 3. Xu et al. [62] produced a nanocomposite hydrogel with CNCs and Ch as a carrier of theophylline. The cumulative drug release at pH 1.5 was significantly greater than at pH 7.4. When rifampicin was loaded in alginate-CNC particles [63], the release was greater at pH 7.4 (100% in 12 h) than at pH 1.2 (10–15% in 2 h), evidencing different profiles depending on the polymer used for the hybrid system.

Elfeky et al. [64] evaluated the photodegradation of the CNC/ZnO/CuO hybrid system. The hybrid particle was more efficient than the pure CNC and Zn/CuO in dye removal. They also studied the antibacterial activity, and the authors concluded that hybrid is more effective against Gram-positive and Gram-negative bacteria.

Some authors synthesized hybrid materials to reinforce mechanical properties that compare the material without CNC. Xie et al. [65] used CNCs and reactive polyhedral oligomeric silsesquioxane and concluded that the hybrid material increased the elastic proprieties of nanocellulose; however, these materials decreased the thermal degradation temperature. Liang et al. [66] produced microcrystalline cellulose hybridized with nano-ZnO. These hybrid systems were added with rubber, silica, and others components to make a sheet by vulcanization. The sheet with hybrid composites improved mechanical properties. Scanning electron microscope micrographs showed that hybrid composites were improved interfacial bonding within a rubber matrix. Salama [67] reviewed hybrids materials using cellulose/calcium phosphate for biomedical and environmental application and concluded that the hybrid materials had the flexibility and functionality of cellulose and the heat resistance and stability of calcium phosphate.

As can be seen, these nanocomposites are potential materials to be added in the production of hydrogels or aerogels to improve their mechanical and physicochemical properties.

4.4 ChNP-CNC-based cryogels: a case study

There are three steps that can produce the nanocellulose aerogels. First, disperse the cellulose; second, form a gel by the sol–gel process; and third, drying. This results in a 3D porous structure. The nanocellulose aerogel can be prepared by dispersing nanocellulose in water using mechanical methods such as ultrasonic, followed by drying with or without solvent exchange [68]. The sol–gel techniques embrace the colloidal suspension into a solid network transformation. By including some components, it is possible to reach ordered porous structures [69,70]. For simplicity, freeze-dried porous cellulose material will be identified as "cryogel," whereas those for which dried process occurred in supercritical conditions will be called "aerogel."

Li et al. [71] demonstrated that atmospheric dried cellulose aerogels can be produced with cross-linking using glycidoxypropyltrimethoxysilane (GPTMS) and branched polyethylenimine (b-PEI). They obtained an improvement on the mechanical strength of the pore walls with the cross-linkers (GPTMS and b-PEI) and freeze-thawing method, in which they maintained a similar density that had a specific area higher (22.4 m^2/g) when compared to those prepared by freezing-drying [71].

In our preliminary study, cryogels were produced using the hybrid system, ChNP-CNC, described by Okamoto-Schalch et al. [20]. The hybrid system suspension was mixed with chitosan solution (2.0 g/L of Ch in 1% acetic acid solution) at different concentrations (Table 4.3). After continuous stirring (10 min), samples were freeze-dried. The image of cryogels can be observed in Fig. 4.3.

Micrographs were made of the samples to evaluate the structure created in the Fig. 4.4. Highly porous structures were observed in all of the samples. The sample with

Table 4.3 Compositions of cryogels contend with ChNP-CNC and Ch in different concentrations.

Sample	ChNP-CNC suspension (% v/v)	Ch solution (% v/v)
A	0	100
B	25	75
C	50	50
D	75	25
E	100	0

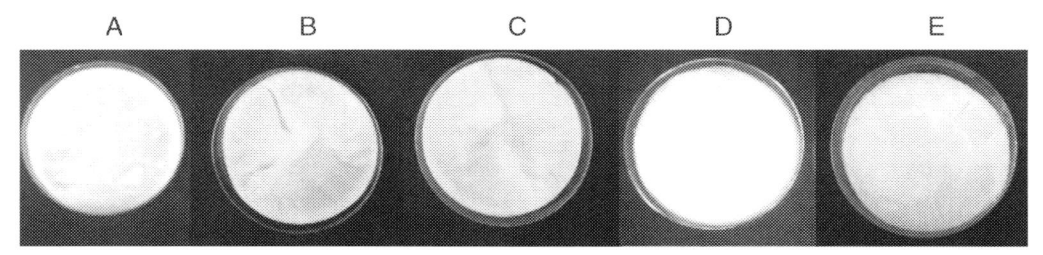

Fig. 4.3 Images of samples with ChNP-CNC hybrid system and Ch solution according to the composition described in Table 4.3.

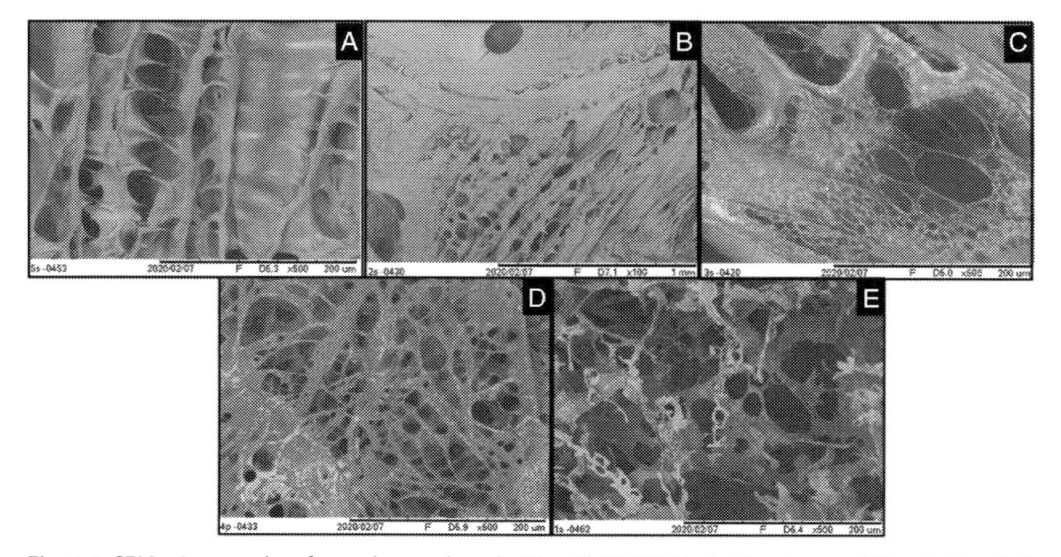

Fig. 4.4 SEM micrographs of samples produced with a ChNP-CNC hybrid system and Ch solution. A, B, C, D, and E is according to their composition (Table 4.3).

ChNP-CNC (sample E) showed greater porosity, larger pores, and hollow cavities. Sample A (100% Ch solution) showed less porosity and a greater amount of material. According to the decrease in the concentration of Ch (B, C, D), there is an increase in the pores.

Kathauria et al. [72] produced cryogels with chitosan and gelatin, the micrographs are shown in Fig. 4.5. They observed that the samples with chitosan–gelatin were of a similar pattern of pore and the sizes range from 30 to 100 μm and presented thicker walls than gelatin cryogels.

Zhang et al. [73] studied the preparation of chitosan cryogels with controllable pore size and the morphology can be observed in Fig. 4.6. They also have similar uniform

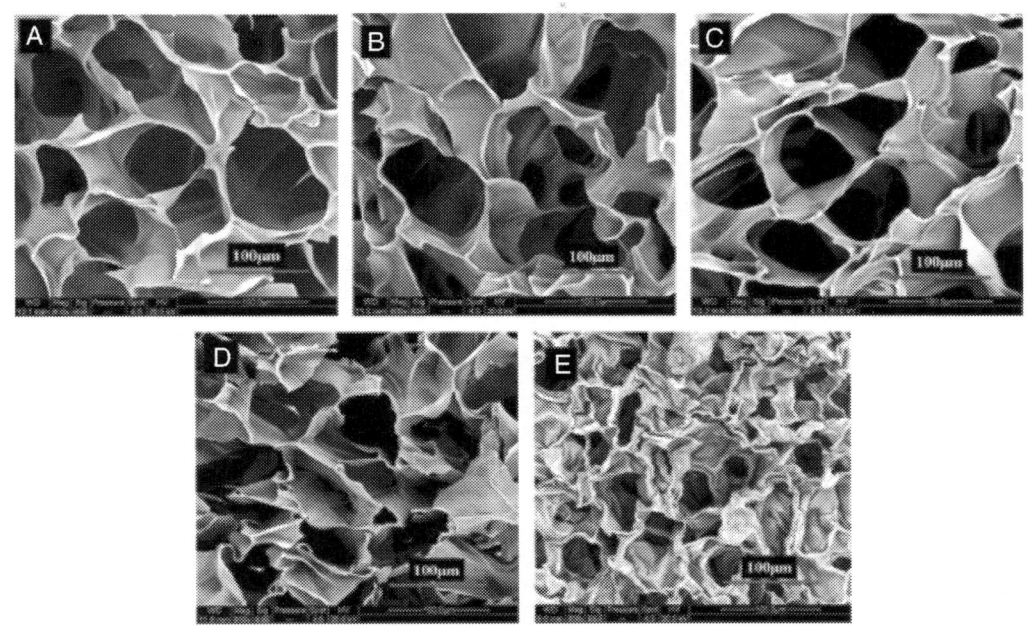

Fig. 4.5 SEM micrographs of cryogels of chitosan and gelatin. (A) 4% chitosan-gelatin, (B) 5%, (C) 6%, (D) 2% chitosan cryogel, and (E) 4% gelatin cryogel. Microscopy images were taken from the work of Kathauria et al. (2009).

pore honeycomb matrices, the difference was the size pore. The lower the temperature and time, the smaller the pore size. The pore of chitosan cryogel is in the range of 60–240 μm.

4.5 Conclusion and future challenges

Cellulose-based nanocomposites exhibit greater mechanical properties than those without nanocelluloses. Acid or enzymatic treatments combined with mechanical processes are normally used for the production of CNCs or nanofibrils, respectively. Many studies have been working on the production of these materials and also on the association of nanocelluloses to other polymers, such as polyacrylamide, starch, chitosan, protein, and others. The network formation can improve the mechanical properties (tensile strength, elongation, and toughness) or pore sizes depending on the materials (films, aerogels, hydrogels, cryogels). Other positive effects that are normally highlighted are the protection of active compounds or improvement on their controlled release. A case study was presented concerning the production of ChNPs and nanocellulose hybrid system (ChNP-CNC) and their application in chitosan cryogels. Highly porous structures were

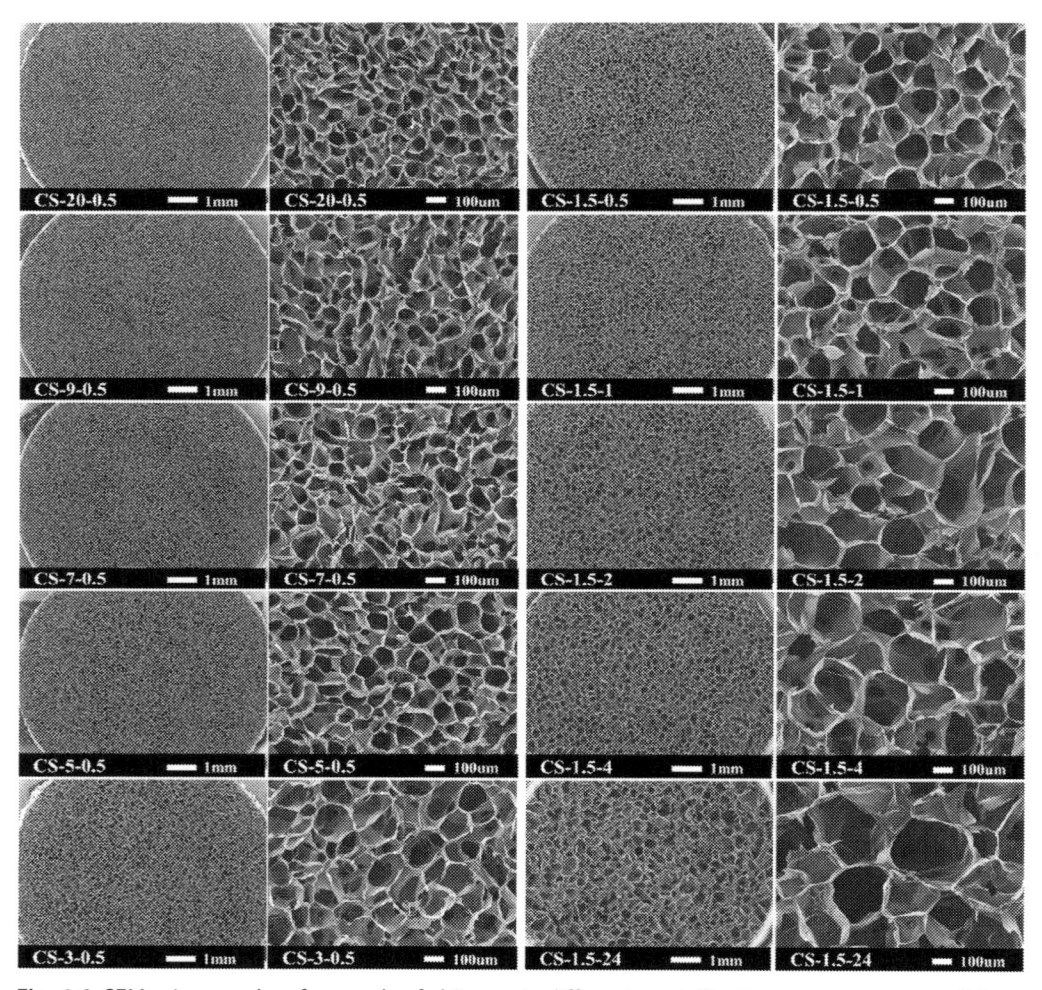

Fig. 4.6 SEM micrographs of cryogels of chitosan in different crystallization temperature conditions. The samples were named CS for chitosan, the first number is the crystal growth temperature (°C), the second number is the crystal growth time (h). Microscopy images were taken from the work of Zhang et al. [76].

observed in all of the samples tested, and they were more evidenced when the concentration of ChNP-CNC increased. The studies of the porous structure of cellulose-based materials are key factors to increase understanding of both the nanocellulose network with the other polymers and their influences on the properties of the materials. It is also worth highlighting that nontoxic and low-cost treatments are needed to make these materials versatile and competitive for a wide range of food, medical, or cosmetics applications.

References

[1]   S. Damodaran, K.L. Fennema, Fennema's Food Chemistry, CRC Press Taylor & Francis Group, New York, 2008.

[2]   M.P. Ansell, L.Y. Mwaikambo, The structure of cotton and other plant fibres, in: S.J. Eichhorn, J.W.S. Hearle, M. Jaffe, T. Kikutani (Eds.), The structure of cotton and other plant fibres, Handbook of Textile Fibre Structure 2009, pp. 62–94.

[3]   H. Kargarzadeh, M. Mariano, J. Huang, N. Lin, I. Ahmad, A. Dufresne and S. Thomas, Recent developments on nanocellulose reinforced polymer nanocomposites: A review. Polymer, 132 (2017) 368–393.

[4]   M. Martelli-Tosi, O.B.G. Assis, N.C. Silva, B.S. Esposto, M.A. Martins, D.R. Tapia-Blácido, Chemical treatment and characterization of soybean straw and soybean protein isolate/straw composite films, Carbohydr. Polym. 157 (2017) 512–520.

[5]   M. Martelli-Tosi, M.M. Masson, N.C. Silva, B.S. Esposto, T.T. Barros, O.B.G. Assis, D.R. Tapia-Blácido, Soybean straw nanocellulose produced by enzymatic or acid treatment as a reinforcing filler in soy protein isolate films, Carbohydr. Polym. 198 (2018) 61–68.

[6]   FAOSTAT, Food and agriculture organization, 2020. http://www.fao.org/faostat/en/#home. Accessed 01 Oct 2020.

[7]   M.L.V. Bose, J.G. Martins Filho, O papel dos resíduos agroindustriais na alimentação de ruminantes, Informe Agropecuário (1984) 3–7.

[8]   E. Cabrera, M.J. Muñoz, R. Martín, I. Caro, C. Curbelo, A.B. Díaz, Comparison of industrially viable pretreatments to enhance soybean straw biodegradability, Bioresour. Technol. 194 (2015) 1–6.

[9]   C. Wan, Y. Zhou, Y. Li, Liquid hot water and alkaline pretreatment of soybean straw for improving cellulose digestibility, Bioresour. Technol. 102 (10) (2011) 6254–6259.

[10] A. Alemdar, M. Sain, Isolation and characterization of nanofibers from agricultural residues – wheat straw and soy hulls, Bioresour. Technol. 99 (6) (2008) 1664–1671.

[11] S.P. Akhlaghi, D. Tiong, R.M. Berry, K.C. Tam, Comparative release studies of two cationic model drugs from different cellulose nanocrystal derivatives, Eur. J. Pharma. Biopharma. 88 (1) (2014) 207–215.

[12] P. Chauhan, N. Yan, Nanocrystalline cellulose grafted phthalocyanine: a heterogeneous catalyst for selective aerobic oxidation of alcohols and alkyl arenes at room temperature in a green solvent, RSC Adv. 5 (47) (2015) 37517–37520.

[13] J.K. Jackson, K. Letchford, B.Z. Wasserman, L. Ye, W.Y. Hamad, H.M. Burt, The use of nanocrystalline cellulose for the binding and controlled release of drugs, Int. J. Nanomed. 6 (2011) 321–330.

[14] R. Kolakovic, L. Peltonen, A. Laukkanen, J. Hirvonen, T. Laaksonen, Nanofibrillar cellulose films for controlled drug delivery, Eur. J. Pharma. Biopharma. 82 (2) (2012) 308–315.

[15] W. Qing, Y. Wang, Y. Wang, D. Zhao, X. Liu, J. Zhu, The modified nanocrystalline cellulose for hydrophobic drug delivery, Appl. Surf. Sci. 366 (2016) 404–409.

[16] M.N.V.R. Kumar, R.A.A. Muzzarelli, C. Muzzarelli, H. Sashiwa, A.J. Domb, Chitosan chemistry and pharmaceutical perspectives, Chem. Rev. 104 (12) (2004) 6017–6084.

[17] N. Devi, J. Dutta, Preparation and characterization of chitosan-bentonite nanocomposite films for wound healing application, Int. J. Biol. Macromol. 104 Part B (2017) 1897–1904.

[18] K.A. Janes, M.J. Alonso, Depolymerized chitosan nanoparticles for protein delivery: preparation and characterization, J. Appl. Polym. Sci. 88 (12) (2003) 2769–2776.

[19] M.R. Moura, F.A. Aouada, L.H.C. Mattoso, Preparation of chitosan nanoparticles using methacrylic acid, J. Colloid Interface Sci. 321 (2) (2008) 477–483.

[20] N.O. Okamoto-Schalch, S.G.B. Pinho, T.T Barros-Alexandrino, G.C. Dacanal, O.B.G. Assis, M. Martelli-Tosi, Production and characterization of chitosan-TPP/cellulose nanocrystal system for encapsulation: a case study using folic acid as active compound, Cellulose 27 (2020) 5855–5869.

[21] N. Durán, L.H.C. Mattoso, P.C. Morais, Nanotecnologia: introdução, preparação e caracterização de nanomateriais e exemplos de aplicação, Artliber, São Paulo, 2006.

[22] Y. Chen, C. Liu, P.R. Chang, D.P. Anderson, M.A. Huneault, Pea starch-based composite films with pea hull fibers and pea hull fiber-derived nanowhisker, Polym. Eng. Sci. (2009) 360–378.

[23] J.I. Morán, V.A. Alvarez, V.P. Cyras, A. Vázquez, Extraction of cellulose and preparation of nanocellulose from sisal fibers, Cellulose 15 (1) (2008) 149–159.

[24] E.M. Teixeira, T.J. Bondancia, K.B.R. Teodoro, A.C. Corrêa, J.M. Marconcini, L.H.C. Mattoso, Sugarcane bagasse whiskers: extraction and characterizations, Ind. Crops Prod. 33 (1) (2011) 63–66.

[25] S. Eichhorn, Raman spectroscopy and x-ray scattering for assessing the interface in natural fibre composites, in: N. Zafeiropoulos (Ed.), Elsevier, Amsterdam, 2011, pp. 379–400.

[26] H. Kargarzadeh, M. Ioelovich, I. Ahmad, S. Thomas, A. Difresne, Methods for extraction of nanocellulose from various sources, in: H. Kargarzadeh, I. Ahmad, S. Thomas, A. Dufresne (Eds.), Methods for extraction of nanocellulose from various sources, Handbook of Nanocellulose and Cellulose Nanocomposites, first ed., 2017, pp. 1–49.

[27] G.K. Dutta, N. Karak, Waste brewed tea leaf derived cellulose nanofiber reinforced fully bio-based waterborne polyester nanocomposite as an environmentally benign material, RSC Adv. 9 (36) (2019) 20829–20839.

[28] K.S. Prado, M.A.S. Spinacé, Isolation and characterization of cellulose nanocrystals from pineapple crown waste and their potential uses, Int. J. Biol. Macromol. 122 (2019) 410–416.

[29] H. Tibolla, F.M. Pelissari, J.T. Martins, E.M. Lanzoni, A.A. Vicente, F.C. Menegalli, R.L. Cunha, Banana starch nanocomposite with cellulose nanofibers isolated from banana peel by enzymatic treatment: in vitro cytotoxicity assessment, Carbohydr. Polym. 207 (2019) 169–179.

[30] S. Widiarto, S.D. Yuwono, A. Rochliadi, I.M. Arcana, Preparation and characterization of cellulose and nanocellulose from agro-indutrial waste – Cassava peel, IOP Conference Series: Materials Science and Engineering, (2016) 176.

[31] A. Perzon, B. Jørgensen, P. Ulvskov, Sustainable production of cellulose nanofiber gels and paper from sugar beet waste using enzymatic pre-treatment, Carbohydr. Polym. 230 (2020) 115581.

[32] P. Kumari, G. Pathak, R. Gupta, D. Sharma, A. Meena, Cellulose nanofibers from lignocellulosic biomass of lemongrass using enzymatic hydrolysis: characterization and cytotoxicity assessment. DARU, J. Pharma. Sci. 27 (2) (2019) 683–693.

[33] J. Marett, A. Aning, E.J. Foster, The isolation of cellulose nanocrystals from pistachio shells via acid hydrolysis, Ind. Crops Prod. 109 (2017) 869–874.

[34] do A F.M. Pelissari, M.M. Andrade-Mahecha, P.J. Sobral, F.C. Menegalli, Nanocomposites based on banana starch reinforced with cellulose nanofibers isolated from banana peels, J. Colloid Interface Sci. 505 (2017) 154–167.

[35] Z. Wang, Z.J. Yao, J. Zhou, Y. Zhang, Reuse of waste cotton cloth for the extraction of cellulose nanocrystals, Carbohydr. Polym. 157 (2017) 945–952.

[36] L.A. Camargo, S.C. Pereira, A.C. Correa, C.S. Farinas, J.M. Marconcini, L.H.C. Mattoso, Feasibility of manufacturing cellulose nanocrystals from the solid residues of second-generation ethanol production from sugarcane bagasse, Bioenergy Res. 9 (3) (2016) 894–906.

[37] E. Fortunati, P. Benincasa, G.M. Balestra, F. Luzi, A. Mazzaglia, D. Del Buono, D. Puglia, L. Torre, Revalorization of barley straw and husk as precursors for cellulose nanocrystals extraction and their effect on PVA_CH nanocomposites, Ind. Crops Prod. 92 (2016) 201–217.

[38] A. Campos, A.C. Correa, D. Cannella, E.M. Teixeira, J.M. Marconcini, A. Dufresne, L.H.C. Mattosos, P. Cassland, A.R. Sanadi, Obtaining nanofibers from curauá and sugarcane bagasse fibers using enzymatic hydrolysis followed by sonication, Cellulose 20 (3) (2013) 1491–1500.

[39] P.B. Filson, B.E. Dawson-Andoh, D. Schwegler-Berry, Enzymatic-mediated production of cellulose nanocrystals from recycled pulp, Green Chem. 11 (2009) 1808–1814.

[40] M. Pääkkö, M. Ankerfor, H. Kosonen, A. Nykänen, S. Ahola, M. Österberg, J. Ruokolainen, J. Laine, P.T. Larsson, O. Ikkala, T. Lindstrom, Enzymatic hydrolysis combined with mechanical shearing and high-pressure homogenization for nanoscale cellulose fibrils and strong gels, Biomacromolecules 8 (6) (2007) 1934–1941.

[41] P. Satyamurthy, P. Jain, R.H. Balasubramanya, N. Vigneshwaran, Preparation and characterization of cellulose nanowhiskers from cotton fibres by controlled microbial hydrolysis, Carbohydr. Polym. 83 (1) (2011) 122–129.

[42] M.M. Campo, M. Darder, P. Aranda, M. Akkari, Y. Huttel, A. Mayoral, J. Bettini, E. Ruiz-Hitzky, Functional hybrid nanopaper by assembling nanofibers of cellulose and sepiolite, Adv. Funct. Mater. 28 (27) (2018) 1–13.

[43] H. Du, W. Liu, M. Zhang, C. Si, X. Zhang, B. Li, Cellulose nanocrystals and cellulose nanofibrils based hydrogels for biomedical applications, Carbohydr. Polym. 209 (2019) 130–144.

[44] R.M.A. Domingues, M.E. Gomes, R.L. Reis, The potential of cellulose nanocrystals in tissue engineering strategies, Biomacromolecules 15 (7) (2014) 2327–2346.

[45] H. Zhang, S. Lyu, X. Zhou, H. Gu, C. Ma, C. Wang, T. Ding, Q. Shao, H. Liu, Z. Guo, Super light 3D hierarchical nanocellulose aerogel foam with superior oil adsorption, J. Colloid Interface Sc. 536 (2019) 245–251.

[46] I. Doench, M.E.W. Torres-Ramos, A. Montembault, P.N. de Oliveira, C. Halimi, E. Viguier, A. Osorio-Madrazo, Injectable and gellable chitosan formulations filled with cellulose nanofibers for intervertebral disc tissue engineering, Polymers 10 (11) (2018) 1202.

[47] N.K. Dehkordi, M. Minaiyan, A. Talebi, V. Akbari, A. Taheri, Nanocrystalline cellulose-hyaluronic acid composite enriched with, GM-CSF loaded chitosan nanoparticles for enhanced wound healing, Biomed. Mater. (2019) 1–51.

[48] X. Li, Y. Liu, Y. Yu, W. Chen, Y. Liu, H. Yu, Nanoformulations of quercetin and cellulose nanofibers as healthcare supplements with sustained antioxidant activity, Carbohydr. Polym. 207 (2019) 160–168.

[49] J.P. Oliveira, G.P. Bruni, S.L.M. el Halal, F.C. Bertoldi, A.R.G. Dias, E.R. Zavareze, Cellulose nanocrystals from rice and oat husks and their application in aerogels for food packaging, Int. J. Biol. Macromol. 124 (2019) 75–184.

[50] T. Paulraj, A.V. Riazanova, A.J. Svagan, Bioinspired capsules based on nanocellulose, xyloglucan and pectin – the influence of capsule wall composition on permeability properties, Acta Biomater. 69 (2018) 196–205.

[51] C. Chen, Y. Wang, T. Meng, Q. Wu, L. Fang, D. Zhao, D. Li, Electrically conductive polyacrylamide/carbon nanotube hydrogel: reinforcing effect from cellulose nanofibers, Cellulose 26 (16) (2019) 8843–8851.

[52] Z. Zhang, X. Wang, M. Gao, Y. Zhao, Y. Chen, Sustained release of an essential oil by a hybrid cellulose nanofiber foam system, Cellulose 27 (5) (2020) 2709–2721.

[53] Z. Jaafar, B. Quelennec, C. Moreau, D. Lourdin, J.E. Maigret, B. Pontoire, A. D'orlando, T. Coradin, B. Duchemin, F.M. Fernandes, B. Cathala, Plant cell wall inspired xyloglucan/cellulose nanocrystals aerogels produced by freeze-casting, Carbohydr. Polym. 247 (2020) 116642.

[54] X. Gong, Y. Wang, H. Zeng, M. Betti, L. Chen, Highly porous, hydrophobic, and compressible cellulose nanocrystals/poly(vinyl alcohol) aerogels as recyclable absorbents for oil-water separation, ACS Sustain. Chem. Eng. 7 (13) (2019) 11118–11128.

[55] J. Supramaniam, R. Adnan, N.H.M. Kaus, R. Bushra, Magnetic nanocellulose alginate hydrogel beads as potential drug delivery system, Int. J. Biol. Macromol. 118 (2018) 640–648.

[56] S.Y. Ooi, I. Ahmad, M.C.I.M. Amin, Cellulose nanocrystals extracted from rice husks as a reinforcing material in gelatin hydrogels for use in controlled drug delivery systems, Ind. Crops Prod. 93 (2016) 227–234.

[57] L. Bugnicourt, C. Ladavière, Interests of chitosan nanoparticles ionically cross-linked with tripolyphosphate for biomedical applications, Prog. Polym. Sci. 60 (2016) 1–17.

[58] D. Britto, M.R. Moura, F.A. Aouada, L.H.C. Mattoso, O.B.G. Assis, N,N,N-trimethyl chitosan nanoparticles as a vitamin carrier system, Food Hydrocoll. 27 (2) (2012) 487–493.

[59] D. Britto, M.R. Moura, F.A. Aouada, F.G. Pinola, L.M. Lundstedt, O.B.G. Assis, L.H.C. Mattoso, Entrapment characteristics of hydrosoluble vitamins loaded into chitosan and N,N,N-trimethyl chitosan nanoparticles, Macromol. Res. 22 (12) (2014) 1261–1267.

[60] A.S.L. Iida, K.N. Luz, T.T. Barros-Alexandrino, C.S. Favaro-Trindade, S.C. Pinho, O.B.G. Assis, M. Martelli-Tosi, Investigation of TPP-chitosomes particles structure and stability as encapsulating agent of cholecalciferol, Polímeros: Ciência e Tecnologia 29 (2019) 2019049.

[61] S.P. Akhlaghi, R.M. Berry, K.C. Tam, Modified cellulose nanocrystal for vitamin c delivery, AAPS PharmSciTech 16 (2015) 306–314.

[62] Q. Xu, Y. Ji, Q. Sun, Y. Fu, Y. Xu, L. Jin, Fabrication of cellulose nanocrystal/chitosan hydrogel for controlled drug release, Nanomaterials 9 (2) (2019) 252.

[63] D. Thomas, M.S. Latha, K.K. Thomas, Synthesis and in vitro evaluation of alginate-cellulose nanocrystal hybrid nanoparticles for the controlled oral delivery of rifampicin, J. Drug Deliv. Sci. Technol. 46 (2018) 392–399.

[64] A.S. Elfeky, S.S. Salem, A.S. Elzaref, M.E. Owda, H.A. Eladawy, A.M. Saeed, M.A. Awad, R.E. Abou-Zeid, A. Fouda, Multifunctional cellulose nanocrystal/metal oxide hybrid, photo-degradation, antibacterial and larvicidal activities, Carbohydr. Polym. 230 (2020) 115711.

[65] K. Xie, X. Gao, W. Zhao, Thermal degradation of nano-cellulose hybrid materials containing reactive polyhedral oligomeric silsesquioxane, Carbohydr. Polym. 81 (2) (2010) 300–304.

[66] Y. Liang, X. Liu, I. Wang, J. Sun, The fabrication of microcrystalline cellulose-nanoZnO hybrid composites and their application in rubber compounds, Carbohydr. Polym. 169 (2017) 324–331.

[67] A. Salama, Cellulose/calcium phosphate hybrids: New materials for biomedical and environmental applications, Int. J. Biol. Macromol. 127 (2019) 606–617.

[68] D.A. Gopakumar, S. Thomas, F.A.T. Owolabi, S. Thomas, A. Nzihou, S. Rizal, H.P.S.A. Khalil, Nanocellulose based aerogels for varying engineering applications. In: Hashmi, S., Choudhury, I.A. (Eds.), Encyclopedia of Renewable and Sustainable Materials, 2 (2020) 155-165. https://doi.org/10.1016/B978-0-12-803581-8.10549-1.

[69] C. Darpentigny, G. Nonglaton, J. Bras, B. Jean, Highly absorbent cellulose nanofibrils aerogels prepared by supercritical drying, Carbohydr. Polym. 229 (2019) 115560.

[70] X. Zhang, M. Cresswell, Inorganic Controlled Release Technology: Materials for Inorganic Controlled Release Technology, (2016) 1–16.

[71] Y. Li, N. Grishkewich, L. Liu, C. Wang, K. Tam, S. Liu, M. Mao Zhiping, X. Sui, Construction of functional cellulose aerogels via atmospheric drying chemically cross-linked and solvent exchanged cellulose nanofibrils, Chem. Eng. J. 366 (2019) 531–538.

[72] N. Kathuria, A. Tripathi, K.K Kar, A. Kumar, Synthesis and characterization of elastic and macroporous chitosan–gelatin cryogels for tissue engineering, Acta Biomater. 5 (1) (2009) 406–418.

[73] H. Zhang, C. Liu, L. Chen, B. Dai, Control of ice crystal growth and its effect on porous structure of chitosan cryogels, Chem. Eng. Sci. 201 (2019) 50–57.

CHAPTER 5

Recent developments of bacterial nanocellulose porous scaffolds in biomedical applications

Swaminathan Jiji[a], Kannan Maharajan[b], Krishna Kadirvelu[a]
[a]DRDO-BU Center for Life Sciences, Bharathiar University Campus, Coimbatore, Tamil Nadu, India
[b]Biology Institute, Qilu University of Technology (Shandong Academy of Sciences), Jinan, Shandong Province, China

5.1 Introduction

Bacterial nanocellulose (BNC) production was first discovered by A.J. Brown in 1886 from *Acetobacter xylinum* [1]. Later, Louis Pasteur described the cellulose produced by microorganism as "a sort of moist skin, swollen, gelatinous and slippery material" [2]. BNC has been used as traditional dessert called Nata-de-coco by people of countries like Philippines, that is, thick gel sheets fermented from coconut water. Bacterial cellulose is produced by many bacterial genera such as *Acetobacter*, *Agrobacterium*, *Azotobacter*, *Pseudomonas*, *Salmonella*, *Rhizobium*, *Sarcina,* and *Alcaligenes.*

In the 20th century, *A. xylinum* was extensively explored as a model species for bacterial cellulose production, which was initiated by Hestrin in 1947 [3,4]. BNC is polysaccharide with β-1-4 glucan chains with the molecular formula of $(C_6H_{10}O_5)n$ and single cell of *A. xylinum* can polymerize about 2,00,000 glucose molecules/second [5,6]. BNC is synthesized by a mechanism involving multiple enzymes and regulatory proteins. The steps involved in the synthesis are: (1) phosphorylation of glucose, (2) isomerization of glucose–6-phosphate to glucose 1-phosphate, (3) synthesis of uridine diphosphate glucose from glucose-1-phosphate, and (4) synthesis of cellulose from uridine diphosphate glucose (UDP)-glucose that are catalyzed by the enzymes glucokinase, phosphoglucomutase, UDP-glucose pyrophosphorylase, and cellulose synthase, respectively. The process of synthesis and purification of BNC is outlined in Fig. 5.1.

5.2 Properties of BNC

BNC has received extensive interest among scientific communities due to its unique structural features and physico-mechanical properties [7]. The properties of BNC are as follows:
- High chemical purity [8,9].
- High crystallinity [10].

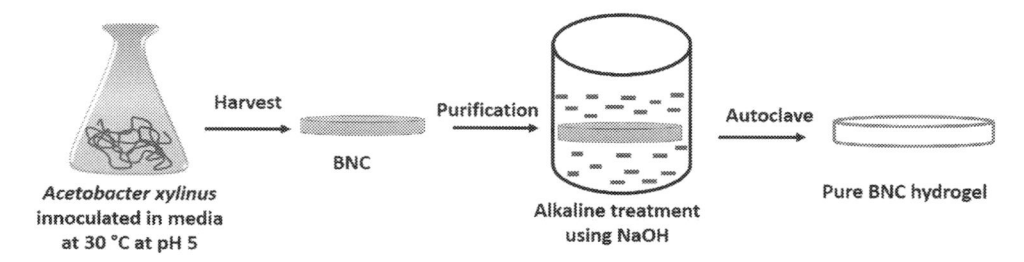

Fig. 5.1 Synthesis and purification of BNC.

- High water absorption capacity makes it hydrophilic (up to 400 times its dry weight) [9,11,12].
- High porosity with a large specific surface area and very narrow fiber diameter in range of 30–40 nm and 3D architecture [13–15].
- Young's modulus and tensile strength are in the range of 15–35 GPa and 200–300 Mpa, respectively [16,17].
- Controllable shape and thickness by varying culture time and culture flasks shape [18].
- Higher thermal stability makes it as sterilizable biomaterial [18].
- Environmental biodegradability (Eco-friendly product) [7,10].
- Excellent biocompatibility [10].

5.3 Importance of BNC in biomedical field

BNC has been widely used in biomedical application either individually or combined with different natural or chemical components such as biopolymers, nanoparticles, etc. [19]. In recent days, BNC production is increasingly important due to the environmentally friendly approach [10,20]. An ideal 3D scaffolds would mimic the native extracellular matrix, be biocompatible, and possess characteristics such as interconnected fibers, porosity, and mechanical properties [21]. The porosity and interconnectivity of fibers are necessary for the cellular movement and proliferation [22]. The important mechanical properties of soft tissues such as skin, muscle, bladder, and blood vessel are the toughness and elasticity [23]. Due to the superior properties of BNC as an effective biopolymer, it has been widely explored in biomedical applications such as scaffolds for tissue engineering, regenerative medicine, drug delivery systems, vascular grafts, dental therapy, artificial urethra, artificial skin, artificial cornea, urethral conduits and blood vessels, and heart valves regeneration (Fig. 5.2) [24–30]. Food and Drug Administration has approved the BNC membranes for tissue engineering applications like wound dressing and cartilage repair [31].

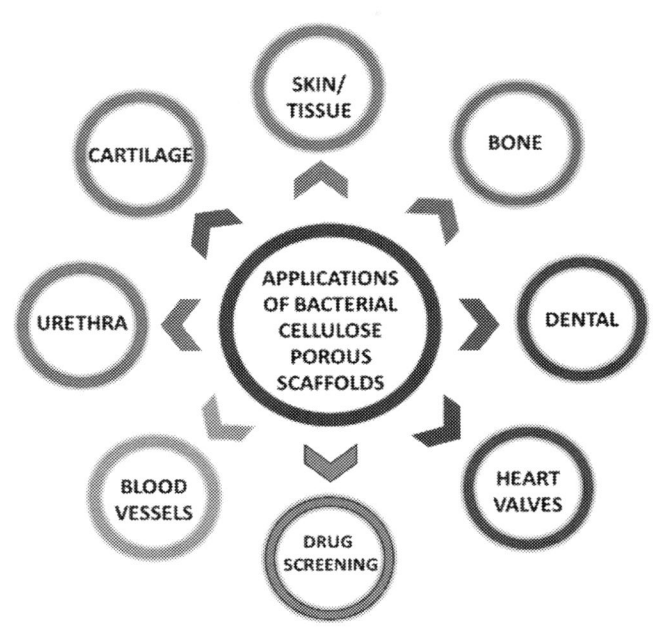

Fig. 5.2 Overview of bacterial nanocellulose in biomedical applications.

5.4 Synthesis and fabrication: development of porous scaffolds

The development of porous structured scaffolds is an essential part for tissue engineering to permit the host cell movement, nutrients, and new tissue establishment [30]. In tissue engineering, the combination of functional cells and a 3D scaffold material is essential for cell attachment, proliferation, and differentiation to support the necessary microenvironment for growth [29]. The densely packed nanofibrils network is one of the challenges of BNC in utilizing as scaffold in tissue engineering, as it impedes their performance and lowers the cell infiltration. In this context, many methods have been reported for the development of pores in the scaffolds (Fig. 5.3).

5.4.1 Phase separation

Phase separation is one of the most versatile fabrication methods to produce highly porous interconnected BNC scaffolds (Fig. 5.3B). For instance, BNC and collagen porous microspheres have been prepared by the reverse-phase suspension method for bone tissue engineering applications [32].

5.4.2 Electrospinning

Electrospinning/electrospraying technique has been widely used for the development of porous scaffolds (Fig. 5.3A). Azarniya et al. [33] have prepared BNC/keratin nanofibrous

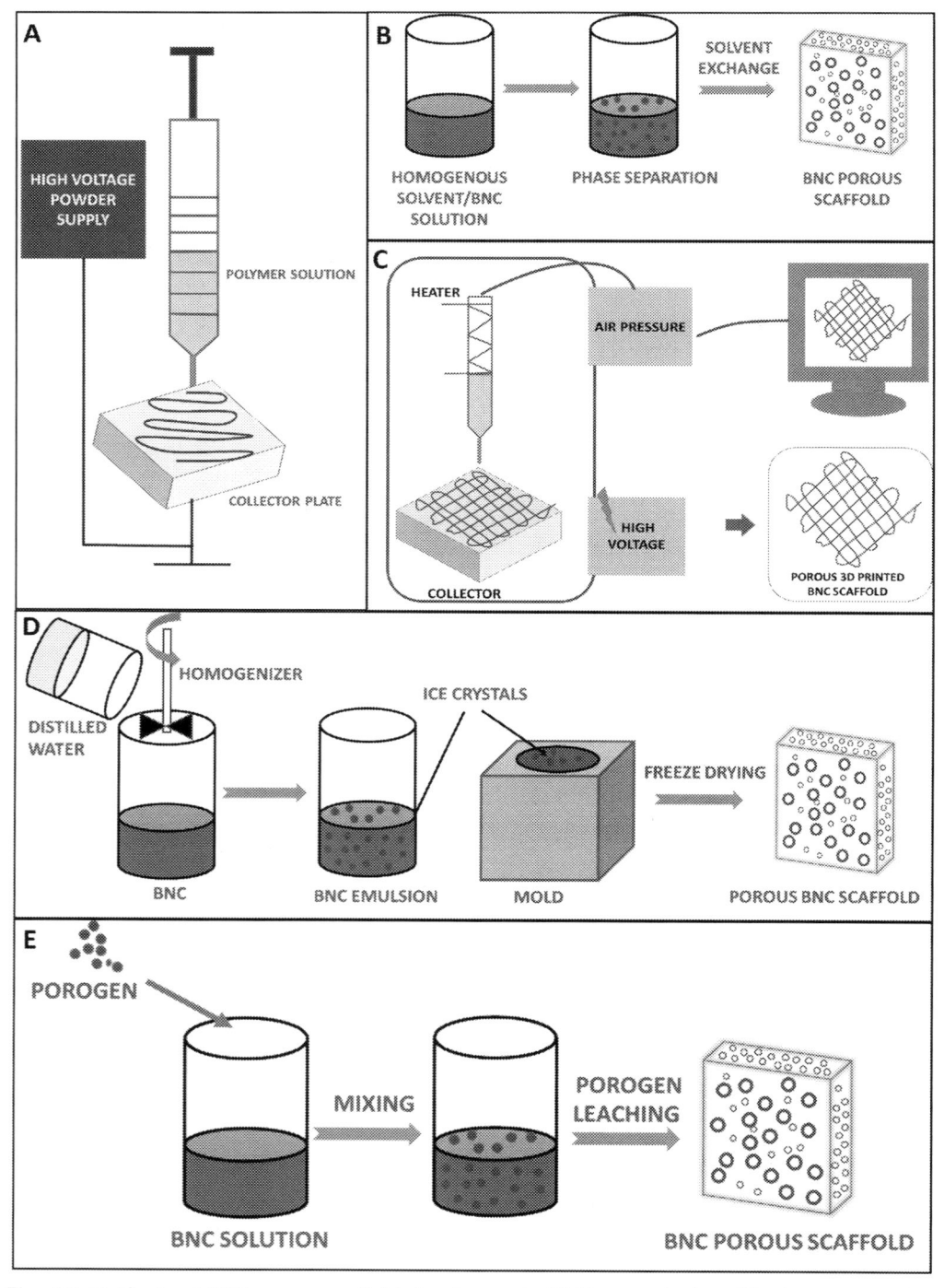

Fig. 5.3 *Methods of BNC porous scaffold development.* (A) Electrospinning, (B) phase separation, (C) 3D printing, (D) emulsion-freeze drying, and (E) particle-leaching techniques.

mats by electrospinning method for wound dressing application. Furthermore, graphene oxide–incorporated cellulose scaffolds were produced by electrospinning for drug screening and cancer biology studies [34].

5.4.3 3D printing

Recently, 3D bioprinting techniques are emerging in medical fields (Fig. 5.3C). For instance, Huang et al. [35] have reported the application of bacterial cellulose for 3D printing of a soft tissue model. Further, alginate/bacterial cellulose composite with *in-situ* synthesized copper nanostructures has been developed by 3D printing and studied their antimicrobial properties for medical application [36].

5.4.4 Emulsion freeze-drying

Emulsion freeze–drying technique is also widely studied in the preparation of porous scaffolds for tissue engineering applications [37,38]. The preparation method of porous scaffold by emulsion freeze–drying is shown in Fig. 5.3D. The porous scaffold of nanocellulose can be obtained from the mixture of homogenized cellulose cross–linked by alginate and freeze–dying process [39]. In addition, Lv et al. [40] have developed silk fibroin/bacterial cellulose scaffolds by the freeze–drying and self–assembling method.

5.4.5 Particle-leaching technique

In particle-leaching technique, the foreign substrates, such as potato starch or paraffin wax were introduced into the BNC matrices during cultivation (Fig. 5.3E) and subsequently extracted out from the matrix by leaching process [24]. For instance, paraffin wax has been utilized to form the highly porous BNC in a bioreactor set up [41]. Khan et al. [29] have used sodium chloride (NaCl) crystals as porogens to fabricate the 3D bacterial cellulose-gelatin composite scaffolds for tissue engineering application.

5.4.6 Solvent casting

Solvent casting is a common technique for preparing porous scaffolds. In this context, Basu et al. [25] have developed inorganic calcium–filled BNC scaffold by solvent casting method for bone tissue regeneration application. Furthermore, polylactic acid/polyethylene glycol copolymer reinforced with BNC nanofibers was developed by solvent casting and particulate leaching for soft tissue engineering application [23].

5.4.7 Irreversible electroporation

Irreversible electroporation is one of the biofabrication methods to generate pores in the BNC scaffold. BNC porous scaffolds produced by the irreversible electroporation (IRE) method could be used in orthopedic applications such as osteochondral defects [42].

5.4.8 Microfluidics

Microfluidics approach has also been reported to develop the porous BNC scaffold. For instance, Yu et al. [43] have developed hollow BNC microspheres porous scaffold by microfluidic process for the wound healing application. Yin et al. [44] have demonstrated the production of porous BNC scaffolds by in the presence of agarose microparticles (porogen).

5.4.9 Laser perforation

In recent days, laser perforation technique is employed to fabricate BNC scaffold for biomedical applications. For instance, various porous BNC scaffolds were reported by laser perforation method for tissue engineering, orthopedic applications, and cartilage implant [26,28,30].

5.5 Bacterial porous scaffold in various biomedical applications

5.5.1 Tissue engineering and wound healing application

Tissue engineering is used to develop the artificial constructs (scaffolds) or biological replacements to repair and retain the normal functions of the organ or tissues from cultured cells [45,46]. Biocompatibility, capability for cell adhesion, migration and proliferation, and low immunogenic potential make BNC a suitable biomaterial for tissue engineering applications [18]. Various studies have reported that BNC helps in survival and differentiation of mesenchymal stem cells [35]. BNC–alginate composite scaffold has supported cell adhesion, migration, and proliferation of human gingival fibroblast cells on the surface [21]. Yin et al. [44] have reported the improved human P1 chondrocytes growth in agarose particle–templated porous BNC scaffold.

Hydroxyapatite (HAp)–bacterial cellulose composite porous scaffolds has improved the osteoblastic differentiation and mineralization in human osteoblast-like cells (SaOs-2 cells) [22]. Bacterial cellulose-gelatin composite scaffolds have promoted the proliferation of fibroblast cells (NIH/3T3), with higher expression of metalloproteases required for extracellular matrix production [29]. Ran et al. [47] have reported that bacterial cellulose/HAp composite has showed better attachment, proliferation, and differentiation of rat bone marrow–derived mesenchymal stem cells (rBMSCs). Electrospinning or electrospraying technique has been widely used to prepare hybrid scaffolds such as drug-eluting nanofibers, hierarchical honeycomb-like structures, and HAp-modified nanofibers [33,48]. For instance, bacterial cellulose/keratin nanofibrous mats modified by a tragacanth gum–conjugated hydrogel were prepared by electrospinning techniques, and the biocompatibility of fibrous composite was determined in fibroblast cells (L929) [33]. Table 5.1 shows the various BNC scaffolds in wound healing/tissue engineering applications.

Table 5.1 BNC scaffolds in wound healing/tissue engineering applications.

S. No.	Material	Preparation method	In-vitro/in-vivo studies	Application	References
1.	Polylactic acid/polyethylene glycol copolymer reinforced with bacterial cellulose nanofibers	Solvent casting and particulate leaching	–	Soft tissue engineering application	Ghalia and Dahman [23]
2.	Bacterial cellulose sponges	Emulsion freeze-drying technique	Fibrous synovium derived mesenchymal stem cells	Tissue engineering application	Gao et al. [37]
3.	Macroporous bacterial cellulose/collagen scaffold	Infrared laser perforation and immersion	Breast cancer cell lines, MDA-MB-231 and MCF-7 cells	Tissue engineering	Xiong et al. [30]
4.	Bacterial cellulose–gelatin composite scaffolds	Casting and leaching approach with porogen	Fibroblast cells (NIH/3T3)	Tissue regeneration	Khan et al. [29]
5.	Bacterial cellulose–alginate composite scaffold	Freeze-drying and Ca^{2+} cross-linking	Human gingival fibroblasts	Tissue engineering	Kirdponpattara et al. [21]
6.	Gelatin–bacterial cellulose sponge scaffold	Freeze-drying and thermal cross-linking	Vero cells	Tissue engineering	Kirdponpattara et al. [49]
7.	Bacterial cellulose/silk fibroin sponge scaffold	Soaking and freeze-drying	V79 cell line	Tissue regeneration	Barud et al. [50]
8.	Hollow bacterial cellulose microspheres porous scaffold	Microfluidics	Human primary epidermal keratinocytes and Sprague-Dawley (SD) rat skin model	Wound healing/tissue regeneration	Yu et al. [43]
9.	Poly(3-hydroxubutyrate-co-4-hydroxubutyrate)/bacterial cellulose composite porous scaffold	Freeze-drying	Chinese Hamster Lung cells	Wound dressing and tissue engineering applications	Zhijiang et al. [51]

(continued)

Table 5.1 Cont'd

S. No.	Material	Preparation method	In-vitro/in-vivo studies	Application	References
10.	Bacterial cellulose/polycaprolactone scaffolds	Electrohydrodynamic (EHD)-3D-bioprinting	Mouse embryo fibroblast (NIH/3T3) cell line	Tissue engineering	Altun et al. [52]
11.	Bacterial cellulose/keratin nanofibrous mats	Electrospinning	Fibroblast cells (L929)	Wound dressing and soft tissue engineering	Azarniya et al. [33]
12.	Bacterial cellulose/poly(methyl methacrylate)	Pressurized gyration	SaOs-2 cell line	Bandage materials	Altun et al. [53]
13.	Bacterial cellulose-poly (acrylic acid) hybrid hydrogels	Immersion and freeze-drying	–	Wound dressings	Chuah et al. [54]
14.	Bacterial cellulose sponges	Immersion, freeze-drying, silanization	RAW 264.7 murine macrophages	Tissue engineering	Frone et al. [55]
15.	Silver nanoparticles/bacterial cellulose	*In-situ* impregnation	Human peripheral blood mononuclear cell	Wound dressings	Tabaii and Emtiazi, [56]
16.	Bacterial cellulose–based hydrogel	Freeze-drying	Human epidermal keratinocytes and dermal fibroblasts and athymic mice	Wound healing	Loh et al. [57]

5.5.2 Bone

Bone is made of osteoblasts, osteocytes, and osteoclasts that provide the mechanical support to the human body by musculoskeletal system, while bone-related diseases or fractures affect their mechanical strength [58]. Osteoporosis has become major problem worldwide with more than 8.9 million fractures annually [25]. Generation of bone extracellular matrix that mimics the native bone with the lamellar structure and providing the microenvironment is a challenging task in bone regeneration [59]. In recent days, the application of bioactive scaffolds in bone tissue engineering is increasing because of their osteoconductive and osteoinductive properties as well as mechanical property and cost-effective production (Table 5.2).

Table 5.2 BNC scaffolds in bone tissue engineering application.

S. No.	Material	Preparation method	In-vitro/in-vivo studies	Application	References
1.	HAp coated bacterial cellulose scaffold	Coating/immersing	Rat calvarial defect model	Bone tissue engineering	Ahn et al. [60]
2.	Inorganic calcium–filled bacterial cellulose scaffold	Solvent casting method	Human diploid fibroblast (Lep–3) cells	Bone tissue regeneration	[25]
3.	Bacterial cellulose scaffold loaded with fisetin	Freeze-drying method	Bone marrow mesenchymal stem cells	Osteogenesis	Kheiry et al. [61]
4.	Bacterial cellulose/silk fibroin HAp composite scaffold	In-situ hybridization method (soaking and self-assembly)	Osteoprogenitor cells (MC3T3-E1)	Bone tissue engineering	Jiang et al. [62]
5.	Biomimetic bacterial cellulose bilayer hydrogel with HAp scaffold	Cross-linking method (bionics principle)	New Zealand White Rabbits osteochondral defect model	Osteochondral tissue engineering	Zhu et al. [63]
6.	Bacterial cellulose membrane conjugated osteopontin	Freeze-drying method	Human periodontal ligament stem cells	Bone tissue regeneration	Klinthoopthamrong et al. [31]
7.	Bacterial cellulose-collagen composite scaffolds	Freeze-drying method	Human mesenchymal stem cells and immunodeficient mice	Osteogenic effect	Noh et al. [64]
8.	HAp–bacterial cellulose composite scaffolds	Powder and shredded agar technique	Human osteoblast-like cells (SaOs–2 cells)	Bone tissue engineering	Bayir et al. [22]
9.	Silk fibroin/BNC nanoribbon composite scaffolds	Multistaged freeze-drying technique	In-vitro bone bioactivity assay	Bone regeneration	Chen et al. [59]

(continued)

Table 5.2 (Cont'd)

S. No.	Material	Preparation method	*In-vitro*/*in-vivo* studies	Application	References
10.	HAp/BNC nanocomposite scaffolds	Biomimetics	Stromal cells derived from human bone marrow (hBMSCs)	Bone tissue engineering	Fang et al. [65]
11.	Micro–nano structured porous bacterial cellulose scaffold with gelatin and HAp coating	Emulsion freeze-drying technique	hBMSCs, nude mice and New Zealand rabbits	Bone tissue engineering	Huang et al. [38]
12.	Bacterial cellulose/HAp composite	Immersion and freeze-drying	rBMSCs	Bone tissue engineering	Ran et al. [47]
13.	Microporous bacterial cellulose	Paraffin wax (porogen)	Osteoprogenitor cells (MC3T3-E1)	Bone regeneration	Zaborowska et al. [66]
14.	Bacterial cellulose-reinforced calcium phosphate cement composite	Molding	Mouse embryonic osteoblast precursor cells (MC3T3-E1)	Bone replacement	Zhang et al. [67]
15.	Bacterial cellulose and collagen porous microspheres	Template method combined with reverse–phase suspension	Mouse embryonic osteoblast precursor cells (MC3T3-E1)	Bone tissue engineering	Zhang et al. [32]
16.	Titanium–aluminum–niobium bone scaffold implants with bacterial cellulose	Selective laser melting and immersion	Osteoblast and fibroblast cell cultures	Orthopedic applications	Dydak et al. [26]

The human bone matrix mainly consists of collagen nanofibers HAp ($Ca_{10}(PO_4)_6(OH)_2$) [38]. HAp has excellent biocompatibility, osteoconductive, or osteoinductive properties, while their application in bone defects is limited due to its brittle nature and lower mechanical stability [65]. To develop a mimic of the natural bone, many researchers have studied HAp in combination with organic nanofibers for suitable composites [38]. HAp/BNC nanocomposite scaffolds have promoted the adhesion, proliferation, and differentiation of stromal cells derived from human bone marrow (hBMSCs) [65]. Huang et al. [38] have reported that micro–nano structured porous BNC scaffold coated with gelatin and HAp improved the osteoinductive properties of hBMSCs and promoted the bone formation in nude mice and rabbits. Further, Jiang et al. [62] have studied the use of BNC/silk fibroin/HAp composite scaffold in osteoprogenitor cells (MC3T3-E1) for bone regeneration.

Microporous bacterial cellulose scaffolds seeded with osteoprogenitor cells (MC3T3-E1) have shown denser mineral deposits on surfaces, which revealed their promising nature in bone tissue engineering applications [66]. In a study, Basu et al. [25] have found that hydrogel scaffolds prepared with inorganic calcium–filled bacterial cellulose has showed possible application in metaphyseal bone regenerations. BNC scaffold loaded with fisetin promoted the osteogenic differentiation from bone marrow–derived mesenchymal stem cells by the adhesion, growth, and biosynthesis of bone-specific matrix [61]. Furthermore, osteochondral regeneration was studied in New Zealand white rabbits by using biomimetic bacterial cellulose bilayer hydrogel with HAp scaffold [63]. Klinthoopthamrong et al. [31] have studied the BNC membrane conjugated osteopontin by *in-vitro* calcification assay in human periodontal ligament stem cells for bone tissue regeneration.

5.5.3 Cartilage

Cartilage damage is one of the common traumas in developed countries. Chondrocytes are the major components necessary for new cartilage generation during cartilage damage [68]. Articular cartilage is a thin layer of connective tissue involved in the prevention of friction within synovial joints [69]. As cartilage lacks the blood vessels, it limits the self-repair by chondrocytes. In this context, cartilage tissue engineering plays a major role in cell growth and differentiation. In recent days, BNC-based scaffolds are widely used for potential cartilage tissue regeneration (Table 5.3). Bacterial cellulose/silk fibroin double network hydrogel has been evaluated for artificial cartilage showed good tensile strength and biocompatibility in preosteoblast cells (MC3T3-E1) [70]. In a study, agarose particle–templated porous BNC has been employed in cartilage growth using human P1 chondrocytes [44]. Besides, Horbert et al. [28] have demonstrated the laser perforated and cell-seeded BNC cartilage implant in neonatal articular chondrocytes of German Holstein–Friesian Cattle by bovine cartilage punch model. Jacek et al. [71] have revealed the cell viability and glycosaminoglycan synthesis by bacterial cellulose in mouse chondrogenic cell line (ATDC5) and RBL-2H3 mast cells degranulation. Methacrylated gelatin/bacterial cellulose composite promotes cell proliferation and

Table 5.3 BNC scaffolds in cartilage tissue engineering application.

S. No.	Material	Preparation method	In-vitro/in-vivo studies	Application	References
1.	Agarose particle–templated porous bacterial cellulose	Agarose particle porogen—microfluidic approach	Human P1 chondrocytes	Cartilage growth	Yin et al. [44]
2.	Bacterial cellulose/chitosan porous scaffold	Freezing and lyophilization	–	Cartilage tissue engineering	Nge et al. [46]
3.	Engineered porous BNC scaffolds	Paraffin wax (porogen)	Neonatal articular chondrocytes	Regenerative cartilage applications	Andersson et al. [41]
4.	Laser perforation and cell-seeded BNC	Laser perforation	German Holstein–Friesian Cattle	Cartilage implant	Horbert et al. [28]
5.	Methacrylated gelatin/bacterial cellulose composite	Immersion and photo cross-linking	Human articular chondrocytes	Cartilage tissue engineering	Gu et al. [69]
6.	Bacterial cellulose/silk fibroin double network hydrogel	Immersion	Preosteoblast cells (MC3T3-E1)	Artificial cartilage	Wang et al. (2020a)
7.	Dialdehyde BNC with DL-allo-hydroxylysine and chitosan composite	Cross-linking, emulsion freeze-drying technique	Bone marrow–derived mesenchymal stem cells and Sprague–Dawley rats	Cartilage regeneration	Wang et al. [72]
8.	Bacterial cellulose–modified lotus root starch and HAp scaffold	Agarose porogen templating	Human P1 chondrocytes	Articular cartilage tissue engineering	Wu et al. [73]

maintains the phenotype of human articular chondrocytes which could be used in cartilage tissue engineering [69]. Wang et al. [72] have studied the dialdehyde BNC with
dextro and levo (DL)-allo-hydroxylysine and chitosan composite by which functional
microtissue was developed and demonstrated the cartilage repair through implantation
in a rat femoral trochlear cartilage defect model.

5.5.4 Urothelium/urethral regeneration

The hollows organs or structures such as esophagus, intestines, blood vessels, urinary bladder,
and ureter comprise epithelial and/or endothelial cells in the lumen enclosed by muscles
and connective tissues [45]. Congenital and acquired pathological condition such as bladder

calculi, sepsis and fistulas can affect the normal functioning of urethra [40]. The multifaceted process of urethral regeneration can be achieved by the growth and proliferation of endothelial, muscle, and epithelial cells [74]. The application of the conventional tissue-engineered grafts limits the angiogenesis and epithelialization in urethral regeneration. The normal functions of urethra can be restored by developing a biomaterial mimic as the native urethral structure with extracellular matrix through the tissue engineering approaches (Table 5.4). In

Table 5.4 BNC scaffolds in urethral regeneration and angiogenesis.

S. No.	Material	Preparation method	In-vitro/in-vivo studies	Application	References
1.	Porous bacterial cellulose	Tubular fermentation with sterile paraffin particles	Urine-derived stem cells and athymic mice	Urinary conduit	Bodin et al. [75]
2.	Porous bacterial cellulose 3D scaffold	Fermentation-gelatin sponge interference	Lingual keratinocytes, New Zealand white male rabbits	Urethral tissue engineering	Huang et al. [76]
3.	Bacterial cellulose/ potato starch nanofibrous scaffolds	Culture in starch-enriched medium	Lingual keratinocytes and muscle cells and beagle dog urethral defect models	Urinary tract reconstruction	Lv et al. [45]
4.	Silk fibroin/ bacterial cellulose scaffolds	Freeze-drying and self-assembling method	Lingual keratinocytes and muscle cells and beagle dog urethral defect models	Urethral reconstruction	Lv et al. [40]
5.	Bacterial cellulose and bladder acellular matrix scaffold	TEMPO-oxidation, freeze-drying	Human umbilical vein endothelial cells (HUVEC) and New Zealand white rabbits urethral defect model	Urethral regeneration	Wang et al. [74]
6.	Bacterial cellulose/ gelatin porous scaffold loaded with vascular endothelial growth factor (VEGF)-silk fibroin nanoparticles	Freezing lyophilization immersion technique	Pig iliac endothelium cells and Dog skin defect model	Angiogenesis for tissue regeneration	Wang et al. [77]
7.	VEGF-loaded heparinized bacterial cellulose/ gelatin scaffold	Freeze-drying immersion technique	Endothelial cells ex-ovo and female rabbits	Angiogenesis for tissue regeneration	Wang et al. [78]

this regard, porous BNC scaffolds seeded with human urine–derived stem cells were indiced to differentiate into urothelial and smooth muscle cells that develops tissue-engineered urinary conduit for urinary diversion application [75]. The application of BNC/potato starch nanofibrous scaffolds has provided preclinical evidence of the urinary tract reconstruction in the beagle dog urethral defect models [45]. In addition, silk fibroin/bacterial cellulose scaffolds supported keratinocytes and muscle cells have evidenced their promising role in urethral regeneration in beagle dogs [40]. Moreover, BNC and bladder acellular matrix scaffold has supported the angiogenesis, epithelialization, and quicker urethral regeneration in a rabbit urethral defect model ([4]).

5.5.5 Angiogenesis or vascular networks

Vascular network supplies oxygen and nutrients to the cells as well as removing the metabolites from the body [79]. The luminal side of blood vessels is lined by endothelial cells that maintain the integrity through the regulation of permeability, vascular tone, inflammation, and prevention of thrombus formation [80]. The pathological conditions in blood vessels lead to atherosclerosis, heart disease, metastasis, etc. In tissue engineering, establishment of the native vascular networks and maintaining them in long term is a challenging part. Several attempts have been made for the development of vascular tissues such as endothelial cells–based biomimetic vascular scaffolds (Table 5.4). For instance, Wang et al. [77] have studied that BNC/gelatin porous scaffold loaded with VEGF-silk fibroin nanoparticles could improve the angiogenesis in pig iliac endothelium cells and beagle dog skin defect model for tissue regeneration. BNC and bladder acellular matrix scaffold have supported the angiogenesis by aiding HUVEC and capillary tube formation [74]. Lei et al. [79] have developed a biodegradable hierarchical microchannel network using bacterial cellulose and demonstrated their perfusable and permeable ability in vasculatures. Further, *in-vitro* and *in-vivo* studies has showed that the construct maintained normal metabolic functions, angiogenesis, and tissue integration in heart cells. Moreover, Wang et al. [78] have developed VEGF-loaded heparinized BNC/gelatin scaffold and demonstrated the angiogenesis for tissue regeneration in chick chorioallantoic membrane assay and female rabbit incision model.

5.5.6 3D bioprinting

In recent years, natural material–based 3D bioprinting techniques have gained significant attention in tissue engineering to construct the tissue or organ for replacing the lost or damaged tissues. BNC has promoted the fidelity and mechanical properties of composite scaffolds for 3D printing of soft tissue model [35]. Alginate/BNC composite with *in-situ* synthesized copper nanostructures has been developed by 3D printing to study their antimicrobial behavior for medical application [36]. Altun et al. [52] have developed BNC/polycaprolactone scaffolds by electrohydrodynamic 3D bioprinting that exhibited biocompatibility through the proliferation of mouse embryo fibroblast (NIH/3T3) cell line. While, Lei et al. [79] has developed a biodegradable perfusable hierarchical microchannel networks by 3D printing technology using bacterial cellulose and demonstrated their perfusable and permeable ability in vasculatures both *in vitro* and *in vivo*. Further, Wei et al. [81] produced the green nanocomposite printing ink to develop 3D printed hydrogel using 2,2,6,6-tetramethylpiperidinyl-1-oxyl (TEMPO) oxidized BNC, sodium alginate, and laponite nanoclay and demonstrated that the developed hydrogel could be used in drug release by their structural stability and sustained release behavior of protein.

5.5.7 Drug delivery

Highly selective, biocompatible, and controlled drug release is very important in drug delivery systems. However, low drug adsorption, restricted bioactivity, and release

Table 5.5 Bacterial cellulose scaffolds in other biomedical application.

S. No.	Material	Preparation method	In-vitro/in-vivo studies	Application	References
1.	Bacterial cellulose and chitosan scaffolds	One-step *ex-situ* solution impregnation	Ovarian cancer cells, A2780 cell lines	Diagnosis of ovarian cancer *in vivo*	Ul-Islam et al. [82]
2.	Oxidized BNC scaffolds	Oxidation using sodium periodate and lyophilization	Schwann cells and red blood cells	Peripheral nerve repair	Hou et al. [83]
3.	Collagen/bacterial cellulose porous microspheres	Template method combined with inverse suspension regeneration	Mouse MC3T3-E1 cells	Drug adsorption and release	Zhang et al. [84]
4.	Polyethylenimine-functionalized bacterial cellulose	TEMPO-oxidation and freeze-drying	Mouse breast tumor cell (4T1), liver hepatocellular carcinoma (HepG2), kidney epithelial cell (293T), and HUVEC	Drug carrier	Chen et al. [85]
5.	Cellulose-binding biosensor scaffolds	Immersion	Human colon cancer cells and primary mouse intestinal organoid model	pH- and Ca^{2+}- sensitive biosensor	O'Donnell et al. [86]
6.	Graphene-oxide–incorporated cellulose scaffolds	Electrospinning and *in-situ* biosynthesis	Human breast cancer cell line (MCF-7)	Drug screening and Cancer biology	Wan et al. [34]
7.	Periodate-oxidized bacterial cellulose	Freeze-drying	Mouse fibroblasts (L929)	Drug delivery in dental therapies	Weyell et al. [87]
8.	Famotidine-loaded bacterial cellulose	Immersion, oven and freeze drying	*In vitro*	Drug delivery	Badshah et al. [88]

specificity in the stomach or intestine are the challenging parts in developing sustainable drug effect [85]. In a view to overcome these pitfalls, BNC-based material is widely reported nowadays (Table 5.5). For example, Zhang et al. [84] have reported the controlled drug adsorption and release technology by the application of collagen/BNC porous microspheres. In addition, Chen et al. [85] have reported the maximum adsorption capacity as 345 mg/g of aspirin at equilibrium with controlled drug release by the polyethylenimine-functionalized bacterial cellulose.

5.5.8 Nerve injury repair

Peripheral nerve injury leads to delayed healing which became serious medical problem and it can be restored by the structural and functional nerve grafts [83]. Oxidized BNC scaffolds developed by Hou et al. [83] have showed potential peripheral nerve repair ability with interconnected pores and mechanical and biodegradable properties.

5.5.9 Cancer diagnosis

Ul-Islam et al. [82] have found that bacterial cellulose and chitosan scaffolds could be used for the diagnosis of ovarian cancer as revealed by the deep infiltration of ovarian cancer cells (A2780 cells) into the matrix and decreased mRNA expression of Notch receptors.

5.6 Conclusion and future perspectives

Bacterial porous scaffolds are the potential biomaterial with tremendous advantages in various biomedical applications, due to its unique properties and biocompatible nature. Different fabrication approaches with other biomaterials, nanocomposites, polymers, etc., have resulted in the development of novel scaffolds, which provides insights into the future application of BNC in medical fields. Scaffold preparation and *in-situ* and *ex-situ* modifications are based on the medical applications, while appropriate monitoring is necessary for the timely biodegradation of BNC implant materials. Overall, the potential of BNC in a wide range biomedical applications is continuously being explored.

References

[1] A.J. Brown, XLIII.—On an acetic ferment which forms cellulose, J. Chem. Soc., Trans. 49 (1886) 432–439.
[2] M. Iguchi, S. Yamanaka, A. Budhiono, Bacterial cellulose—a masterpiece of nature's arts, J. Mater. Sci. 35 (2) (2000) 261–270.
[3] S. Bielecki, A. Krystynowicz, M. Turkiewicz, H. Kalinowska, Bacterial cellulose. In: E.J. Vandamme, S. De Baets, A. Steinbüchel (Eds.), Biopolymers, Wiley-VCH, Weinheim, 5 (2005) 37–90. https://doi.org/10.1002/3527600035.bpol5003.
[4] J. Wang, J. Tavakoli, Y. Tang, Bacterial cellulose production, properties and applications with different culture methods–a review, Carbohydr. Polym. 219 (2019) 63–76.

[5] Y. Huang, C. Zhu, J. Yang, Y. Nie, C. Chen, D. Sun, Recent advances in bacterial cellulose, Cellulose 21 (1) (2014) 1–30.

[6] P. Ross, R. Mayer, M. Benziman, Cellulose biosynthesis and function in bacteria, Microbiol. Mol. Biol. Rev. 55 (1) (1991) 35–58.

[7] N. Shah, M. Ul-Islam, W.A. Khattak, J.K. Park, Overview of bacterial cellulose composites: a multi-purpose advanced material, Carbohydr. Polym. 98 (2) (2013) 1585–1598.

[8] S.M. Keshk, Bacterial cellulose production and its industrial applications, J. Bioprocess. Biotech. 4 (150) (2014) 2.

[9] W. Liu, H. Du, M. Zhang, K. Liu, H. Liu, H. Xie, X. Zhang, C. Si, Bacterial cellulose-based composite scaffolds for biomedical applications: a review, ACS Sustain. Chem. Eng. 8 (20) (2020) 7536–7562.

[10] P. Lv, Y. Yao, D. Li, H. Zhou, M.A. Naeem, Q. Feng, J. Huang, Y. Cai, Q. Wei, Self-assembly of nitrogen-doped carbon dots anchored on bacterial cellulose and their application in iron ion detection, Carbohydr. Polym. 172 (2017) 93–101.

[11] R. Mangayil, S. Rajala, A. Pammo, E. Sarlin, J. Luo, V. Santala, M. Karp, S. Tuukkanen, Engineering and characterization of bacterial nanocellulose films as low cost and flexible sensor material, ACS Appl. Mater. Interfaces 9 (22) (2017) 19048–19056.

[12] Z. Shi, Y. Zhang, G.O. Phillips, G. Yang, Utilization of bacterial cellulose in food, Food Hydrocoll. 35 (2014) 539–545.

[13] Y. Ge, S. Chen, J. Yang, B. Wang, H. Wang, Color-tunable luminescent CdTe quantum dot membranes based on bacterial cellulose (BC) and application in ion detection, RSC Adv. 5 (69) (2015) 55756–55761.

[14] H. Ullah, H.A. Santos, T. Khan, Applications of bacterial cellulose in food, cosmetics and drug delivery, Cellulose 23 (4) (2016) 2291–2314.

[15] Z. Yang, L. Ren, L. Jin, L. Huang, Y. He, J. Tang, W. Yang, H. Wang, In-situ functionalization of poly (m-phenylenediamine) nanoparticles on bacterial cellulose for chromium removal, Chem. Eng. J. 344 (2018) 441–452.

[16] M.L. Cacicedo, M.C. Castro, I. Servetas, L. Bosnea, K. Boura, P. Tsafrakidou, A. Dima, A. Terpou, A. Koutinas, G.R. Castro, Progress in bacterial cellulose matrices for biotechnological applications, Bioresour. Technol. 213 (2016) 172–180.

[17] D.R. Ruka, G.P. Simon, K.M. Dean, Bacterial cellulose and its use in renewable composites, in: V.K. Thakur (Ed.), Nanocellulose Polymer Nanocomposites: Fundamentals and Applications, John Wiley and sons, NJ, 2014, pp. 89–130.

[18] K. Qiu, A.N. Netravali, A review of fabrication and applications of bacterial cellulose based nanocomposites, Polym. Rev. 54 (4) (2014) 598–626.

[19] M. Moniri, A.B. Moghaddam, S. Azizi, R.A. Rahim, S.W. Zuhainis, M. Navaderi, R. Mohamad, In vitro molecular study of wound healing using biosynthesized bacteria nanocellulose/silver nanocomposite assisted by bioinformatics databases, Int. J. Nanomed. 13 (2018) 5097.

[20] A.F. Jozala, L.C. de Lencastre-Novaes, A.M. Lopes, V. de Carvalho Santos-Ebinuma, P.G. Mazzola, A. Pessoa-Jr, D. Grotto, M. Gerenutti, M.V. Chaud, Bacterial nanocellulose production and application: a 10-year overview, Appl. Microbiol. Biotechnol. 100 (5) (2016) 2063–2072.

[21] S. Kirdponpattara, A. Khamkeaw, N. Sanchavanakit, P. Pavasant, M. Phisalaphong, Structural modification and characterization of bacterial cellulose–alginate composite scaffolds for tissue engineering, Carbohydr. Polym. 132 (2015) 146–155.

[22] E. Bayir, E. Bilgi, E.E. Hames, A. Sendemir, Production of hydroxyapatite–bacterial cellulose composite scaffolds with enhanced pore diameters for bone tissue engineering applications, Cellulose 26 (18) (2019) 9803–9817.

[23] M.A. Ghalia, Y. Dahman, Fabrication and enhanced mechanical properties of porous PLA/PEG copolymer reinforced with bacterial cellulose nanofibers for soft tissue engineering applications, Polym. Test. 61 (2017) 114–131.

[24] Z. Ashrafi, L. Lucia, W. Krause, Bioengineering tunable porosity in bacterial nanocellulose matrices, Soft Matter 15 (45) (2019) 9359–9367.

[25] P. Basu, N. Saha, P. Saha, Inorganic calcium filled bacterial cellulose based hydrogel scaffold: novel biomaterial for bone tissue regeneration, Int. J. Polym. Mater. Polym. Biomater. 68 (1-3) (2019) 134–144.

[26] K. Dydak, A. Junka, P. Szymczyk, G. Chodaczek, M. Toporkiewicz, K. Fijałkowski, B. Dudek, M. Bartoszewicz, Development and biological evaluation of Ti6Al7Nb scaffold implants coated with gentamycin-saturated bacterial cellulose biomaterial, PloS One 13 (10) (2018) e0205205.

[27] L. Fu, J. Zhang, G. Yang, Present status and applications of bacterial cellulose-based materials for skin tissue repair, Carbohydr. Polym. 92 (2) (2013) 1432–1442.

[28] V. Horbert, J. Boettcher, P. Foehr, F. Kramer, U. Udhardt, M. Bungartz, O. Brinkmann, R.H. Burgkart, D.O. Klemm, R.W. Kinne, Laser perforation and cell seeding improve bacterial nanocellulose as a potential cartilage implant in the in vitro cartilage punch model, Cellulose 26 (1) (2019) 647–664.

[29] S. Khan, M. Ul-Islam, M. Ikram, M.W. Ullah, M. Israr, F. Subhan, Y. Kim, J.H. Jang, S. Yoon, J.K. Park, Three-dimensionally microporous and highly biocompatible bacterial cellulose–gelatin composite scaffolds for tissue engineering applications, RSC Adv. 6 (112) (2016) 110840–110849.

[30] G. Xiong, H. Luo, C. Zhang, Y. Zhu, Y. Wan, Enhanced biological behavior of bacterial cellulose scaffold by creation of macropores and surface immobilization of collagen, Macromol. Res. 23 (8) (2015) 734–740.

[31] N. Klinthoopthamrong, D. Chaikiawkeaw, W. Phoolcharoen, K. Rattanapisit, P. Kaewpungsup, P. Pavasant, V.P. Hoven, Bacterial cellulose membrane conjugated with plant-derived osteopontin: preparation and its potential for bone tissue regeneration, Int. J. Biol. Macromol. 149 (2020) 51–59.

[32] W. Zhang, X.C. Wang, X.Y. Li, F. Jiang, A 3D porous microsphere with multistage structure and component based on bacterial cellulose and collagen for bone tissue engineering, Carbohyd. Polym. 236 (2020) 116043.

[33] A. Azarniya, E. Tamjid, N. Eslahi, A. Simchi, Modification of bacterial cellulose/keratin nanofibrous mats by a tragacanth gum-conjugated hydrogel for wound healing, Int. J. Biol. Macromol. 134 (2019) 280–289.

[34] Y. Wan, Z. Lin, Q. Zhang, D. Gan, M. Gama, J. Tu, H. Luo, Incorporating graphene oxide into bio-mimetic nano-microfibrous cellulose scaffolds for enhanced breast cancer cell behavior, Cellulose 27 (2020) 4471–4485.

[35] L. Huang, X. Du, S. Fan, G. Yang, H. Shao, D. Li, C. Cao, Y. Zhu, M. Zhu, Y. Zhang, Bacterial cellulose nanofibers promote stress and fidelity of 3D-printed silk based hydrogel scaffold with hierarchical pores, Carbohydr. Polym. 221 (2019) 146–156.

[36] E. Gutierrez, P.A. Burdiles, F. Quero, P. Palma, F. Olate-Moya, H. Palza, 3D printing of antimicrobial alginate/bacterial-cellulose composite hydrogels by incorporating copper nanostructures, ACS Biomater. Sci. Eng. 5 (11) (2019) 6290–6299.

[37] C. Gao, Y. Wan, C. Yang, K. Dai, T. Tang, H. Luo, J. Wang, Preparation and characterization of bacterial cellulose sponge with hierarchical pore structure as tissue engineering scaffold, J. Porous Mater. 18 (2) (2011) 139–145.

[38] Y. Huang, J. Wang, F. Yang, Y. Shao, X. Zhang, K. Dai, Modification and evaluation of micro-nano structured porous bacterial cellulose scaffold for bone tissue engineering, Mater. Sci. Eng. C, Mater. Biol. Appl. 75 (2017) 1034.

[39] P. Krontiras, P. Gatenholm, D.A. Hägg, Adipogenic differentiation of stem cells in three-dimensional porous bacterial nanocellulose scaffolds, J. Biomed. Mater. Res. B: Appl. Biomater. 103 (1) (2015) 195–203.

[40] X. Lv, C. Feng, Y. Liu, X. Peng, S. Chen, D. Xiao, H. Wang, Z. Li, Y. Xu, M. Lu, A smart bilayered scaffold supporting keratinocytes and muscle cells in micro/nano-scale for urethral reconstruction, Theranostics 8 (11) (2018) 3153.

[41] J. Andersson, H. Stenhamre, H. Bäckdahl, P. Gatenholm, Behavior of human chondrocytes in engineered porous bacterial cellulose scaffolds, J. Biomed. Mater. Res. A 94 (4) (2010) 1124–1132.

[42] A. Baah-Dwomoh, A. Rolong, P. Gatenholm, R.V. Davalos, The feasibility of using irreversible electroporation to introduce pores in bacterial cellulose scaffolds for tissue engineering, Appl. Microbiol. Biotechnol. 99 (11) (2015) 4785–4794.

[43] J. Yu, T.R. Huang, Z.H. Lim, R. Luo, R.R. Pasula, L.D. Liao, S. Lim, C.H. Chen, Production of hollow bacterial cellulose microspheres using microfluidics to form an injectable porous scaffold for wound healing, Adv. Healthc. Mater. 5 (23) (2016) 2983–2992.

[44] N. Yin, M.D. Stilwell, T.M. Santos, H. Wang, D.B. Weibel, Agarose particle-templated porous bacterial cellulose and its application in cartilage growth in vitro, Acta Biomater. 12 (2015) 129–138.

[45] X. Lv, J. Yang, C. Feng, Z. Li, S. Chen, M. Xie, J. Huang, H. Li, H. Wang, Y. Xu, Bacterial cellulose-based biomimetic nanofibrous scaffold with muscle cells for hollow organ tissue engineering, ACS Biomater. Sci. Eng. 2 (1) (2016) 19–29.

[46] T.T. Nge, M. Nogi, H. Yano, J. Sugiyama, Microstructure and mechanical properties of bacterial cellulose/chitosan porous scaffold, Cellulose 17 (2) (2010) 349–363.

[47] J. Ran, P. Jiang, S. Liu, G. Sun, P. Yan, X. Shen, H. Tong, Constructing multi-component organic/inorganic composite bacterial cellulose-gelatin/hydroxyapatite double-network scaffold platform for stem cell-mediated bone tissue engineering, Mater. Sci. Eng. C 78 (2017) 130–140.

[48] N. Lavielle, A. Hébraud, G. Schlatter, L. Thöny-Meyer, R.M. Rossi, A.M. Popa, Simultaneous electrospinning and electrospraying: a straightforward approach for fabricating hierarchically structured composite membranes, ACS Appl. Mater. Interfaces 5 (20) (2013) 10090–10097.

[49] S. Kirdponpattara, M. Phisalaphong, S. Kongruang, Gelatin-bacterial cellulose composite sponges thermally cross-linked with glucose for tissue engineering applications, Carbohydr. Polym. 177 (2017) 361–368.

[50] H.O. Barud, H.D.S. Barud, M. Cavicchioli, T.S. do Amaral, O.B. de Oliveira Junior, D.M. Santos, A.L.D.O.A. Petersen, F. Celes, V.M. Borges, C.I. de Oliveira, P.F. de Oliveira, Preparation and characterization of a bacterial cellulose/silk fibroin sponge scaffold for tissue regeneration, Carbohydr. Polym. 128 (2015) 41–51.

[51] C. Zhijiang, H. Chengwei, Y. Guang, Poly (3-hydroxubutyrate-co-4-hydroxubutyrate)/bacterial cellulose composite porous scaffold: preparation, characterization and biocompatibility evaluation, Carbohydr. Polym. 87 (2) (2012) 1073–1080.

[52] E. Altun, N. Ekren, S.E. Kuruca, O. Gunduz, Cell studies on electrohydrodynamic (EHD)-3D-bioprinted bacterial cellulose/polycaprolactone scaffolds for tissue engineering, Mater. Lett. 234 (2019) 163–167.

[53] E. Altun, M.O. Aydogdu, F. Koc, M. Crabbe-Mann, F. Brako, R. Kaur-Matharu, G. Ozen, S.E. Kuruca, U. Edirisinghe, O. Gunduz, M. Edirisinghe, Novel making of bacterial cellulose blended polymeric fiber bandages, Macromol. Mater. Eng. 303 (3) (2018) 1700607.

[54] C. Chuah, J. Wang, J. Tavakoli, Y. Tang, Novel bacterial cellulose-poly (acrylic acid) hybrid hydrogels with controllable antimicrobial ability as dressings for chronic wounds, Polymers 10 (12) (2018) 1323.

[55] A.N. Frone, D.M. Panaitescu, C.A. Nicolae, A.R. Gabor, R. Trusca, A. Casarica, P.O. Stanescu, D.D. Baciu, A. Salageanu, Bacterial cellulose sponges obtained with green cross-linkers for tissue engineering, Mater. Sci. Eng. C 110 (2020) 110740.

[56] M.J. Tabaii, G. Emtiazi, Transparent nontoxic antibacterial wound dressing based on silver nano particle/bacterial cellulose nano composite synthesized in the presence of tripolyphosphate, J. Drug Deliv. Sci. Technol. 44 (2018) 244–253.

[57] E.Y.X. Loh, N. Mohamad, M.B. Fauzi, M.H. Ng, S.F. Ng, M.C.I.M. Amin, Development of a bacterial cellulose-based hydrogel cell carrier containing keratinocytes and fibroblasts for full-thickness wound healing, Sci. Rep. 8 (1) (2018) 1–12.

[58] P. Basu, N. Saha, P. Saha, Swelling and rheological study of calcium phosphate filled bacterial cellulose-based hydrogel scaffold, J. Appl. Polym. Sci. 137 (14) (2020) 48522.

[59] J. Chen, A. Zhuang, H. Shao, X. Hu, Y. Zhang, Robust silk fibroin/bacterial cellulose nanoribbon composite scaffolds with radial lamellae and intercalation structure for bone regeneration, J. Mater. Chem. B 5 (20) (2017) 3640–3650.

[60] S.J. Ahn, Y.M. Shin, S.E. Kim, S.I. Jeong, J.O. Jeong, J.S. Park, H.J. Gwon, Y.C. Nho, S.S. Kang, C.Y. Kim, J.B. Huh, Characterization of hydroxyapatite-coated bacterial cellulose scaffold for bone tissue engineering, Biotechnol. Bioprocess Eng. 20 (5) (2015) 948–955.

[61] E.V. Kheiry, K. Parivar, J. Baharara, B.S.F. Bazzaz, A. Iranbakhsh, The osteogenesis of bacterial cellulose scaffold loaded with fisetin, Iran. J. Basic Med. Sci. 21 (9) (2018) 965.

[62] P. Jiang, J. Ran, P. Yan, L. Zheng, X. Shen, H. Tong, Rational design of a high-strength bone scaffold platform based on in situ hybridization of bacterial cellulose/nano-hydroxyapatite framework and silk fibroin reinforcing phase, J. Biomater. Sci. Polymer edition 29 (2) (2018) 107–124.

[63] X. Zhu, T. Chen, B. Feng, J. Weng, K. Duan, J. Wang, X. Lu, Biomimetic bacterial cellulose-enhanced double-network hydrogel with excellent mechanical properties applied for the osteochondral defect repair, ACS Biomater. Sci. Eng. 4 (10) (2018) 3534–3544.

[64] Y.K. Noh, A.D.S. Da Costa, Y.S. Park, P. Du, I.H. Kim, K. Park, Fabrication of bacterial cellulose-collagen composite scaffolds and their osteogenic effect on human mesenchymal stem cells, Carbohydr. Polym. 219 (2019) 210–218.

[65] B. Fang, Y.Z. Wan, T.T. Tang, C. Gao, K.R. Dai, Proliferation and osteoblastic differentiation of human bone marrow stromal cells on hydroxyapatite/bacterial cellulose nanocomposite scaffolds, Tissue Eng. A 15 (5) (2009) 1091–1098.

[66] M. Zaborowska, A. Bodin, H. Bäckdahl, J. Popp, A. Goldstein, P. Gatenholm, Microporous bacterial cellulose as a potential scaffold for bone regeneration, Acta Biomater. 6 (7) (2010) 2540–2547.

[67] Q. Zhang, Z. Lei, M. Peng, M. Zhong, Y. Wan, H. Luo, Enhancement of mechanical and biological properties of calcium phosphate bone cement by incorporating bacterial cellulose, Mater. Technol. 34 (13) (2019) 800–806.

[68] S. Gea, R.M. Sari, A.F. Piliang, D.P. Indrawan, Y.A. Hutapea, Study of bacterial cellulose as scaffold on cartilage tissue engineering, AIP Conference Proceedings, 2049, AIP Publishing LLC, 2018 020061.

[69] L. Gu, T. Li, X. Song, X. Yang, S. Li, L. Chen, P. Liu, X. Gong, C. Chen, L. Sun, Preparation and characterization of methacrylated gelatin/bacterial cellulose composite hydrogels for cartilage tissue engineering, Regen. Biomater. 7 (2) (2020) 195–202.

[70] K. Wang, Q. Ma, Y.M. Zhang, G.T. Han, C.X. Qu, S.D. Wang, Preparation of bacterial cellulose/silk fibroin double-network hydrogel with high mechanical strength and biocompatibility for artificial cartilage, Cellulose 27 (4) (2020) 1845–1852.

[71] P. Jacek, M. Szustak, K. Kubiak, E. Gendaszewska-Darmach, K. Ludwicka, S. Bielecki, Scaffolds for chondrogenic cells cultivation prepared from bacterial cellulose with relaxed fibers structure induced genetically, Nanomaterials 8 (12) (2018) 1066.

[72] Y. Wang, X. Yuan, K. Yu, H. Meng, Y. Zheng, J. Peng, S. Lu, X. Liu, Y. Xie, K. Qiao, Fabrication of nanofibrous microcarriers mimicking extracellular matrix for functional microtissue formation and cartilage regeneration, Biomaterials 171 (2018) 118–132.

[73] J. Wu, N. Yin, S. Chen, D.B. Weibel, H. Wang, Simultaneous 3D cell distribution and bioactivity enhancement of bacterial cellulose (BC) scaffold for articular cartilage tissue engineering, Cellulose 26 (4) (2019) 2513–2528.

[74] B. Wang, X. Lv, Z. Li, M. Zhang, J. Yao, N. Sheng, M. Lu, H. Wang, S. Chen, Urethra-inspired biomimetic scaffold: a therapeutic strategy to promote angiogenesis for urethral regeneration in a rabbit model, Acta Biomater. 102 (2020) 247–258.

[75] A. Bodin, S. Bharadwaj, S. Wu, P. Gatenholm, A. Atala, Y. Zhang, Tissue-engineered conduit using urine-derived stem cells seeded bacterial cellulose polymer in urinary reconstruction and diversion, Biomaterials 31 (34) (2010) 8889–8901.

[76] J.W. Huang, X.G. Lv, Z. Li, L.J. Song, C. Feng, M.K. Xie, C. Li, H.B. Li, J.H. Wang, W.D. Zhu, S.Y. Chen, Urethral reconstruction with a 3D porous bacterial cellulose scaffold seeded with lingual keratinocytes in a rabbit model, Biomed. Mater. 10 (5) (2015) 055005.

[77] B. Wang, X. Lv, S. Chen, Z. Li, J. Yao, X. Peng, C. Feng, Y. Xu, H. Wang, Bacterial cellulose/gelatin scaffold loaded with VEGF-silk fibroin nanoparticles for improving angiogenesis in tissue regeneration, Cellulose 24 (11) (2017) 5013–5024.

[78] B. Wang, X. Lv, S. Chen, Z. Li, J. Yao, X. Peng, C. Feng, Y. Xu, H. Wang, Use of heparinized bacterial cellulose based scaffold for improving angiogenesis in tissue regeneration, Carbohydr. Polym. 181 (2018) 948–956.

[79] D. Lei, Y. Yang, Z. Liu, B. Yang, W. Gong, S. Chen, S. Wang, L. Sun, B. Song, H. Xuan, X. Mo, 3D printing of biomimetic vasculature for tissue regeneration, Mater. Horiz. 6 (6) (2019) 1197–1206.

[80] S.L. Arias, A. Shetty, J. Devorkin, J.P. Allain, Magnetic targeting of smooth muscle cells in vitro using a magnetic bacterial cellulose to improve cell retention in tissue-engineering vascular grafts, Acta Biomater. 77 (2018) 172–181.

[81] J. Wei, B. Wang, Z. Li, Z. Wu, M. Zhang, N. Sheng, Q. Liang, H. Wang, S. Chen, A 3D-printable TEMPO-oxidized bacterial cellulose/alginate hydrogel with enhanced stability via nanoclay incorporation, Carbohydr. Polym. 238 (2020) 116207.

[82] M. Ul-Islam, F. Subhan, S.U. Islam, S. Khan, N. Shah, S. Manan, M.W. Ullah, G. Yang, Development of three-dimensional bacterial cellulose/chitosan scaffolds: analysis of cell-scaffold interaction for potential application in the diagnosis of ovarian cancer, Int. J. Biol. Macromol. 137 (2019) 1050–1059.

[83] Y. Hou, X. Wang, J. Yang, R. Zhu, Z. Zhang, Y. Li, Development and biocompatibility evaluation of biodegradable bacterial cellulose as a novel peripheral nerve scaffold, J. Biomed. Mater. Res. A 106 (5) (2018) 1288–1298.

[84] W. Zhang, X.C. Wang, J.J. Wang, Drugs adsorption and release behavior of collagen/bacterial cellulose porous microspheres, Int. J. Biol. Macromol. 140 (2019) 196–205.

[85] X. Chen, X. Xu, W. Li, B. Sun, J. Yan, C. Chen, J. Liu, J. Qian, D. Sun, Effective drug carrier based on polyethylenimine-functionalized bacterial cellulose with controllable release properties, ACS Appl. Biomater. 1 (1) (2018) 42–50.

[86] N. O'Donnell, I.A. Okkelman, P. Timashev, T.I. Gromovykh, D.B. Papkovsky, R.I. Dmitriev, Cellulose-based scaffolds for fluorescence lifetime imaging-assisted tissue engineering, Acta Biomater. 80 (2018) 85–96.

[87] P. Weyell, U. Beekmann, C. Küpper, M. Dederichs, J. Thamm, D. Fischer, D. Kralisch, Tailor-made material characteristics of bacterial cellulose for drug delivery applications in dentistry, Carbohydr. Polym. 207 (2019) 1–10.

[88] M. Badshah, H. Ullah, A.R. Khan, S. Khan, J.K. Park, T. Khan, Surface modification and evaluation of bacterial cellulose for drug delivery, Int. J. Biol. Macromol. 113 (2018) 526–533.

CHAPTER 6

Characteristic features and functions of nanocellulose for its feasible application in textile industry

P. R Sreeraj[a], Santosh Kr. Mishra[a], Purushottam Kumar Singh[b]
[a]Department of Production Engineering, National Institute of Technology Tiruchirappalli, Tamil Nadu, India
[b]Department of Mechanical Engineering, BIT Sindri, Jharkhand, India

6.1 Introduction

Textile materials play a vital role in the human life from time to time. It is developed from natural fibers that are extensively used because of their sole properties that provide good quality textile materials. Cellulose is abundantly available naturally and there are various naturally accessible sources such as vegetable biomass, cotton, wood, etc. from which it can be retrieved. Cellulose is widely available in natural form on this planet. The powdered nature of cellulose also referred as nanocellulose or microfibrillated cellulose is highly regarded as a utility product. Moreover, apart from textile industries, nanocellulose is a natural, renewable, and biocompatible, which make it a necessary material in the fields of biomedical, food industries, etc. [1–3].

The natural fibers with small principal lengths do not have the capability to be used to spin yarns because of the variations in the principle lengths. Due to this, the natural fibers such as cotton, wool, hemp, or silk are lost in the course of processing stages and finishing usages. A novel method of recycling these fibers in the form of nanocellulose has large marketing benefits due to its outstanding inherent characteristics. Nanowhiskers are very fine powders usually made from cellulose fibers or proteins that provide numerous functional characteristics not only in the textile industry, but also in the field of medicines and biotechnology, electronics, composite materials, etc. The other applications lie in the domain of paper industries used as a surface strengthening agent or dry strengthening agent, in the field of nanobarriers or nanocoatings, food industries, cosmetics, skin creams, bionanocomposites, medical fields, spongy products, emulsive agents, and oil recycling applications. The research are still progressing on the novel methods for synthesis of nanocellulose and extending their use for diverse fields such as biotechnology and biomedicines [4–7].

Nanocellulose nanotechnology happens to be a promising domain especially in the areas dealing with training, research, and development and can be considered as a viable option for long-term durable practical applications. Their applicability can be observed in diverse sectors ranging from food to textile, ceramics, paints, and several different industries. This variability in application can be contributed mainly as a result of them possessing distinct features such as particle sizes having nanosize with considerable surface areas along with owing different inorganic as well as inorganic materials.

Nanoproducts are basically the byproducts extracted from naturally available nanocellulose. The generalized categorization for the generation of nanocellulose can be grouped under the two approaches: bottom-up and top-down. Natural fibers can be utilized to get nanocellulose through acid hydrolysis leading to the generation of firm nanoparticles having crystalline nature usually known as nanowhiskers having short geometry in the range of 100s to 1000 nm in contrast to nanofibrils that are obtained by means of homogenization process. The resultant material obtained is called nanocrystalline cellulose (NCC).

The general lateral dimensions range from 5 to 20 nm and the variation in the longitudinal direction ranges from 10 nm up to micrometers. It behaves as a pseudoplastic by exhibiting the features similar to gels or certain fluids having sufficiently high viscous behavior during normal operating conditions but exhibits less viscous nature after certain duration of time under the conditions of agitation, shaking, or else when being highly stressed. This feature is referred as thixotropy [8].

The cellulose materials projected to be utilized as nanofillers for its use in composites are generally exposed to hydrolysis via acids such as H_2SO_4 or HCl subsequently leading to the selective degrading of amorphous portions pertaining to cellulose and ultimately resulting in the breaking of microfibril beams. The fragmentation in its advanced structure basically results from cellulose hydrolysis leading to the formation of crystalline nanofibers that are commonly discussed as nanowhiskers in the literature. The parameters which determine the structure of obtained nanocrystals are the concentration of acid, correct proportion of acid to cellulose, reaction time as well as operating temperature conditions utilized for hydrolysis of cellulose.

The utilization of enhanced nanomaterials resulted in providing great flexibility especially in the area of textile industry. Nanotechnology plays a major contributing factor in enhancing the fibers employed in textile firms. In this connection, nanocellulose can be quoted as a very good example that possesses the features of having lighter weight with relatively lesser cost as well as having excellent conductivity for electricity; most importantly retrieved from natural sources, therefore offers a plenty of avenues and scopes for its diverse applications.

There are few nanofabrics that possess excellent features of cleaning themselves by utilizing the benefits of the lotus effect, or might have capability of UV blocking, inertness to flame, or might be having antiwearing properties. Under any situation, it is likely that they will enter in the form of fabric in common man's lives and they could finally be utilized in the task of wireless biomonitoring concerning critical functions from remote distance. Textiles with antibacterial features enabled by embedded nanoparticles will be having a major contribution especially in hospitals where the issue of cross-contamination by bacteria can be a critical issue particularly for the elderly people. Proper control measures need to be put in place to restrict the increase in bacterial populations under these environments to have minimum infection.

Nanotechnology usually deals with materials from 1 to 100 nm in dimensions. The basics of nanotechnology lies within the principle that the characteristics of materials significantly vary when their magnitudes are decreased to nanometer ranges. Currently, the textile industries are making most use of the potentials of nanotechnology. Thus, nanotechnology in textile can be considered as the knowledge, manipulation, and mechanism of the matter at the nanoscale ranges so that the chemical, physical, and biological properties of the materials (bulk matter, individual atoms, and molecules) can be engineered, synthesized, or altered to develop the next generation of enhanced devices, materials, systems, and structures. It can be employed to provide preferred textile properties such as increased durability, tensile strength, water repellency, exceptional surface structure, antimicrobial properties, soft hand, fire retardancy, and the same.

Nanotechnologies can be effectively used for customized applications and thus opening a wide spectrum of avenues in case of textile firms. They are major contributing factors in enhancing the properties of textile materials for their effective utilization by both common as well as professional people, wherein nano-enhanced materials embedded within the matrix of the polymer or sometimes coated upon the top surface depending upon fibers, for example, cotton fibers that are basically naturally available fibers of a varying length. Acid hydrolysis technique is used to form very minute-sized tiny cellulose utilizing short-length fibers that have been retrieved from long fibers under the conditions of spinning the systems. The developed powder is coated on the fabrics of different natures, be it natural or synthetic. They have superior qualities such as improvement in the durability, dirt-repelling capabilities, etc.

By manufacturing e-textiles, the clothing enabled with nanotechnology will over-protect the person wearing it from various toxic as well as hazardous entities, pathogens, facilitating medical services along with the military and can finally help in constantly monitoring from a remote location the various complex and critical functions of patients who are bedridden or ill since long duration. Nanofibers possess the capability of absorbing the products meant for plant production and thus monitoring their slow release, which depicts their vital application in the domain of agriculture and optimizing the production of crops.

In this chapter, the various sources and preparation techniques of nanocellulose are discussed followed by other general techniques that include mechanical, chemical, and biological methods. The produced nanocelluloses were characterized for determining shape, size, and composition. The effect of nanoscale cellulose on the properties of the

Table 6.1 The common sources of cellulose.

Sources	Features
Agro residues and plants Green algae	Plants and agro residues are the most dominant sources of cellulose. The extraction from the resources is less expensive due to its large availability. Green algae are also one among the suitable material resources for cellulose production. Red algae and yellow algae are also considered as best species suitable for cellulose extraction. The advantage with algae is that the cellulose obtained from cladophorales and siphonocladales resulted in a 95% crystalline in nature.
Tunicates	Tunicates are a type of invertebrate species found in water, especially in seas and oceans. There is an enzyme complex in their epidermal layer that is responsible for the secretion of cellulose. The features of the cellulose depend largely on the properties of the concerned species.
Bacteria	The major bacterial species that is usually used for the preparation of cellulose is *Komagataeibacter xylinus*. The cellulose fibers obtained from its evolution medium show the best characteristic features compared to other sources.

textiles was discussed extensively. The advantages and limitations along with current trends and future scopes are also described in this chapter.

6.2 Sources of nanocellulose

The cellulose is one of the abundant materials available in the nature. The typical sources of cellulose [9] are shown in Table 6.1.

6.3 Classification of nanocellulose structures

The different types of nanocelluloses can be classified into various subdivisions based on their technique of preparation, dimensions, profile, and utility. The sources and processing conditions are dependent on the variations of the above characteristics. The nomenclature of the nanocelluloses based on their size is shown in Fig. 6.1. The terminologies of the various types of nanocelluloses are based on their characteristics.

The comparisons of the various nanocelluloses based on their sources and sizes [10] are shown in Table 6.2.

6.4 Preparation of nanocellulose

The following are the various methods used for the preparation of nanocellulose.

6.4.1 Experimental procedure

Nanocelluloses are usually made by stirring unwanted viscose rayon fibers with sodium zincate solution.

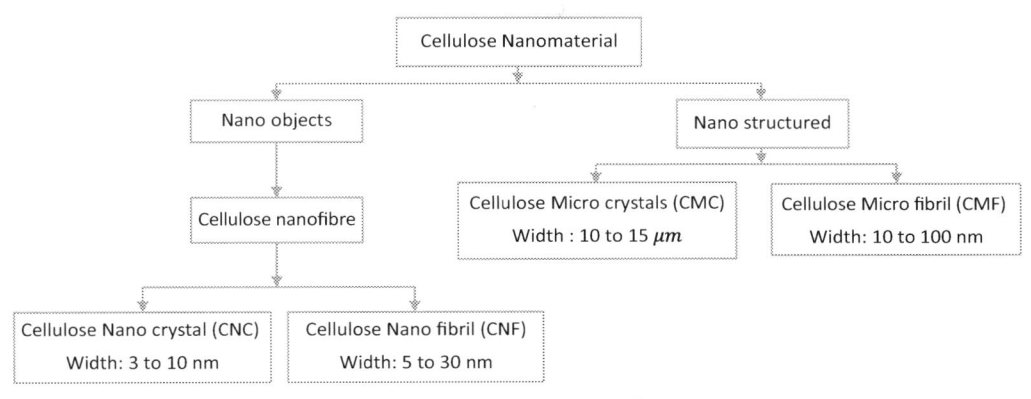

Fig. 6.1 The terminologies of the various nanocelluloses based on size.

Table 6.2 Comparison of different nanocelluloses based on their sources and sizes.

Classifications of nanocellulose	Various sources of each nanocelluloses	Mean lengths and diameters of each nanocelluloses
Bacterial nanocellulose (BNC)	Sugars and alcohols with less molar weights	Length is in micrometer range with diameter varying from 20 nm to 100 nm
Nanofibrillated cellulose (NFC)	Wood, hemp, potato tuber, sugar beet, flax	Length is in micrometer range with diameter varying from 5 nm to 60 nm
Nanocrystalline cellulose (NCC)	Rice straw, wheat straw, cotton, flax, wood, tunicin, hemp, Avicel, BC mulberry bark, microcrystalline cellulose (MCC), ramie, algae	Length varies from 100 nm to 250 nm from the plant to some micrometers with diameter ranges from 5 nm to 70 nm

6.4.2 Preparation of solution of sodium zincate

Sodium zincate solution is made by mixing NaOH and water in the ratio 9:10 and then steadily adding ZnO and constantly stirring it. After placing the solution idle for one day, filter the solution using Whatman No.1 filter paper to obtain the solution of sodium zincate.

6.4.3 Preparation of nanocellulose

The viscose rayon fibers are initially grinded to minor sized powder particles. Then it is stirred with prepared sodium zincate solution in the ratio of 1 g: 9 mL. The temperature is maintained at 50 °C for diffusing the sodium zincate solution into the amorphous part of the fibers. Hence the glycosidic bonds formed will undergo cleavage and the particles undergo neutralization with glacial acetic acid solution. The suspended particles are eroded away. Whatman No.1 filter paper is used to filter those suspended particles. The colloidal solution undergoes evaporation and gets converted into powder size particles. This powder is cleaned using purified water and finally dried to obtain nanocellulose [11].

6.5 General techniques used (specific techniques)

The initial stage of the preparation of the nanocellulose includes separation of cellulose from the cellulose pulp that is a source of lignin and hemicellulose. Cellulose nanofiber (CNF) and Cellulose nanocrystals (CNC) are prepared by additional processing on the purified cellulose pulp.

6.5.1 Top-down process

Mostly all CNFs and CNCs are produced by this method that involves splitting the cellulose fibers into nano-sized particles.

6.5.2 Bottom-up processes

The bottom–up processes use bacteria for bacterial cellulose (BCs) and an electrospinning technique for electrospun CNF (ECNF), respectively [12].

The processes involved in the production techniques determine the crystallinity, morphology, and surface chemistry of the different types of CNC/CNF [13]. The techniques for preparing nanocellulose include chemical and biological pretreatments, mechanical disintegration, etc.

6.5.3 Mechanical disintegration

Mechanical disintegration is one of the most commonly used methods for the preparation of CNFs and posttreatment and purification methods for producing CNCs [14]. The cellulose pulp will be broken down into smaller particles. The delamination is mostly preferred instead of fiber shredding for the efficient mechanical disintegration because delamination can produce nanocellulose with better mechanical characteristics. The delaminations can also be improved by incorporating an aqueous medium for mechanical disintegration that will evade accumulation of fibers or inverse coalescence and loosen the hydrogen bonding within fibers. The usually applied methods for the effective delamination of cellulose fibers are grinding, homogenization, and refining [15,16]. The normally used practices for efficient delamination of cellulosic fibers comprises grinding, homogenization, and refining. Mechanically disintegrated nanocellulose produces large-dimensional clusters compared to those produced by chemical means [17].

Park et al. [18], discussed the dependence of the composition of wood-based CNF on the defibrillation effectiveness in wet-disk milling. The results showed improvement in the defibrillation in the absence of hemicellulose and lignin in cellulose. The thermal stability, viscosity, and crystallinity undergo reduction as we move from microdisintegration range to nano-scale range. The mostly preferred mechanical treatments include ball milling, aqueous counter collision, electrospinning, blending, extrusion, cryocrushing, grinding, steam explosion, refining, ultrasonication, homogenization, etc. A combination of the above methods is also preferred for enhanced properties of the fibers.

6.5.4 Chemical reaction

Chemical reaction for production of nanocellulose is mostly preferred for large-scale requirement. The chemicals added can also considerably aid in the defibrillation processes.

The various types of chemical techniques applied for separating cellulose fibers are carboxymethylation, acid hydrolysis, quaternization, carboxylation, ionic liquids, sulfonation, and solvent–aided pretreatments [19-22].

Acid hydrolysis is mainly employed to obtain nanocelluloses from cellulose fibers. This method will deplete the amorphous sections, but the crystalline areas remain unchanged. The method of sulfuric acid hydrolysis has higher adsorption capacity compared to mechanical grinding. It is also used to manufacture bacterial nanocrystals from the BC microfibrils [23].

The procedure of surface charge modification by sulfonation, carboxylation, and carboxymethylation uses anionic property for defibrillation of cellulose. It is centered on the principle of electrostatic repulsion of same charges. Carboxylation is able to produce nanocellulose with minimum agglomeration and improved delamination characteristics. Oxidative sulphonation of cellulose using bisulphate and periodate compounds is one of the environmentally stable methods to promote nanofibrillation of cellulose.

Sulfonation of cellulose consumes harmless and less costly chemicals during recycling that reduces formation of halogen wastes [24]. During processes such as quaternization that provides cationic property to nanocellulose and results in accumulation of pollutants that clot kaolin colloids has been discussed by Liimatainen et al. [24].

6.5.5 Biological reaction

Enzymatic hydrolysis is an added pretreatment phase mainly employed for reducing chemical waste and energy requirement and thereby reducing the cost and environmental problems. This method is usually practiced before mechanical breakdown of cellulose (blending or refining) to support delaminations of the cell wall [25].

The commonly used enzymes for the manufacture of nanocellulose include pectinases, cellulases, xylanases, and ligninases [26-27]. The use of cellulose as an enzyme substrate along with acetate buffer produces CNCs with higher dimensions and low sulfur content. Hemicellulose existing in sources of cellulose obstructs the division of cellulose nanofibrils.

Xylanase and lytic polysaccharide mono-oxygenases (LPMO) known as "accessory enzymes" can increase the hydrolytic characteristics of cellulose, whereas endoglucanase assists the purifying and fibrillation of cellulose.

The synergistic effect of the enzymes aids the overall accessibility of the substrate, whereas the rheological characteristics of CNFs prepared using enzyme pretreatment provides better flocculation property apart from chemical pretreatment.

6.6 Pretreatment of nanocellulose

The major problems that exist during fibrillation of cellulose (commonly in mechanical techniques) are aggregation of fibers when the slurry is forced into the disintegration mechanism and large amount of energy requirement for effective fiber delamination due to large number of passes. This large quantity of energy is essential to separate the fibers and to oppose the hydrogen bonding within fibers [28]. An effective pretreatment method is necessary for reducing energy requirement. The suitable pretreatment method is selected based on the sources of cellulose and its morphological characteristics. The pretreatment provides accessibility, changes its crystalline structure, breakage of hydrogen bonds, etc. [29,30].

The partial separation of nanocellulose materials such as lignin, hemicellulose, etc. and the parting of each fiber can be achieved by the pretreatment of plant constituents. Pretreatment also aids in the exclusion of the protein matrix and separation of the fibers and mandrel for tunicates [31]. Pretreatment for algae usually helps in the exclusion of the matrix material in their cell wall [32], while for bacterial nanocellulose pretreatment facilitates the exclusion of the bacterial contaminants. Pretreatment modifies the structure, polymorphism, and crystallinity of the cellulose. The various most efficient pretreatment techniques are discussed next.

6.6.1 Pulping processes

Pulping is a method employed to separate fibers from wood and plants. There are two types of applying pulping: mechanical and chemical. Mechanical application technique consumes high energy and use almost entire wood material. The wooden logs are ground using a rotating cylinder made of sandstone to cut out the fibers or the wooden chips that can be fed into the center of a rotating disc with water supply [33]. But the drawback is that these mechanical methods may create harm to the size and form of the wooden fibers, and thus diminishing their relevant properties.

Kraft pulping is one of the chemical methods in which the plant resources are mixed with a hot solution of NaOH and sulfides in a digester. One part of the wood is transformed into pulp, whereas the other part will get dissolved in the solution. Soda cooking uses a mixture of NaOH and sulfite acid or its salts [33]. Solution of sodium chlorite for lignin oxidation provides a greater amount of delignified fibers compared to the kraft method. It is also to be understood that when the amount of hemicelluloses in the fibers is more, the nanofibrillation will also be more. The strong bond between amorphous and water absorbent hemicelluloses prevents the accumulation and ease the fibrillation.

6.6.2 Bleaching

Bleaching method uses elimination of lignin residues and other contaminations of the pulp without affecting the polymorphism or crystalline characteristics of the cellulose to gain better aging resistance. The commonly used bleaching agents are oxygen, hydrogen peroxide, sodium chlorite, ozone, chlorine, peracetic acid, and chlorine dioxide [34–36].

6.6.3 Alkaline-acid-alkaline pretreatment

This pretreatment process consists of soaking the fibers in 12–17.5 wt% sodium hydroxide for 2 h thus increasing the fiber surface area hydrolysis process, followed by mixing the fibers with 1M Hyaluronic acid (HA) at 60–80 °C for hydrolyzing the hemicelluloses and finally mixing the fibers with 2 wt% sodium hydroxide solution for 2 h at 60–80 °C to dislocate the lignin structure [37,38].

6.6.4 Enzymatic pretreatment

The use of enzymes for pretreatment imparts environmental stability and harmless substitute of chemical pretreatment with less energy requirement. The commonly used enzymes are xylanases and ligninases. They have the ability to degrade hemicelluloses and maintaining cellulose. The cellular enzymes known as cellulases assist hydrolysis [39].

Cellulose nanofibrils are made from the bleached wood by an amalgamation of enzymatic hydrolysis using endoglucanase and high-pressure disintegration [40]. The results show that there is an increase in bulging of the pulp fibers in liquid water and assists in its breakdown, thus inhibiting the microfluidizer from blockage. Enzymatic process gives identical CNFs with good aspect ratios.

The endoglucanase enzymatic pretreatment promotes the creation of a mixture of CNF and stiffer rod-shaped nanoparticles, while exoglucanase conserves the web-shaped form of the CNF. The enzymatic method has only less influence on the final size of the nanocellulose.

6.6.5 Ionic liquids

Ionic liquids are a class of organic salts that exists in liquid form below 100 °C. The characteristic features of ionic liquids are their low vapor pressure, chemical and thermal stability, and nonflammability [41,42]. The suspension of cellulose in particular hydrophilic ionic liquids uses environmentally harmless solvents and biorenewable sources.

The mechanism of dissolution is that the cations of ionic liquids react with the oxygen atoms, while the anions react with the protons of the OH groups in cellulose hence helps in breaking of hydrogen bonds. Cellulose generation can be made faster by the adding acetone, ethanol, or water. High solubility of the cellulose in the ionic liquids is obtained at a dissolution temperature of 130 °C. The major advantages with ionic liquids are that it will not be consumed in the process and can be restored by ion exchange, evaporation, or reverse osmosis and finally reused [43].

6.6.6 Oxidation

(2,2,6,6-Tetramethylpiperidin-1-yl)oxyl (TEMPO) is a commonly used sublime solid for the pretreatment of cellulose in the laboratories to decrease the energy requirement for mechanical disintegration. It serves as a catalyst for oxidizing primary alcohols to aldehydes and carboxylic acids. The side reactions results in tough cellulose depolymerization, decrease in thermal stability, and discoloration after TEMPO oxidation of cellulose under alkaline conditions [44]. To inhibit these side reactions, the TEMPO pretreatment must be performed in mere acidic environments with pH 6.0–6.5 and temperatures of 50–60 °C. When the carboxyl amount increases, the cellulose starts diffusing at the amorphous regions, but not in the crystalline regions.

The oxidation rate is significant in decreasing the energy requirement and enhancing the process of nanofibrillation. Preparation of nanocellulose fibers may be optimized by controlling the time and degree for oxidation. TEMPO oxidation results in the formation of small and uniform nanocelluloses. The studies of nanocelluloses reveal that the oxidation process does not change the form and surface area of the nanocellulose particles. The separation of nanocellulose fibers from cellulose that has undergone high oxidation can be easily attained without any further mechanical treatments except stirring during chemical methods.

6.6.7 Steam explosion

Steam explosion pretreatment technique used for the removal of cellulose fibers from plant biomass can be applied individually or can be combined with elevated pressure disintegration. This pretreatment involves heating the vapor phase at 180 to 210 °C under a steam pressure in the range of 1–3.5 MPa. This pretreatment is completed by an explosive decompression, which results in the breaking of the material by a flash evaporation of water at a very large force.

This method can cause hemicellulose hydrolysis and fiber fibrillation. The nanocellulose fibers prepared by this method have large aspect ratios [45]. This method requires only low energy and low intake of chemicals, which results in lower environmental impact and low cost. But for more efficient fibrillation, this method must be repeated more times [46].

6.6.8 Other pretreatments

There are also other methods used in limited cases such as carboxymethylation and acetylation [47]. Carboxymethylation provides high charge celluloses that encourage the separation of nanocellulose fibers. The disintegration energy requirement is also less with carboxymethylation. Acetylation involves imposing acetyl groups that helps in breaking of hydrogen bonding within nanofibrils and thus promoting dispersion.

Currently researchers are concentrating on the isolation methods that are environmentally stable, less expensive, and provide highly efficient nanocellulose fibers. A combination of various individual treatment methods together is giving better results and can be considered as a developing domain for future.

6.7 Characterization and thermal analysis of nanocellulose particles

The experimental procedure for characterization and thermal analysis of nanocellulose [48] is as follows.

6.7.1 Characterization of nanocellulose particles

The size of the particles and their distribution in the nanocellulose can be investigated using a particle size analyzer. Scanning electron microscope (SEM) technique is commonly employed for analyzing the morphology of nanocellulose and the chemical composition of the prepared nanocellulose can be determined by Spectrometer (Thermo Scientific).

6.7.2 Characterization of prepared nanocellulose

The nanocellulose is made into the form of dried powder and is examined by dispersing the powder in water using a particle size analyzer.

The shape of the prepared nanocellulose particles is observed to be in the form of small rods. Whiskers are usually these rod-like structured particles formed by the breakage of the cellulose chains, which contain a large order of crystalline locations, linked with lower order amorphous locations. The absorption in the crystallinity band indicates a decline in the intensity resulting in the variation of crystalline property of the samples.

6.7.3 Characterization of polyester/nanocellulose composite

Image analyzer is used to investigate the distribution of nanoparticles on the fiber surface and their penetration in the polymer matrix. The preparation of the longitudinal and the cross-sectional sections of the sample was carried out using the American Association of Textile Chemists and Colorists (AATCC) test.

The cellulose materials are initially applied on a polyester fabric and a direct dye is used to spot the cellulose particles. Deposition and dispersion of the cellulose particles can be seen on the polyester fiber surface by the longitudinal and cross-sectional views. There are no substantial variations in the spectrum detected after the treatment of nanocellulose over polyester fibers, which appears over the course.

6.7.4 Thermal analysis

The thermal analysis focuses on the thermal reactivity of the nanopowders, treatment of the polyester fabric materials, and untreated and the onset temperatures. Differential scanning calorimeter is used for measurement of the onset temperatures of the samples. It can be found that there is a small rise in onset temperature because of the amalgamation of the nanocellulose into the fiber matrix; but, the effect is insignificant.

During investigation on thermal degradation of polyester fabric, the maximum temperature of the nanocellulose-treated sample is shifted to higher values when the degree of crystallinity drops. The reason for this is by the principle that the beginning of thermal degradation takes place in the amorphous region in the cellulose. The amorphous will be very low in pure polyester and increases with the addition of nanocellulose.

6.8 Effects of nanocellulose on the properties of the textiles

The nanocellulose has a wide variety of applications in the textile industries. It is because the application of nanocellulose can enhance the characteristics and properties of the textiles. The physical properties of the polyester fabrics were found to change with breaking load and crease recovery angle.

6.8.1 Effect on tensile strength

The results obtained when treated and untreated polyester samples were evaluated for the effect of tensile strength indicate that the use of nanocellulose in polyester fibers enhances their load-bearing capacity. The amount of diffusion of nanocellulose into the polymer matrix is an important factor for the tensile strength improvement.

6.8.2 Effect on crease recovery

There is a slight improvement in the crease recovery angle after the addition of nanocellulose into the fibers. The nanocellulose particles have smaller size that helps them to enter in between the polymer particles and function as a cross-linking agent. As more and more nanoparticles are added, the crease recover angle was found to be increased. Hence, the mechanical interlocking of the nanocellulose inside the pores results in the improvement in physical properties of the fibers.

6.8.3 Effect of nanocellulose on water absorbency of polyester fabric

Drop test and Wicking test are mostly used to study the water absorbency of nanocellulose-added polyester fabric. It is found that polyester fabric added with less concentration of nanocellulose takes lower time for water absorption. But as more and more nanocellulose are added, more time will be required for water absorption because it obstructs the diffusion of the water molecules by the formation of nanowhiskers that will not permit the water drop to be accommodated inside the polymer spaces. The hydrophilicity exhibited by the polyester is due to less quantity of nanocellulose treatment.

6.8.4 Effect of nanocellulose on water permeability of polyester fabric

The water permeability through the nanocellulose-added polyester fibers reduces compared to that of the normal fibers. It is because of the resistance applied by the nanoparticle in the polymer matrix against the movement of water within the fabrics. It can also be understood that as the amount of nanoparticle in fabrics was more, the permeability of water was found to be minimized.

6.8.5 Effect of nanocellulose on air permeability of polyester fabric

The air permeability of nanocellulose-added polyester fiber is lower compared to the normal fiber because of the resistance developed in the polymer matrix against the stream of air.

6.8.6 Effect of nanocellulose on dyeing of treated fabric with a direct dye

The procedures used for dyeing nanocellulose-added polyester fiber are pad-dry-cure method followed by a direct dye (Congo red BDC) by the pad-dry-cure method. The results show that as the amount of dye increases, there is an increase in the K/S value of the sample successively applied with a direct dye. The dyed nanocellulose-added polyester fiber samples were washed using nonionic detergent for 15 min at a temperature of 70 °C. This experimental outcome indicates that the proportionate decrease in K/S values exists higher in the samples dyed without the nanocellulose addition.

6.9 Applications, advantages, and limitations of nanocellulose

The nanocelluloses we used commonly have a large number of applications. The various applications of different types of nanocelluloses [48] are shown in Table 6.3.

Nanocellulose provides several benefits for textiles such as lighter fiber mass without losing strength and high filler content. It also improves multiple and cross-functionality in different industrial applications. In crystal-like structure, it has eight times the tensile strength compared to steel. The higher stiffness extends their functionality to a wide range of applications. Nanocellulose is a renewable, environmentally compatible,

Table 6.3 The recent applications of various nanocelluloses in different sectors.

Types of nanocelluloses	Applications
Bacterial nanocellulose (BNC)	• Exclusion of dense metals from impure water • Cleaning of dyes
Nanofibrillated cellulose (NFC)	• Biopackaging of foods • Removal of contaminants from water • Exclusion of vanadium metal in water • Preparing mix of aerogels for the redevelopment of bone • The cleaning of injuries in medical fields • Supply of drugs in medical fields • Packaging or covering of food products in food sector • Cartilage and bone tissue implants in health sector • Implants of blood vessels • Control of mobility of proteins and enzymes • Preparation of coatings • Maintenance of super capacitors • Improvement of paper board barrier characteristics • Development of complex membrane in fuel cell (Proton-exchange membrane fuel cells (PEMFC)) • Production of oil with improved properties • Improvement in the performance of lithium–sulfur battery
Bacterial nanocellulose (BNC)	• Developing good support material for different catalytic systems • Papermaking industries for preparation of additives • Increased mechanical strength applications • Excites enlargement of mesenchymal stem cells and formation of collagens in tissues • Used for supply of drugs and creating a base for various medicines in medical field • Applied as an additive for wheat bread in food industries

sustainable, and biodegradable material. It is formed out of cellulose that is one of the most available polymers.

Even though nanocellulose is considered as one of the assured environmental friendly materials, there still exist many limitations and challenges for the use of this nanomaterial. The use of nanocellulose must be having low cost and easily scalable to promote the commercialization and marketability of nanocellulose. Most of the studies on the production of nanocellulose are carried out only on the laboratory scale. The difficulty is in the environmental impact, scalability, and the cost for the choice of treatment methods for production of nanocellulose in industrial applications. The mechanical disintegration method is commonly used for upscaling, but consumes more energy. Chemical pretreatment of nanocellulose results in a larger impact on the environment as it generates acidic and solvent waste even though consumption of energy is lower. Carboxymethylation chemical process requires higher cumulative energy for cellulose-treated fiber due to the larger usage of chemicals such as alcohols for preparing CNF. The challenges also lie during production of BC and ECNF, as it depends on the growth of nanofibers from less molecular weight sugars by bacteria and dissolved cellulose by electrospinning. The preparation techniques of nanocellulose always need regular examination for producing environmentally stable and less economical nanocellulose. Moreover, the economic and regulatory limitations due to the toxicity of different nanocellulose materials to environment are a greatest obstacle for its use and commercialism.

6.10 Current trends and future scopes

Even though nanocellulose is having several desirable characteristics, their high hydrophilicity, hygroscopicity, and affinity to form groups can significantly constrain their uses. The above characteristics lead to the formation of agglomerates in polymers because of the large intermolecular hydrogen bonding between nanocellulose molecules. Accumulation of nanocellulose particles can considerably obstruct the performance of the matrix. Hence, one of the scopes for the development lies in resolving the agglomeration of nanocellulose for improved performance during applications. But, its water-absorbing ability can also play a vital role in improving hybrid systems. Hence, the characteristics of nanocellulose that are considered as undesirable can be beneficial for some other specific applications.

However, the characteristics can be altered to decrease or entirely eradicate its effects. But the modifications are to be done only to the exterior surface for maintaining its morphological and structural integrity. Most of the modifications comprise violent oxidization that increases the concentration of OH groups on its surface and it agitates further modifications. The mechanical properties are also improved by these changes due to the variations in its crystallinity and structural integrity.

Although nanocellulose has good thermal stability in comparison to other fibers, the concern is also with the long durability of its products in their service life. The anticipated increase in the nanocellulose production in future will definitely give rise to the more industrial applications.

The wide variety of nanocellulose function ability will drive the future of biocompatible, sustainable, and biodegradable materials for different uses over the materials used currently that generates greenhouse gases and petroleum products that are harmful to the environment. Hence, it is likely that in the near future, nanocellulose will be commonly used in the electronics, biotechnology, medical, food, oil and gas industries. Nanocelluloses have the capability to substitute synthetic materials for various applications.

The research on nanocellulose is getting continuous consideration throughout the world. Keeping this in mind, with further extended knowledge in application point of view with improvised technologies for preparation can utilize nanocellulose from lab to industrial scale.

6.11 Conclusions

- The results of fourier transform infrared spectroscopy (FTIR) spectral analysis indicate that nanocellulose is formed from the waste viscose rayon fiber by the addition of sodium zincate.
- The crystals of the cellulose prepared are appeared to be in the shape of rods.
- The addition of nanocellulose increases the rigidity of the material, changes the thermal property, and increases the onset temperature.
- Nanocellulose also decreases air and water permeability of polyester fibers and increases the absorbing capacity.
- The addition of nanocellulose can enhance the color strength of direct dyed polyester and also improve soaping properties.

References

[1] K. Havancsak, Nanotechnology at present and its promise for the future, Mater. Sci. Forum, 414, 2003 85–94.
[2] J. Hengstentaerg, H. Mark, Crystalline materials, J. Crystallogr 69 (1929) 271.
[3] D.J. Gardner, G.S. Oporto, R. Mills, M.A.S.A Samir, Adhesion and surface issues in cellulose and nanocellulose, J. Adhes. Sci. Technol. 22 (5-6) (2008) 545–567.
[4] R.H. Atalla, Structures of cellulose. MRS Online Proceedings Library 197 (1987) 89–98 .
[5] M. Plunguian, Cellulose Chemistry, Chemical publishing Company Co. Inc., New York, 1943.
[6] B.G. Rånby, Fibrous macromolecular systems. Cellulose and muscle. The colloidal properties of cellulose micelles, Discuss. Faraday Soc. 11 (1951) 158–164.
[7] D. Ciolacu, F. Ciolacu, V.I. Popa, Amorphous cellulose—structure and characterization, Cellul. Chem. Technol. 45 (1) (2011) 13.
[8] S.M. Salah, Application of nano-cellulose in textile, J. Textile Sci. Eng. 3 (2013) 142.
[9] A. Sharma, M. Thakur, M. Bhattacharya, T. Mandal, S. Goswami, Commercial application of cellulose nano-composites–a review, Biotechnol. Rep. 21 (2019) e00316.
[10] P. Thomas, T. Duolikun, N.P. Rumjit, S. Moosavi, C.W. Lai, M.R.B. Johan, L.B Fen, Comprehensive review on nanocellulose: recent developments, challenges and future prospects, J. Mechan. Behav. Biomed. Mater. (2020) 103884.
[11] D.P. Chattopadhyay, B.H. Patel, Synthesis, characterization and application of nano cellulose for enhanced performance of textiles, J. Text. Sci. Eng. 6 (248) (2016) 2.
[12] M. Rajinipriya, M. Nagalakshmaiah, M. Robert, S. Elkoun, Importance of agricultural and industrial waste in the field of nanocellulose and recent industrial developments of wood based nanocellulose: a review, ACS Sustain. Chem. Eng. 6 (3) (2018) 2807–2828.
[13] O. Nechyporchuk, M.N. Belgacem, J. Bras, Production of cellulose nanofibrils: a review of recent advances, Ind. Crops Prod. 93 (2016) 2–25.

[14] A. García, A. Gandini, J. Labidi, N. Belgacem, J. Bras, Industrial and crop wastes: a new source for nanocellulose biorefinery, Ind. Crops Prod. 93 (2016) 26–38.

[15] S.S. Nair, J.Y. Zhu, Y. Deng, A.J. Ragauskas, Characterization of cellulose nanofibrillation by micro grinding, J. Nanopart. Res. 16 (4) (2014) 1–10.

[16] M. He, G. Yang, J. Chen, X. Ji, Q. Wang, Production and characterization of cellulose nanofibrils from different chemical and mechanical pulps, J. Wood Chem. Technol. 38 (2) (2018) 149–158.

[17] I.A. Sacui, R.C. Nieuwendaal, D.J. Burnett, S.J. Stranick, M. Jorfi, C. Weder, J.W. Gilman, Comparison of the properties of cellulose nanocrystals and cellulose nanofibrils isolated from bacteria, tunicate, and wood processed using acid, enzymatic, mechanical, and oxidative methods, ACS Appl. Mater. Interfaces 6 (9) (2014) 6127–6138.

[18] C.W. Park, S.Y. Han, H.W. Namgung, P.N. Seo, S.Y. Lee, S.H. Lee, Preparation and characterization of cellulose nanofibrils with varying chemical compositions, BioResources 12 (3) (2017) 5031–5044.

[19] O.M. Vanderfleet, D.A. Osorio, E.D. Cranston, Optimization of cellulose nanocrystal length and surface charge density through phosphoric acid hydrolysis, Philos. Trans. R. Soc. A Math. Phys. Eng. Sci. 376 (2112) (2018) 20170041.

[20] K.P.Y. Shak, Y.L. Pang, S.K Mah, Nanocellulose: recent advances and its prospects in environmental remediation, Beilstein J. Nanotechnol. 9 (1) (2018) 2479–2498.

[21] A.J. Onyianta, M. Dorris, R.L. Williams, Aqueous morpholine pre-treatment in cellulose nanofibril (CNF) production: comparison with carboxymethylation and TEMPO oxidisation pre-treatment methods, Cellulose 25 (2) (2018) 1047–1064.

[22] D. Santos, D. M., I.S. Leite, A. de Lacerda Bukzem, R.P. de Oliveira Santos, E. Frollini, N.M. Inada, S.P Campana-Filho, Nanostructured electrospun nonwovens of poly (ε-caprolactone)/quaternized chitosan for potential biomedical applications, Carbohydr. Polym. 186 (2018) 110–121.

[23] M. Börjesson, G. Westman, Crystalline nanocellulose—preparation, modification, and properties, in: Matheus Poletto (Ed.), Intechopen, London, UK, 2015, pp. 159–191. DOI: 10.5772/61899. https://www.intechopen.com/books/cellulose-fundamental-aspects-and-current-trends/crystalline-nano-cellulose-preparation-modification-and-properties.

[24] H. Liimatainen, T. Suopajärvi, J. Sirviö, O. Hormi, J. Niinimäki, Fabrication of cationic cellulosic nanofibrils through aqueous quaternization pretreatment and their use in colloid aggregation, Carbohydr. Polym. 103 (2014) 187–192.

[25] M. Pääkkö, M. Ankerfors, H. Kosonen, A. Nykänen, S. Ahola, M. Österberg, T. Lindström, Enzymatic hydrolysis combined with mechanical shearing and high-pressure homogenization for nanoscale cellulose fibrils and strong gels, Biomacromolecules 8 (6) (2007) 1934–1941.

[26] A. Hideno, K. Abe, H. Yano, Preparation using pectinase and characterization of nanofibers from orange peel waste in juice factories, J. Food Sci. 79 (6) (2014) N1218–N1224.

[27] B.M. Cherian, L.A. Pothan, T. Nguyen-Chung, G. Mennig, M. Kottaisamy, S. Thomas, A novel method for the synthesis of cellulose nanofibril whiskers from banana fibers and characterization, J. Agric. Food Chem. 56 (14) (2008) 5617–5627.

[28] S. Kalia, S. Boufi, A. Celli, S. Kango, Nanofibrillated cellulose: surface modification and potential applications, Colloid Polym. Sci. 292 (1) (2014) 5–31.

[29] A. Šturcová, G.R. Davies, S.J. Eichhorn, Elastic modulus and stress-transfer properties of tunicate cellulose whiskers, Biomacromolecules 6 (2) (2005) 1055–1061.

[30] H.A. Khalil, Y. Davoudpour, M.N. Islam, A. Mustapha, K. Sudesh, R. Dungani, M. Jawaid, Production and modification of nanofibrillated cellulose using various mechanical processes: a review, Carbohydr. Polym. 99 (2014) 649–665.

[31] O. Van den Berg, J.R. Capadona, C. Weder, Preparation of homogeneous dispersions of tunicate cellulose whiskers in organic solvents, Biomacromolecules 8 (4) (2007) 1353–1357.

[32] A. Mihranyan, Cellulose from cladophorales green algae: from environmental problem to high-tech composite materials, J. Appl. Polym. Sci. 119 (4) (2011) 2449–2460.

[33] M. Ek, G. Gellerstedt, G. Henriksson, Pulping Chemistry and Technology, Vol. 2, Walter de Gruyter, Berlin, Germany, 2009.

[34] B.H. Fritzvold, U.S. Patent No. 4,278,496, U.S. Patent and Trademark Office, Washington, DC, 1981.

[35] B.H. Fritzvold, N. Soteland, U.S. Patent No. 4,450,044, U.S. Patent and Trademark Office, Washington, DC, 1984.

[36] Z. Yuan, Y. Ni, A.R.P Van Heiningen, Kinetics of peracetic acid decomposition: Part I: spontaneous decomposition at typical pulp bleaching conditions, Can. J. Chem. Eng. 75 (1) (1997) 37–41.

[37] A. Bhatnagar, M. Sain, Processing of cellulose nanofiber-reinforced composites, J. Reinf. Plast. Compos. 24 (12) (2005) 1259–1268.

[38] B. Wang, M. Sain, K. Oksman, Study of structural morphology of hemp fiber from the micro to the nanoscale, Appl. Compos. Mater. 14 (2) (2007) 89–103.

[39] S. Janardhnan, M.M. Sain, Isolation of cellulose microfibrils–an enzymatic approach, Bioresources 1 (2) (2006) 176–188.

[40] M. Henriksson, G. Henriksson, L.A. Berglund, T. Lindström, An environmentally friendly method for enzyme-assisted preparation of microfibrillated cellulose (MFC) nanofibers, Eur. Polym. J. 43 (8) (2007) 3434–3441.

[41] A. Pinkert, K.N. Marsh, S. Pang, M.P. Staiger, Ionic liquids and their interaction with cellulose, Chem. Rev. 109 (12) (2009) 6712–6728.

[42] S. Zhu, Y. Wu, Q. Chen, Z. Yu, C. Wang, S. Jin, G. Wu, Dissolution of cellulose with ionic liquids and its application: a mini-review, Green Chem. 8 (4) (2006) 325–327.

[43] A.N. Frone, D.M. Panaitescu, D. Donescu, Some aspects concerning the isolation of cellulose micro- and nano-fibers, UPB Sci. Bull. B Chem. Mater. Sci. 73 (2) (2011) 133–152.

[44] T. Saito, A. Isogai, TEMPO-mediated oxidation of native cellulose. The effect of oxidation conditions on chemical and crystal structures of the water-insoluble fractions, Biomacromolecules 5 (5) (2004) 1983–1989.

[45] B. Deepa, E. Abraham, B.M. Cherian, A. Bismarck, J.J. Blaker, L.A. Pothan, M. Kottaisamy, Structure, morphology and thermal characteristics of banana nano fibers obtained by steam explosion, Bioresour. Technol. 102 (2) (2011) 1988–1997.

[46] B.M. Cherian, A.L. Leão, S.F. De Souza, S. Thomas, L.A. Pothan, M. Kottaisamy, Isolation of nanocellulose from pineapple leaf fibres by steam explosion, Carbohydr. Polym. 81 (3) (2010) 720–725.

[47] N. Lavoine, I. Desloges, A. Dufresne, J. Bras, Microfibrillated cellulose–its barrier properties and applications in cellulosic materials: a review, Carbohydr. Polym. 90 (2) (2012) 735–764.

[48] P. Thomas, T. Duolikun, N.P. Rumjit, S. Moosavi, C.W. Lai, M.R.B. Johan, L.B Fen, Comprehensive review on nanocellulose: recent developments, challenges and future prospects, J. Mech. Behav. Biomed. Mater. (2020) 103884.

CHAPTER 7

Nanocellulose in plastic industry

Sapna Jain, Bhawna Yadav Lamba, Sanjeev Kumar
Department of Chemistry, School of Engineering, University of Petroleum and Energy Studies, Dehradun, Uttarakhand, India

7.1 Introduction

Cellulose is most pervasive and abundant organic substance in the biosphere. It has been largely used by packaging industry. Cellulosic packaging falls under different categories such as wrapping materials and containers, flexible and rigid packaging, primary and secondary packaging, etc. [1]. About 40% of total packaging material, worldwide is cellulosic [2]. The properties of a cellulosic packaging depend upon the sources of cellulosic fiber, separation process, arrangement of network matrix, nature of nonfibrous chemical additives used, and the various conversion processes.

There is a paradigm shift in the various applications of cellulosic material with the introduction of nano form of cellulose. Nanocellulose is a promising material for development of innovative materials that have applications in different industries. It is also widely used for enhancement of properties of conventional materials. It can attain remarkable and encouraging properties and can be used as filler, material of composites, coatings, thin films, etc. The main application of nanocellulose as plastic is in packaging.

7.2 Plastic in packaging

The food packaging industries are growing in accordance with the demands of consumers, with the help of advancement in materials and technology. Currently, the packaging is not restricted to efficient distribution and preservation of food or other articles but also about its final fate in nature, and communication with end users. The most used material for packaging applications is plastic, owing to their outstanding properties such as light weight, ease of handling, and durability. Globally, the plastic packaging market was valued at USD 345.91 billion in 2019 and is expected to reach a value of USD 426.47 billion by 2025, at a compound annual growth rate (CAGR) of 3.47% over the forecast period 2020–2025 [3]. Plastics are polymeric material that is lightweight, durable, and less expansive and its wide range of applications is not a hidden fact. We all use plastics in our day today life. Plastics are primarily polymerized molecules obtained from petroleum or natural gas kind of substances. The crude petroleum product (left after refining of crude oil) is used as raw material in production of plastics. There are mainly two types of plastics: thermoplastic and thermosetting (Table 7.1).

Table 7.1 Comparison of Thermoplastic anc thermosetting plastics.

Thermoplastic plastic	Thermosetting plastic
It can be reheated, remolded, and cooled, as necessary, without causing any chemical changes.	It is a material that strengthens when heated, but cannot be remolded or heated after the initial forming.
Low melting point	High melting point
Less degree of cross-linkage is present.	High degree of cross-linkage is present.
On heating, the individual polymer chains can undergo major molecular motions and thermoplastic material becomes soft and ultimately melts.	On heating the polymer chains may undergo relatively weak molecular motions and once softened on heating they undergo chemical changes and thus they cannot be softened again by heat.
Example: PET, LDPE, HDPE, PVC, PS PP	Example: Bakelite, polyurethanes, etc.

Plastics are generally strong and lightweight, which facilitate use of less packaging material to transport more products. Another advantage is their bacteria resistance that provides food safety and freshness of foods for a longer period.

A range of plastics, such as polypropylene (PP), polyethylene, polyvinyl chloride, and polyethylene terephthalate (PET), has been in use as a packaging material. The packaging addresses needs of different industries such as food items, electronic items, chemicals, adhesives, coatings etc.

The various polymers used in manufacturing of various products for packaging are summarized in Fig. 7.1.

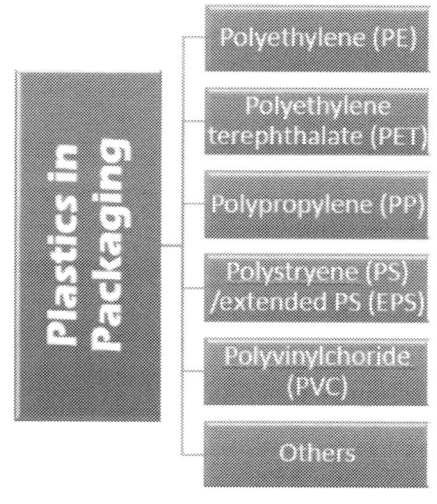

Fig. 7.1 Different polymers used as plastic in packaging.

- *Polyethylene terephthalate (PET)*

It is commonly known as stomach plastic. It is a type of plastic that is suave, clear, and comparatively thin plastic. It is used in packaging industry for different food items (oil, soft drinks, juices, salad dressing), cosmetics, water bottles production, etc. It provides an air-tight packaging [4].

- *High-density polyethylene (HDPE)*

High-density polyethylene (HDPE) is a hard, opaque, lightweight, and strong plastic. It possesses mostly linear structure that imparts 90% crystallinity, lower impact strength, tear strength, higher tensile, and bursting strength. It is stiffer than low-density polyethylene (LDPE) having density 0.92–0.94 g/cm^3. HDPE possesses excellent resistant to oil, grease, and moisture, and is the most prevalent plastic film in use. Sealing in HDPE is comparatively difficult and HDPE are blow molded to form bottles, milk jugs, or opaque, such as the packaging for household detergents or bleaches, plastic bags for carrying food and retail items, reusable shipping containers, and wire and cable sheathing.

- *Low-density polyethylene*

LDPE is tough translucent material generally formed under high pressure that leads to great extent of chain branching with long and short chains. The presence of branch chains prevents the close packaging of the chain that imparts less crystallinity (50–70%) and relatively low density of 0.91–0.94 g/cm^3. It possesses good tensile strength, burst strength, tear strength, and impact strength. It possesses good resistant to water, water vapor, acid, alkali inorganic solution but possesses poor resistant to oil, grease, and hydrocarbon. Chief characteristic is that it can be fusion welded to itself to give good, tough, liquid-tight seals; thus it is ideal for packaging applications involving heat sealing. Also used to protect dry cleaning and as bags for bread. It also coats food cartons, flexible lids, and disposable plates and cups. Recycled LDPE is used in the production of heavy-duty garbage bags, paneling, lawn furniture, trashcans, and floor tile.

- *Polypropylene*

PP is a rigid, tough, and, crystalline thermoplastic hydrocarbon resin. It is cheapest plastic used and possesses density in the range of 0.898 g/cm^3 to 0.920 g/cm^3. PP is highly flammable plastic and possesses excellent resistance to acids, alcohols, bases aldehydes, esters, aliphatic hydrocarbons, ketones. It possesses good resistance to steam sterilization thus it is used to make hospital accessories; it can be molded into bottles, pots, electronic industry, crates, graphic arts applications, disposable diaper tabs, packaging, closures, etc.

- *Polyvinyl chloride*

Polyvinyl chloride (PVC) is the most commonly used plastic having density of 1.3–1.7 g/cm^3. It exists in two general forms: rigid and flexible grades. PVC possesses some extent resistant to acid, base, alcohols, fuels, and some paint thinner. Rigid PVC is mainly used for making pipes used in construction; however, this flexible plastic is also used for coating electrical wires, electrical fittings, shower curtain, etc.

- *Polystyrene*

Polystyrene is a clear and hard aromatic polymer having density in the range of 0.96–1.05 g/cm³. It is available as solid or foamed. The expanded polystyrene is known as styrofoam. It is water-resistant and possesses resistance to some acids and bases. The styrofoam is used for food packaging and for making laboratory equipment. Polystyrene is used for making electrical appliances, toys gardening pots, and automobile parts.

The main issue related to the use of plastics is the nonbiodegradability and thus their disposal is a menace. The most effective way of handling this issue is to recycle the products after use. Recycled plastics find many applications especially where food packaging is not done. Another way is to work with alternatives such as bioplastics.

7.3 Bioplastics

In pursuit of environmental safety and sustainability, advancements are being made to develop bio-based materials. Bioplastics are materials obtained from renewable bio-based resources or biodegradable resources. Both the types of bioplastics ensure more environmental safety and greater sustainability in comparison to conventional plastic materials.

Both the kinds of biopolymers, natural and synthetic, are used in packaging industries (Fig. 7.2).

The synthetic biopolymers are more in use pertaining to freedom of enhancement in desirable properties such as high tensile strength, more flexibility, high gloss, clarity, etc. [5]. Polylactic acid (PLA) is the most common synthetic biopolymer. It is biodegradable and has margin to alter its lifespan [6].

Biopolymers are potent candidate in development of plastic materials and that too for food packaging. However, the biopolymers cannot replace plastics in many

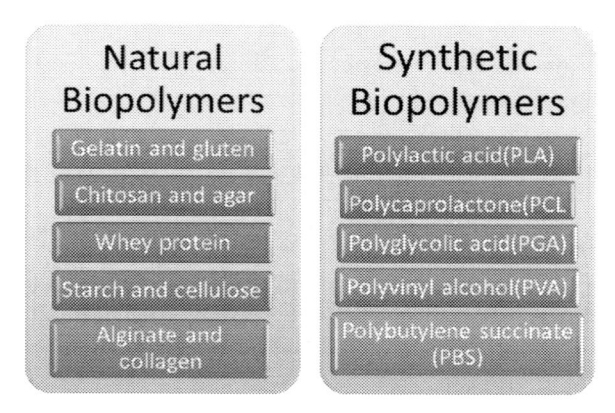

Fig. 7.2 Examples of natural and synthetic biopolymers.

Table 7.2 Different sources of cellulose.

S. No	Source of cellulose	Reference
1	Wheat and cereals	[7]
2	Jute	[8]
3	Soybean husk	[9]
4	Flax fibers and straw	[10]
5	Corn, pineapple, bananas, sorghum etc.	[11]
6	Sugarcane bagasse	[12]

applications such as packaging owing to the less mechanical strength, thermal stability, and solvent impenetrability with respect to conventional plastics.

Studies have shown that the use of nanomaterials may enhance the mechanical and barrier properties of biopolymers [13,14]. Nanocomposites are polymeric materials where most of the dispersed phase components have nanomaterials. Nanomaterials have one or more dimensions of the order of 100 nm or less. In development of nanocomposites, the nanofillers have a crucial role in improving the mechanical and barrier properties as the tension of matrix could be transferred to nanofillers through the interface [15,16]. It is also reported that the presence of nanofillers in the dispersed phase increases the barrier strength of material by controlling the extent of gaseous through it [15].

7.4 Cellulose

As reported by Anselme Payen, a French Chemist, cellulose is a fibrous solid resistant to action of acid and ammonia (during treatment of different plant tissues) [17]. The study showed that the molecular formula of cellulose is $C_6H_{10}O_5$ and it is isomeric to starch. Now we know the exact structure of cellulose as a natural polymer with monomeric unit of β−D-glucose [18]. The natural sources of cellulose are wood, cotton, hemp, sisal, etc. A number of agricultural residues can also act as source of cellulose (Table 7.2).

Cellulose is ubiquitously used in polymeric industries, production of papers, fibers, and films due to its renewable nature and eco-safety. The nano form of cellulose offers more functionalities. One of these applications has been the development of nanocellulose in virtue of its super functionalities, due to its extremely large, active surface area and low cost [37].

7.5 Nanocellulosic composites: potential to replace plastics

Nanotechnology is a revolutionary technique to improve properties of any substance finding application in medicine, sensing, composites, etc. [19]. Nanomaterials are replacing the macromaterials owing to their extraordinary features as mentioned by Salimi and coworkers (Fig. 7.3) [20].

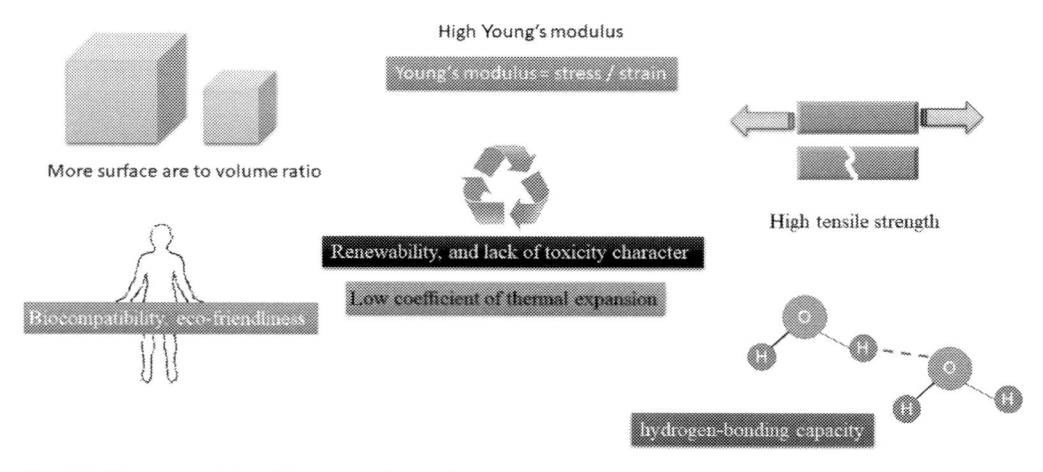

Fig. 7.3 Nanomaterials with extraordinary features.

In the last few years, nanocellulose has also appeared as a promising material in place of cellulose [21]. Nanocellulose can be of different forms based on source (plant origin or synthesized by bacteria) and structure (Fig. 7.4).

Electron microscopic studies have shown that in NFC (nanofibrillated cellulose) the cellulosic fiber is generally of a micrometer length and smaller than 100 nm of diameter. The cellulosic fiber entangles and may have both crystalline and amorphous portions [22]. The electron microscopic studies revealed that the structure of NCC (nanocrystalline cellulose) is crystalline and rod like [23]. The diameter may range from 5 to 40 nm. NCC can be produced by acid hydrolysis of an innate cellulose by removal of

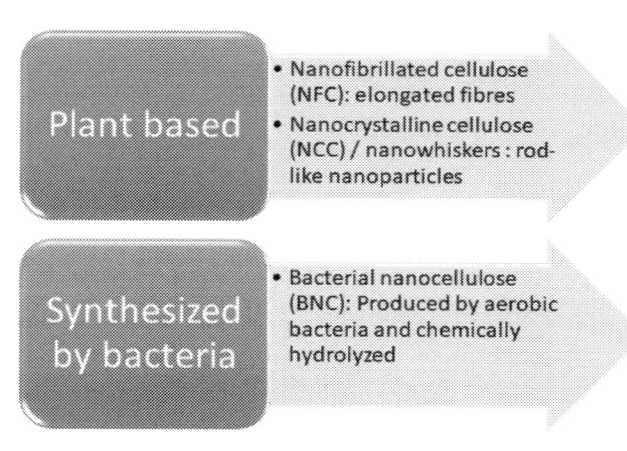

Fig. 7.4 Types of nanocellulose [Khalil et al. [24]].

the amorphous portions. Bacterial nanocellulose is produced by bacteria and generally of remarkable purity, more biocompatible and crystalline as compared to the above two types of nanocellulosic fibers [24,25].

7.5.1 Nanocellulose: enhancement in properties packaging material

The application of nanocellulosic material in plastic industries is primarily as packaging component or as reinforcement material in composites.

Manufacturing of a packaging material is a complex process involving insightful engineering to get a product of desirable properties [26]. The presence of nanocellulose may enhance the various mechanical properties such as tensile strength, Young's modulus, breaking point of a material, etc. There are several reports showing the positive effect of increasing the content of nanoparticle in a composite. Abdollahi et al. showed that the tensile strength of a composite increased by about 33% on increasing the content of high-strength cellulose nanoparticles from zero to five weight percentage. Packaging industry requires a film that can show a better puncture resistance, a study showed that the increase in concentration of NCC can increase the puncture resistance in films of starch/gelatin/NCC [27].

The resistance to sorption and diffusion of air are indicative of barrier properties of a packaging material. The barrier properties directly influence the shelf life of the packaged food item. For packaging materials the barrier properties of major importance are oxygen and water vapor permeability. A study showed that NCC improved barrier property of a nanocomposite film by making path sinuous for moisture reaching the membrane [28]. The water vapor permeability in a PVA (Polyvinyl alcohol) nanocomposite observed a decline by 7%, 20%, and 29% with addition of 3%, 6%, and 9% of 30 oct 2020 NCC as filler, respectively [29]. The addition of NCC or NFC shown to increase the impermeability of membrane for oxygen and water vapor [30,31]. The NCC along with other nanoparticles of other material is used with bioplastics, PLA, to increase the mechanical properties, thermal properties, and resistance toward bacteria [32].

The presence of nanocellulose can also improve the thermal properties of a composite. The analysis of thermal stability of a polymeric material can be done by thermogravimetric analysis (TGA), dynamic mechanical analysis, and differential scanning calorimeters (DSCs). As reported, the blends of PLA/polyhydroxy butyrate (PHB) with nanocellulose exhibit an improved interaction among components shown by one-step degradation as compared to blend of only PLA and PHB, which showed a two-step degradation as studied by TGA [33].

The use of nanocellulosic material for food packaging is also gaining attention as it can help in holding the freshness, decreasing the rate of decay or ripening, and retaining the nutrients such as ascorbic acid of the food item [34].

A study also showed the superiority of nanocellulose-based composites over nylon-based packaging material in increasing the shelf life of ground meat due to decrease in the growth of lactic acid bacteria. The study was conducted at two different temperatures for a storage period of six days [35]. In similar study an improvement in the shelf life and overall quality of fruit strawberry was reported by application of a blend of chitosan with nanocellulose. This can be attributed to the effect of coating of the blend on the decrease in senescence and metabolism rate. The presence of nanocellulose along with chitosan helped to preserve the phenols and anthocyanins in the strawberry making it look fresh and keeping the nutrient value. Chitosan is also a natural polymer and is the second most abundant polysaccharide after cellulose. It is biocompatible, biodegradable, nontoxic, and is used in packaging materials as it also shows antifungal and antimicrobial activities. The NCC is commonly used with chitosan to enhance the tensile strength of the packaging material even at low concentration of 3–5%. It is reported that the two natural polysaccharides form a percolating network and show strong filler–matrix interaction. As mentioned previously, NCC enhances the barrier properties of the chitosan also. NCC gets easily dispersed in the chitosan matrix forming a homogenous structure. Thus NCC-reinforced nanocomposite films have a bright future as a packaging material, replacing the plastic material. In another study a high tensile strength (up to 8.2 MPa) was observed with 6% cellulosic nanocrystals in thermoplastic cassava starch. These films were tested for their prospective use in food packaging [36].

7.6 Conclusions

It is a daunting fact that 50% of the plastic is used only once which includes plastic bottles, coffee cup lids, and food packaging. The amount of plastic packaging that is thrown away every single year is enough in length to circle the globe four times [37]. In nut shell, it is requisite to replace the conventional plastics by other environment-benign materials in different applications such as packaging, etc. The use of bioplastics is an alternative to plastics; however, the drawback with the bioplastics is their poor mechanical strength. This urged scientists to work for bionanomaterials as a promising alternative in food packaging, etc. The studies have shown that the utilization of nanoparticles, such as nanocellulose, enhances the mechanical, thermal, and barrier properties of biopolymers.

The nanocellulosic materials are securing their position as a strong contender for food packaging materials owing to its green nature, environment sustainability, recyclability, and compatibility to form excellent composites with other biopolymers.

The studies have also confirmed the increase in shelf life of food items with the usage of nanocellulose-based composites in packaging. Thus nanocellulose offers an excellent opportunity in development of nanocomposites of biopolymers for development of high-quality packaging material and may help to reduce the burden of plastics on the mother earth.

References

[1] D.S. Lee, K.L. Yam, L. Piergiovanni, Food Packaging Science and Technology, CRC Press, Boca Raton: New York, 2008, pp. 243–274.

[2] L. Piergiovanni, European Union (EU) packaging. In: In The Wiley Encyclopedia of Packaging Technology, John Wiley & Sons, Inc., Hoboken, NJ, 2009, pp. 883–884.

[3] Industry report Mordor Intelligence. https://www.mordorintelligence.com/industry-reports/plastic-packaging-market.

[4] R. Proshad, M.S. Islam, T. Kormoker, M.A. Haque, M.D. Mahfuzur Rahman, et al., Toxic effects of plastic on human health and environment: A consequences of health risk assessment in Bangladesh, Int. J. Health 6 (2018) 1–5.

[5] N.G. Rhim, Jong-Whan, K.W. Perry, Natural Biopolymer-Based Nanocomposite Films for Packaging Applications. Crit. Rev. Food Sci. Nutr. 47 (4) (2007) 411–433. doi:10.1080/10408390600846366. In this issue.

[6] K. Halász, Y. Hosakun, L. Csóka, Reducing water vapor permeability of poly (lactic acid) film and bottle through layer-by-layer deposition of green-processed cellulose nanocrystals and chitosan, Int. J. Polym. Sci. 2015 (2015).

[7] A. Alemdar, M. Sain, Biocomposites from wheat straw nanofibers: morphology, thermal and mechanical properties, Compos. Sci. Technol. 68 (2) (2008) 557–565.

[8] M.S. Jahan, A. Saeed, H. Zhibin, Y. Ni, Jute as raw material for the preparation of microcrystalline cellulose, Cellulose 18 (2011) 451–459.

[9] Y.U. Nelson, A.G. Edgardo, A.W. Ana, Microcrystalline cellulose from soybean husk: effects of solvent treatments on its properties as acetylsalicylic acid carrier, Int. J. Pharm. 206 (2000) 85–96.

[10] A.M. Bochek, I.L. Shevchuk, V.N. Lavrentev, Fabrication of microcrystallinecellulose and powdered cellulose from short flax fiber and flax straw, Russ. J. Appl. Chem. 76 (10) (2003) 1679–1682.

[11] S. Istvan, D. Plackett, Microfibrillated cellulose and new nanocomposite materials: a review, Cellulose 17 (3) (2010) 459–494.

[12] D. Bhattacharya, L.T. Germinario, W.T. Winter, Isolation, preparation and characterization of cellulose microfibers obtained from bagasse, Carbohydr. Polym. 73 (3) (2008) 371–377.

[13] S. Sanuja, A. Agalya, M.J. Umapathy, Studies on magnesium oxide reinforced chitosan bionanocomposite incorporated with clove oil for active food packaging application, J. Polym. Mater. Polym. Biomater 63 (2014) 733–740.

[14] J.P. Reddy, J.W. Rhim, Characterization of bionanocomposite films prepared with agar and paper-mulberry pulp nanocellulose, Carbohydr. Polym 110 (2014) 480–488.

[15] P. Kanmani, J.W. Rhim, Physicochemical properties of gelatin/silver nanoparticle antimicrobial composite films, Food Chem 148 (2014) 162–169, doi:10.1016/j.foodchem.2013.10.047.

[16] E. Trovatti, S.C.M. Fernandes, L. Rubatat, C.S.R. Freire, A.J.D. Silvestre, C.P Neto, Sustainable nanocomposite films based on bacterial cellulose and pullulan, Cellulose 19 (2012) 729–737, doi:10.1007/s.10570-012-9673-9.

[17] A. Payen, C.R. Hebd, Seances, Acad. Sci. (7) (1838) 1052–1056.

[18] C. Doree, The Methods of Cellulose Chemistry Including Methods for the Investigation of Substances Associated with Cellulose in Plant Tissues, Chapman and Hall, London, 1947, p. 543.

[19] A. Arof, N.M. Nor, N. Aziz, M. Kufian, et al. Investigation on morphology of composite poly (ethylene oxide)-cellulose nanofibers. Mater. Today Proc. 17 (2019) 388–393, doi:10.1016/j.matpr.2019.06.265.

[20] S. Salimi, et al., Production of nanocellulose and its applications in drug delivery: a critical review. ACS Sustain. Chem. Eng. 7 (2019) 15800–15827, doi:10.1021/acssuschemeng.9b02744.

[21] L. Bacakova, et al., Versatile application of nanocellulose: from industry to skin tissue engineering and wound healing. Nanomaterials 9 (2019), doi:10.3390/nano9020164.

[22] A. Suzuki, C. Sasaki, C. Asada, Y. Nakamura. Production of cellulose nanofibers from Aspen and Bode chopsticks using a high temperature and high pressure steam treatment combined with milling, Carbohydr. Polym. 194 (2018) 303, doi:10.1016/j.carbpol.2018.04.047.

[23] E. Fortunati, P. Benincasa, G.M. Balestra, F. Luzi, A. Mazzaglia, D. Del Buono, D. Puglia, L. Torre. Revalorization of barley straw and husk as precursors for cellulose nanocrystals extraction and their effect on PVA_CH nanocomposites, Ind. Crops Prod 92 (2016) 201.

[24] H.P.S. Abdul Khalil, A.H. Bhat, A.F. Ireana Yusra. Green Composites from Sustainable Cellulose Nanofibrils: A Review, Carbohydr. Polym. 87 (2012) 963.

[25] A. Müller, M. Zink, N. Hessler, F. Wesarg, F.A. Müller, D. Kralisch, D. Fischer. Bacterial nanocellulose with a shape-memory effect as potential drug delivery system, RSC Adv 4 (2014) 57173, doi:doi.org/10.1039/C4RA09898F.

[26] C.J. Webert, V. Haugaard, R. Festersen, G. Bertelsen, Production and applications of biobased packaging materials for the food industry, Food Addit. Contam. 19 (2002) 172–177, doi:10.1080/026500301 1008748.3.

[27] J.S. Alves, K.C. Reis, E.G.T. Menezes, F.V. Pereira, J. Pereira, Effect of cellulose nanocrystals and gelatin in corn starch plasticized films, Carbohydr. Polym. 115 (2015) 215–222, doi:10.1016/j.carbpol.2014.08.057.

[28] S.A. Paralikar, J. Simonsen, J. Lombardi, Poly (vinyl alcohol)/cellulose nanocrystal barrier membranes, J. Membrane Sci. 320 (2008) 248–258.

[29] H.A. Silvério, W.P.F. Neto, D. Pasquini, Effect of incorporating cellulose nanocrystals from corncob on the tensile, thermal and barrier properties of poly (vinyl alcohol), Nanocompos. J. Nanomater. 2013 (2013) Article ID 289641.

[30] C. Aulin, M. Gallsted, T. Lindstrom, Oxygen and oil barrier properties of microfibrillated cellulose films and coating, Cellulose 17 (2010) 559–574.

[31] S. Belbelchouche, J. Bras, G. Sequeira, C. Chappey, L. Lebrun, B. Khelifi, Water sorption behaviour and gas barrier properties of cellulose whiskers and microfibrils films, Carbohydr. Polym. 83 (2011) 1740–1748.

[32] E. Fortunati, I. Armentano, Q. Zhou, A. Iannoni, E. Saino, L. Visa, L.A. Berglund, J.M. Kenny, Multifunctional bionanocomposite films of poly (lactic acid), cellulose nanocrystals and silver nanoparticles, Carbohydr. Polym. 87 (2012) 1596–1605.

[33] M.P. Arrieta, E. Fortunati, F. Dominici, E. Rayón, J. López, J.M. Kenny (2015).

[34] H.M.C. Azeredo, K.W.E. Miranda, H.L. Ribeiro, M.F. Rosa, D.M Nascimento, Nanoreinforced alginate–acerola puree coatings on acerola fruits, J. Food Eng. 113 (4) (2012) 505–510.

[35] D. Dehnad, H. Mirzaei, Z. Emam-djomeh, S. Jafari, Thermal and antimicrobial properties of chitosan – nanocellulose films for extending shelf life of ground meat, Carbohydr. Polym 109 (2014) 148–154, doi:10.1016/j.carbpol.2014.03.063.

[36] J.A. Piermaria, A. Pinotti, M.A. Garcia, AG. Abraham, Films based on kefiran, an exopolysaccharide obtained from kefir grain: development and characterization, Food Hydrocoll. 2 (2009) 684–690.

[37] Protega global Web source: https://www.protega-global.com/blog/10-daunting-plastic-packaging-statistics. Last access: 30 Oct 2020.

CHAPTER 8

Nanocellulose in the sports industry

Archana Singh[a], Deepak Rawtani[b], Shruti Jha[c]
[a]School of Engineering and Technology, National Forensic Sciences University (Ministry of Home Affairs, GOI), Gandhinagar, Gujarat, India
[b]School of Pharmacy, National Forensic Sciences University (Ministry of Home Affairs, GOI), Gandhinagar, Gujarat, India
[c]Sardar Vallabhbhai National Institute of Technology, Surat, Gujarat, India

8.1 Introduction

The advent of the twenty-first century has established the development of new technologies. Different materials that are still used in daily life somehow manage to harm the environment in different ways. Therefore, studies have been conducted on renewable resources, environmental and waste management strategies that owe to curb down a considerable amount of harm done to the environment. One of the most harmful used materials is plastics, as they are nonbiodegradable (takes around 1000 years to decompose); nevertheless it is an abundant material at present because of its cost effectiveness and its ease of availability. Biodegradable materials are used to overcome such decomposition hurdles. Nanocellulose is a biodegradable material with exclusive characteristics such as recyclable nature, ease in processing, cost effectiveness, lofty strength, and ease in surface modification [1]. Apart from its wide spectrum applications in the sports industry, nanocellulose has also been employed as an adsorbent for toxic metal ions, removal of anionic species, antibiotics, etc., [2].

Sources of nanocellulose include algae, bacteria, plant cell walls, cotton, straw, wood, hemp, agro-industrial waste, and tunicates [3]. Cladophorales and siphonoclades are an order of green algae that serve as a natural source of cellulose. Nanocellulose fibrils with width of 20–30 nm can be extracted from these algae via acid hydrolysis [4]. Algae biomass plays an important role in many industrial sectors as well as in nanocellulose production [5]. Bacterial nanocellulose (BNC) is a class of nanocellulose synthesized by bacteria and it is useful for protection against the chemical environment and UV radiation [6,7]. The dissimilarity between plant cellulose and bacterial cellulose is their chemical and physical properties. The diameter and crystallinity of plant cellulose are approximately 13–22 µm and 45–65%, respectively [8]. For instance, the *Juncus* plant produces cellulose microfibers (CMF) and cellulose nanofibrils (CNF) where the diameter of CMF is around 3.5 µm and the diameter of CNF is 2.5 nm [9]. The primary source of nanocellulose is wood [10]. Nanocellulose is also used for making nanopapers [11]. Tunicates belong to the subphylum "*Tunicata*" that produces cellulose from outer

133

tissue called *tunic* [12]. Cellulose extracted from tunicates shows the best recovery of platinum and palladium [13]. Nontoxicity, renewability, sustainability, high-specificity strength, and biodegradable nature are some of the intrinsic characteristics of hemp [14] and also serve as a source of cellulose [15]. For the extraction of nanocellulose from hemp, chemical and mechanical treatment is performed where chemical treatment involves enzymatic hydrolysis [16], ionic liquid [17], acid hydrolysis [18], or treatment with an alkali [19]. Mechanical treatment of hemp involves steam explosion [20], cryogenic grinding, homogenization [21], and mechanical grinding. Chemical and mechanical treatment strategies are combined to extract nanocellulose. The combination of both treatment strategies is advantageous as it decreases the energy consumption used in the mechanical treatment, additionally this combination of chemical and mechanical treatment also reduces the cost of production [22]. The fabrication of nanocellulose from cellulose involves different methodologies. Enzymatic hydrolysis [23], substrate-allied, and enzyme-allied elements are some of the fabrication techniques from which nanocellulose can be obtained [24]. The intrinsic characteristics of ionic liquid include nonflammability, low vapor pressure, recyclability, chemical, and thermal stability. These ionic liquids are also called "green solvents" [25] and are utilized for fabricating nanocellulose fiber and crystals. It has a low melting point (<100 °C) [26]. Acid hydrolysis is one of the rapid methods that produces nanocellulose of smaller size [27] and eliminates the amorphous region of cellulose to produce nanocellulose crystals [28]. Acid hydrolysis carries properties such as low cost, less time-consuming, and easily accessible [29]. Problems such as low melting point, poor fiber wetting, and adhesion are conquered by alkali treatment and remove the noncellulosic region [19]. The steam explosion process involves high-pressure steaming followed by decomposition [20] used to extract cellulose from feed stocks and biomass. Hence, provides appreciable characteristics such as environmentally friendly, low cost, and great durability [30]. The process where the sample is cooled before grinding and the cryogenic temperature is maintained throughout the process is termed cryogenic grinding [31], which involves crushing with the shear process and high impact [32]. Homogenization is an appreciable approach due to its reproducibility, high efficiency [33], and the environment-friendly process performed to reduce the size of cellulose [34]. Mechanical grinding produces smooth nanocellulose [35]. However, mechanical treatment is preferable over chemical treatment as there is no involvement of chemicals.

Primarily nanocellulose is classified on the basis of their structure into the following important classes represented in Fig. 8.1, namely [2];

- Nanocrystalline cellulose (NCC),
- Nanofibrillated cellulose (NFC),
- Bacterial nanocellulose (BNC).

The basic differences between these categories lie in particle or nanocrystal size, morphology; these are the primary considerations that have been considered corresponding

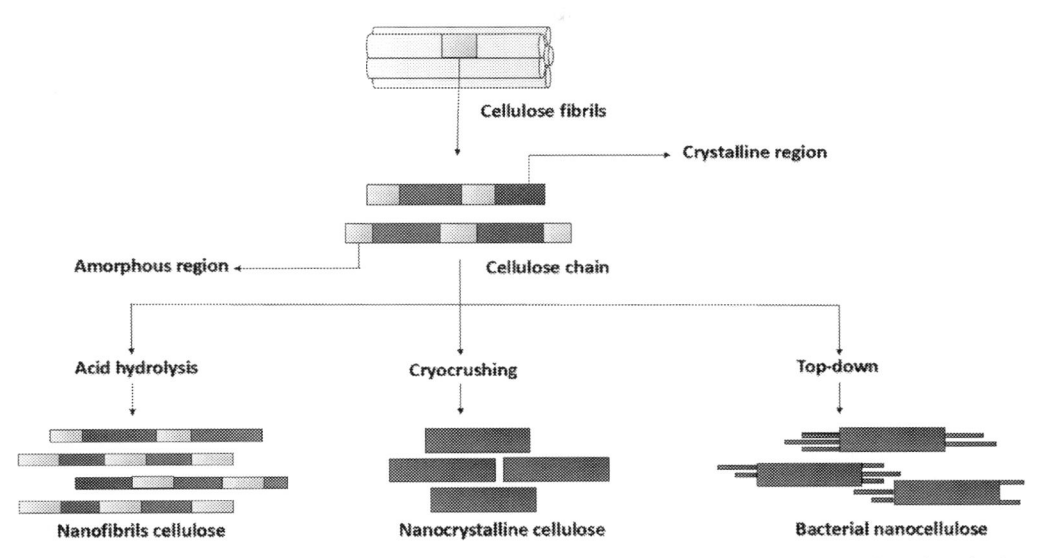

Fig. 8.1 Synthesis of nanocellulose via mechanical and chemical strategies. Reproduced with the permission of [100].

to the varying source and extraction strategies. They are homogeneous in chemical compositions [36]. Nanocrystalline nanocellulose possesses attributes such as biocompatibility, biodegradability, high surface area, lightweight, stiffness, optical properties, feasibility, improved mechanical properties, and nontoxicity. However, the structure differs based on length (5–20 nm) and diameter (100 nm) [37,38]. In the polymer, NCC shows extremely poor dispersion because of a greater number of a surface hydroxyl group [39]. Lignocellulose matrix consists of lignin, cellulose, hemicellulose, and other materials that are disintegrated regarding cellulose extraction to produce NCC [38]. The NFC is fabricated from cellulose by various mechanical and chemical treatments and carries two major limitations—the primary limitation of NFC corresponds to an excessive amount of hydroxyl functional groups and the latter drawback related to NFC is its greater hydrophilicity [40]. However, it has also gained attention due to the presence of abundant sites of reaction and substantial surface area [41]. A. J. Brown in the late nineteenth century reported the production of BNC from the bacteria *Acetobacter xylinum*. The synthesis of BNC from this bacterial strain could be observed in oxygen and glucose-rich environment [42,43]. The sources of BNC may vary, and different species of bacteria and fungi may act as a source for the extraction of BNC. *Rhizobium* and *Agrobacterium* are the most commonly utilized sources for BNC extraction [8]. The biocompatibility of BNC is exceptional, the physiochemical characteristics and sturdiness of BNC are remarkable too. Biodegradable nature [44], augmented tensile strength, and water retention capacity are some other intrinsic characteristics of BNC. The application spectrum of BNC as broad as it can be used for electronics [45], manufacturing cement composites [46], food, cosmetics, medicine [8], textile [47], and many more. The difference between various parameters of the three nanocellulose (BNC, NCC and NFC) have been discussed in Table 8.1.

Nanocellulose obtained the attention of many researchers and scientists because of its excellent intrinsic characteristics and usages in multifarious domains [48]. Nanocellulose has applications in numerous fields such as the sports industry [49], pharmaceutical industry [50], food industry [51], biomedical products [52], paper industry [53], textile industry [54], and so on. The lineage of sports emerged centuries ago and it is celebrated still. Sports involves strategic reproach and thinking, strength, physical and mental activeness to compete with different opponents. Sports brings numerous organizations and communities together, and different activities in sports include different outdoor and indoor games such as hockey, cricket, chess, weightlifting, badminton, boxing, golf, soccer, volleyball, tennis, table tennis, etc. The sports industry utilizes plastic composites for manufacturing different products such as shoes, rackets, footballs, mouth guards, shoulder pads, and different protective gears because of their light weightiness, cost-effectiveness, and easy accessibility. But the consequences of plastics are that they are nonbiodegradable, consume fossil resources, produce harmful gases upon burning or heating, and to overcome such hurdles different alternatives corresponding to biodegradable materials are considered [55]. Nanocellulose is a material that has the potential to replace plastics in the sports industry as they are biodegradable, nontoxic, naturally occurring polymer, abundant in nature, and eco-friendly [56]. Hence, the above properties of nanocellulose are utilized in the sports industry, which assist in manufacturing of great protective gears and other numerous products. The advancements in sports throughout the years have seen a sudden surge of interest of the upcoming generation into sports. Hence, the demand for sports products such as shoes, pads, bands, clothes, balls, bats, etc., is escalating. The production of such products of superior quality is a major challenge. Nanocellulose, hence as a naturally occurring material, contributes majorly to the sports industry and provides the best alternative for plastic- or synthetic-based products Fig. 8.2.

This chapter intends to provide a source of information about the morphology and intrinsic characteristics of nanocellulose. Nanocellulose is an abundant nonsynthetic polymer used in the sports industry due to its outstanding intrinsic characteristics, for example, affordability, easy availability, stability, large surface area, high tensile strength, moldability, and comfortability. Nanocellulose is used in manufacturing shoes, clothes, protective guards, sports vehicles, and sports equipment on the basis of mechanical, thermal, optical, and barrier properties. It will also provide information about the procedure and treatment process that is hinged for the manufacturing of products with nanocellulose as the primary constituent.

Table 8.1 Classification of nanocellulose.

Parameters	Nanocrystalline cellulose	Nanofibrils cellulose	Bacterial nanocellulose
Another entitles	Cellulose nanocrystals, crystals, crystallites, whiskers, microcrystals	Microfibrillated cellulose, nanofibrils, microfibrils, nanofibrillated cellulose	Bacterial cellulose, microbial cellulose, bio-cellulose
Extracted from	Cellulose fibrils	Cellulose fibrils	Lignocellulosic biomass
Treatment	Acid hydrolysis	Cryocrushing, High-pressure homogenization	Top-down
Size	Diameter: 5–70 nm	Diameter: 1–100 nm	Diameter: 20–100 nm
	Length: 100–250	Length: 500–2000 nm	Length: 100 nm
Tensile strength	13–180	7.5	0.12
Young modulus	80	120–143	145
Advantages	Renewable, light in weight	Highly absorbed, nontoxic	Electrically conductive, photonics
Industrial sector	Food agriculture, textiles	Electronics, biomedicines	Sports industry, photonics

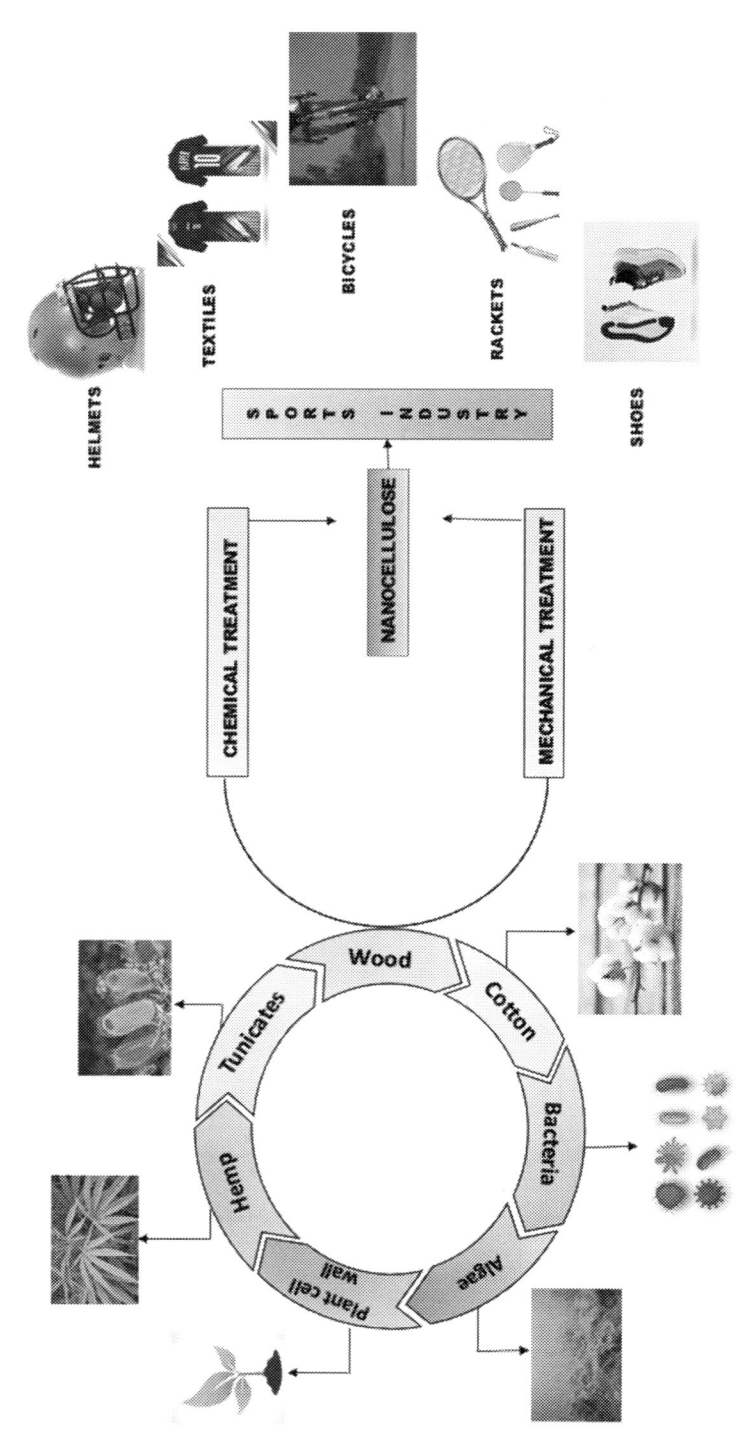

Fig. 8.2 Sources of cellulose extraction, nanocellulose fabrication, and its application in the sports industry.

8.2 Nanocellulose–morphology and characteristics

Cellulose is a naturally occurring abundant polymeric structure with numerous useful properties such as miniature dimensions, greater absorption potential, and heterogeneity in shapes and sizes [57]. The length of wood cellulose fiber lies in the range of 1–3 mm and the size of cotton, ramie, and flax is generally greater, unlike cellulose. The thickness of cellulose fiber lies in the range of 3–6 μm and the width of cellulose fibers lies around 10–30 μm. The fiber of wood is flat, cotton fiber is twisted, and the fiber of flax and ramie is round and straight [58]. Cellulose is a polysaccharide of 1, 4-β glycoside bonds linked with D-glucopyranose. The density of the cellulose crystal morphology and its physical characteristics is determined by the triple hydroxyl (-OH) functional groups, the secondary hydroxyl group present at carbon no. 2 and 3, and the primary hydroxyl group that is present at carbon no. 6. Due to the high molecular mass of cellulose that is, 500.000 Da, and its long-chain structure, cellulose tends to be nonsoluble in water. Nanocellulose is a great alternative to cellulose owing to its ease in modification and water-soluble characteristics [27]. Nanocellulose is synthesized from cellulose by the means of mechanical and chemical processing [59]. Due to its nanoscale size, it has a large surface area which ultimately has more surface atoms, and having different electronic properties leads to a change in optical, catalytic, and various properties [60]. While fabricating nanocellulose by chemical and mechanical treatment, there is a change in the structure and properties of nanocellulose [61]. The development of hydrogen (H-) bonds results in the assembly of the cellulose chain in a highly ordered structure owing to the large number of hydroxyl (OH) group on the surface of cellulose. The cell wall of the fiber consists of polymeric material, and hydrogen bond plays an important role in adhesion between the nanocellulose and polymeric material. Cellulose nanocrystal (CNC) has a cylindrical shape, right-hand chirality, and rigidity [62]. The dissolution of the amorphous region and the elimination of the noncellulosic region lead to changes in the morphological structure of nanocellulose. The thermal properties and structure of the nanocellulose are highly dependable on the sources and extraction process. Cellulose I and cellulose II are present in nanocellulose attributed by X-ray diffraction. The chemical modification of nanocellulose can damage its properties so it has to be lenient. Chemical modification is categorized into three groups: (1) molecules scion on nanocellulose, (2) molecules scion to nanocellulose, (3) replacement of hydroxyl groups [63]. The lower the crystallinity of nanocellulose the lower will be the modulus. The thermal expansion coefficient (TEC) of dried nanocellulose is low due to high crystallinity and strength [64].

Due to nanocellulose tremendous mechanical, optical, thermal properties, its demands are very high in various fields. Nanocellulose is a green material and does not harm the environment, which makes it a crucial material used in various applications such as in manufacturing textiles, in the biomedical field, for packaging and protecting food items, in the sports industry, and so forth. The role of nanocellulose in the sports industry is an important parameter for the manufacturing of sports equipment such as protective shoes, protective clothes, protection pads, helmets, textiles, and many more based on different properties of nanocellulose. Fig. 8.3 illustrates structure of cellulose along with various extraction and enzymatic hydrolytic processes, apart from that it also classifies nanocellulose according to it size.

8.3 Nanocellulose in the sports industry

Nanocellulose is a connate nanomaterial with a 1–3 mm length, ≈30 μm width [36]. Due to the strong interaction between the solvent molecule and hydroxyl groups present on the surface, the nanocellulose can be dissolved in a few strong polar solvents [65]. Nanocellulose is a great alternative to cellulose because of its ease in modification and solubility [27]. Due to its nanoscale size, it has a large surface area which ultimately has more surface atoms, and having different electronic properties leads to a change in optical, catalytic, and various properties Fig. 8.4 [60]. The dissolution of the amorphous region and confiscation of the noncellulosic region lead to changes in the morphological structure of nanocellulose. The thermal properties and structure of the nanocellulose are highly dependable on the sources and extraction process [63]. The lower the crystallinity of nanocellulose the lower will be the modulus. The TEC of dried nanocellulose is low due to high crystallinity and strength. The other properties of nanocellulose are a great surface-to-volume ratio, easy processability, cost-effectiveness, environment friendly, durability, stability, reduced mobility, depletion of penetrant diffusivity, antioxidants, antibacterial, nontoxicity, and biodegradability [66].

The immense properties of nanocellulose make it a demandable material in the sports industry and various equipment such as shoes, textile, protective pads, rackets, cricket bats, and many more are manufactured by utilizing nanocellulose for better results and production [49].

The basic ideology of the sports industry behind the manufacturing of sports textiles is that it provides UV protection, tremendous durability, perspiration control, impermeability toward dirt and water, visual and aesthetic appeal, environment friendly, pathogen resistance [67], incombustible, breathability, biodegradability, electricity resistant [68], and many more. The purpose of protective helmets such as football [69] and bicycle helmets [70], mouth guards [71], shin pads [72], elbow pads [73], shoulder pads [74], sports gloves [75] is to prevent the wearer from severe injuries. Injuries such as brain damage, eye injury, ear bone fracture, internal bleeding, fracture in the spine, etc. Shoes incorporating nanocellulose were designed and manufactured for various sports activities such as volleyball, cricket, badminton, basketball, football, etc. Numerous activities take place under sports and it also involves cars, bikes, and bicycle racing. Games equipment such as cricket bats, hockey, sticks, balls, nets, rackets, sticks, wickets, and so forth play a

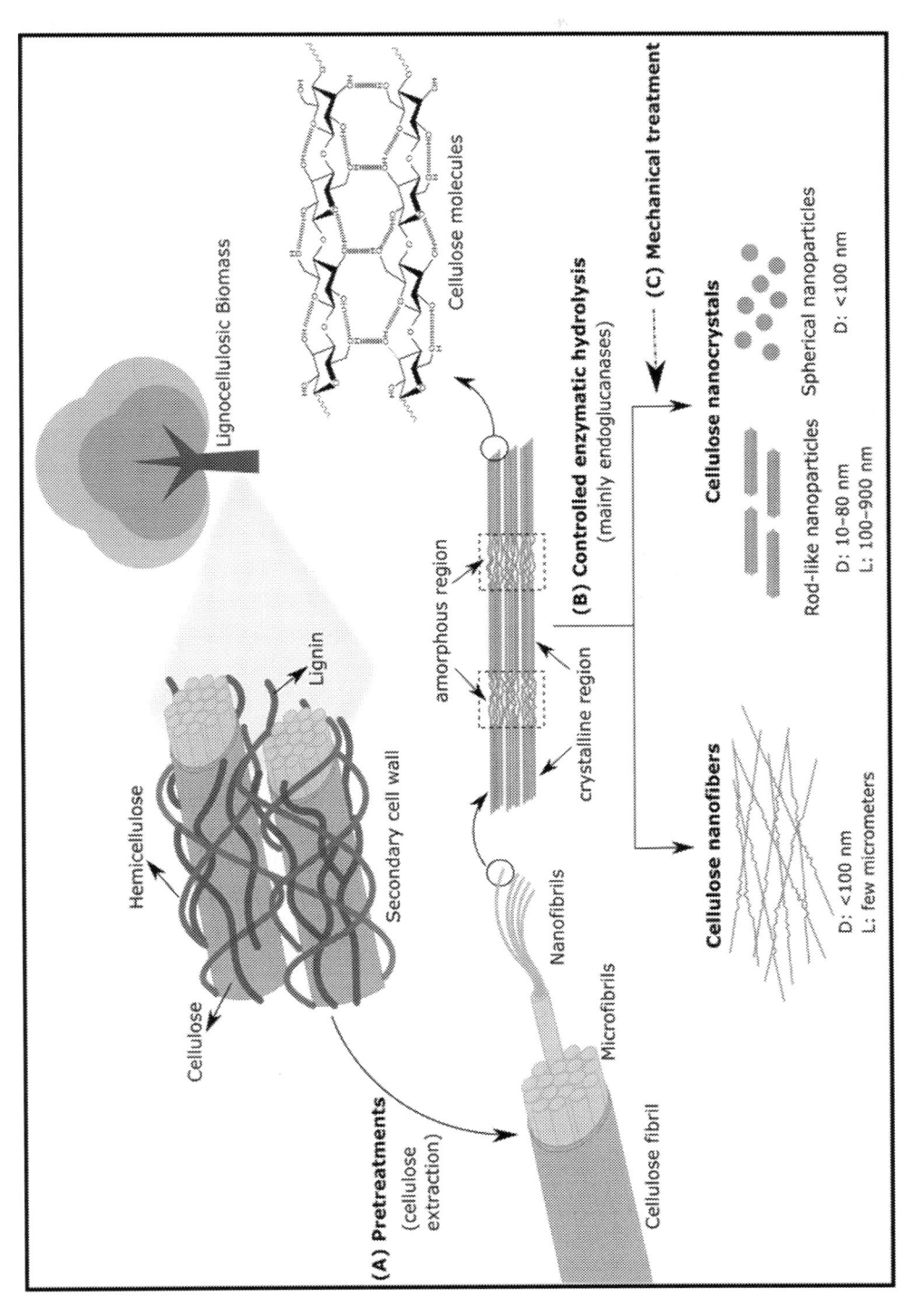

Fig. 8.3 Process for nanocellulose (NC) production through enzymatic hydrolysis. (A) Pretreatments of the lignocellulosic biomass for cellulose extraction; (B) Controlled enzymatic hydrolysis for production of cellulose nanofibers and cellulose nanocrystals (rod-like and spherical) and their respective sizes; (C) Indication of the possible application of mechanical treatment after enzymatic hydrolysis, usually employed to obtain more uniform particles. Reproduced with the permission from [53].

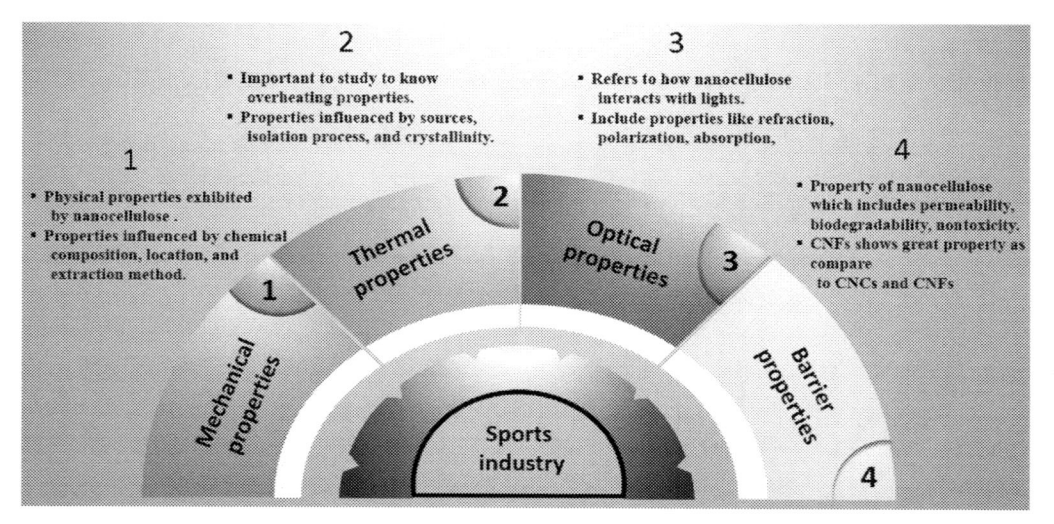

Fig. 8.4 Properties of nanocellulose and applications in the sports industry.

vital role in conducting a sports event. It is responsibility of the sports industry to fulfill the demands of the public along with the quality and durability of the sports products. Another reason for using nanocellulose is to cut down the price of sports products [76]. The sports industry has to meet two crucial aspects that are flexible according to market commute/demands and efficacious products [77].

8.3.1 The mechanical property of nanocellulose

The reaction to different external forces, when applied, upon any substance or material is known as the mechanical or intrinsic physical characteristic of that particular material. It includes properties such as Young's modulus, tensile strength, stiffness, elongation, and fatigue strength, brittleness. These properties can be influenced by factors such as chemical composition and location in plants, separating methods, and many more [78]. The great mechanical properties of nanocellulose make it an ideal material for manufacturing sports products. The first shoe consisting CNF was manufactured by the ASICS company of Japan in 2018 for better performance in long-distance running. For benefaction great mechanical properties, the FlyteFoamLyte1 added along with surface hydrophobized CNF. For the refinement in tensile strength, light-weight, and Young's modulus of the composite, the CNF was added to the matrix. The addition of CNF into the matrix also leads to the strengthening and toughening of foam composites [79]. The various sports activities are easily carried out by lightweight shoes that is why the demand for lightweight shoes is high, and to fulfill this condition nanocellulose is utilized which has low density [80]. The concern of the sports industry is to manufacture sports clothes that are adhesive and mechanically firm throughout wet and dry circumstances. The film degradation and loss of properties occur due to swelling up of sturdy hydrogen-bond of nanocellulose material in aqueous condition and to overcome such problems cationic polyelectrolyte poly(ethylene imine) (PEI), covalent cross-linking, and functional copolymers are used for adhesion and stability. For imparting a physical network of nanocellulose polymer substrate and nanocellulose particles PEI is added in order to form a hydrogen bond with cellulose, affinity with nanocellulose hydrogels, and for cross-linking functional copolymer added and to examine this covalent cross-linking along with polycarboxylic acid is added. All this together improves the adhesiveness and stability of the sports textile [54]. In an experiment the nanocellulose was applied to the polyester fabric by paddling technique and its 2 cm × 8 cm fabric sample was analyzed for the determination of tensile strength. It shows the improvement in bearing capacity and crease recovery angle of fiber or textile [81]. The nanocellulose along with fabric provides great tensile strength and it is used in sports industry to manufacture clothes.

There is a concern in the sports industry to manufacture a helmet of low density and with more impact energy [82]. Chioma E. Njoku in his work extracted *Urena lobata* fiber by water retting method and then treated it with NaOH solution. The tensile property of fiber is better due to more exposure of cellulose on the fiber surface. Hence, it is utilized for the manufacturing of helmets [83]. Aerial roots of the banyan tree treated with alkali results in enhancement of cellulose content, and decrease in hemicellulose, lignin, and wax leads to high tensile strength, rise in density but less than the synthetic fibers. Hence, this is used in helmets due to properties such as high tensile strength and light-weight [84]. Using flex and hemp cellulose in making protective helmets results in high stiffness and strength, easily bendable, and higher vibration damping capacity [85]. The vehicle utilized in sports activities should be lightweight, stable, flexible, and other various properties. Ideas2cycles company in 2018 manufactured a cycle by replacing synthetic fibers with CNF for better quality and properties. Mandrels are utilized to synthesize CNFs which are cut into prescribed length by the companies. These CNFs are further attached to fixture tubes at a particular angle with the help of 3D printing technique [79]. Nanocellulose is used in many different parts of sports cars such as seat, handle, and many more to provide comfort and stability/ balance. Inserting nanocellulose in bicycles consolidates mechanical properties, comfortability, stability, and fortify capacity [86]. In a work, the nanocellulose was extracted, and to decrease the strength between the fibrils they are pretreated with chemical or thermal methods. Fibers are compressed by the homogenizer. This approach is costly; however, development in the neoteric time is feasible to carry out this process at a lower cost. The addition of acid leads to the cancellation of negative charges which in turn allows the fibrils to come close to each other and create a strong bond. Hence, fibrils are miscellaneous with water and sodium chloride and not with water and acid. The fibers of nanocellulose are 1 µm in length and thickness is around 4 and 20 nm so it is passed through a 1.1 mm hole and dried. These obtained fibers are strong and stable, hence used in the manufacturing of bicycles [87].

Darker (the company of Japan) developed a racket with lightweight and high strength by incorporating a layer of nanocellulose to eliminate the repercussion of adhesiveness [79]. Nanocellulose plays a vital role in the sports industry due to its great mechanical properties, which makes it a demandable and interesting material.

8.3.2 Nanocellulose and its intrinsic thermal characteristics

An important dynamic of nanocellulose to focus on is its thermal characteristics. The thermal characteristics of nanocellulose play a significant role in determining its applications in the sports industry. The thermal characteristics of any substance include thermal stability, thermal behavior, thermal conductivity, melting point, glass transition temperature, crystallization temperature, and thermal degradation property. The thermal stability of nanocellulose is influenced by factors such as sources of cellulose, the isolation process of nanocellulose, the crystallinity of nanocellulose. After the incorporation of nanocellulose as a nucleating agent, there is an increase in the thermal behavior of the composites. The thermal conductivity of nanocellulose in the sports industry is important to study to know the overheating properties [88]. Thermal analysis of sports textiles is observed by incorporation of nanocellulose inside the polyester fiber matrix. The decrease in crystallinity degree leads to the shift in temperature of nanocellulose-treated polyester fiber matrix. Thermal degradation reactions start in the amorphous region of nanocellulose material [81]. Nanocellulose undergoes thermal degradation, which involves depolymerization, dehydration, decomposition, and charred residue that occurs between 250 °C and 350 °C. The thermal stability of the cotton textile depends on the thermal stability of nanocellulose used [89]. Thermogravimetric analysis is used to determine the thermal properties such as temperature and the amount of moisture absorbed by nanocellulose [90]. In general, the thermal comfort of the helmets used in sports activities is in a narrow range that is within 34–35 °C. According to the study, there is a slight increase in head temperature of the wearer between 1 and 2°C which may cause discomfort to the wearer [91]. Hence, nanocellulose is utilized in manufacturing helmets for different types of sports. The thermal stability of nanocellulose requires a thorough examination when it comes to the applications of nanocellulose in different products. The nanocellulose has poor thermal stability as compared to the matrix polymer. The nanocellulose in a study showed lower thermal stability for textiles but their degradation temperature is not a limiting factor in the sports industry. Nanocellulose can be used for manufacturing textiles because of other properties such as mechanical and barrier properties [92].

8.3.3 Barrier properties of nanocellulose

Nanocellulose carries great barrier properties, which makes it an important material or polymer in the sports industry. CNFs are a strong gas barrier material as compared to CNCs and BNCs. Hence, it is used as a filler for coating film surface to increase barrier properties and to obtain nanocomposite material [93]. The greenhouse gases are produced in both the process of manufacturing and discarding shoes as they contain synthetic rubbers, chemical-based adhesive, and chromium-tanned leather. Hence, it is toxic and not environment friendly. Nanocellulose is an eco-friendly material as it does not bring out any toxicity, greenhouse gases, and sustains expeditious biodegradation [94]. Polyamide (PA11) along with CNC provides various advantages such as biodegradability, nontoxicity, cost-effectiveness, and energy consumption. Hence, it is used in manufacturing high staging athletic shoes or sports shoes [95]. The removal of hazardous liquid waste such as hydroxide, salts, dyes in a colossal amount, and utilizing a large amount of water during the manufacturing of sports clothes ultimately increase water pollution level. A wood mash added to the homogenizer converts cellulose fibrils into nanocellulose fibrils (NFC) at high shear forces. The diameter of the NFC is around 10–50 nm. The association of NFC with dye at upraised temperature results in colored NFC-dye pigment of nanocellulose. Owing to NFC hydrogels there is an increase in surface to volume ratio. Nanocellulose fibrils overcome the problem of water pollution as it requires six times less water and hazardous substances such as salts, hydroxide, and alkali [96]. During any sports activity, it is natural to sweat, and to maintain the property of the clothes sports industry combines hydrophilic cellulose along with the hydrophobic polyethylene terephthalate (PET) to obtain great infrared transmittance and satisfactory thermal-wet condition. The textile is designed in such a way so that the PET layer could touch the human skin surface and can absorb sweat/water, whereas nanocellulose is a double-layer material that absorbs sweating from PET due to its tremendous hydrophilicity property [97]. The sports industry manufactured clothes are also antifungal and antimicrobial. Here cellulose is attached with an enzyme and an amine-allicin combination is made because the attachment of allicin is not easy with other chemical molecules. Allicin-conjugated nanocellulose and lysozyme-conjugated nanocellulose are used to make antimicrobial and antifungal sports textiles [98]. With the help of the padding approach, nanocellulose incorporated into the textile fiber results in great color strength and soap resistance [81].

The quantity of UV light reaching the earth's surface is determined by time, altitude, latitude, season, etc. It carries a positive and negative effect on human beings by providing vitamin-D, treatment of skin diseases, skin cancer, photosensitivity disorder, etc. Hemp fiber is extricated and used in making UV protection textiles without any involvement of chemical treatment such as bleaching, dying. It is an environmentally friendly, cost-effective, less energy-consumption process [99].

8.3.4 Optical characteristics

The optical characteristics of a material can be defined on the basis of, how a material interacts with lights. It includes properties such as refraction, polarization, absorption, dispersion, diffraction, photosensitivity, etc. The CNF- and CNC-based films are optically transparent when nanocellulose fiber is densely packed and the intervening space between the fibers is small, which avoids scattering of light. CNCs are aligned to form a chiral nematic liquid crystalline phase in equilibrium along with the isotropic phase in the aqueous medium [64]. In one study, the measurement of absorption of polyester fabric treated with nanocellulose was done by drop test and wicking behavior test. Owing to the inherent hydrophilicity of cellulose, there is an improvement in hydrophilicity of polyester treatment when the low dosage of nanocellulose used, but as soon as nanocellulose dosage was increased is starts acting as nanowhiskers and avoid the accommodation of water droplets within the interpolymeric spaces [81]. Table 8.2 represents properties and application of nanocellulose in sports industry.

8.4 Conclusion and future prospects

In recent years, nanocellulose as a polymer material has gained primary focus from different fields of science and industries for the manufacture of different products that are used in day-to-day life. The abundance and the application spectrum of nanocellulose are well established. The real challenge lies in the extraction of nanocellulose. The intrinsic characteristics of nanocellulose-based material make it an impeccable replacement for other plastic-based materials and goods. The benefits of nanocellulose do not go unrecognized and it has been utilized for the manufacturing of various sports products such as shoes, textiles, protective pads, vehicles, gloves, and other miscellaneous equipment whose utilization is evident in the sports industry. A comment upon the durability and physical strength of nanocellulose has to be made as it is used extensively used in the sports industry. Nanocellulose has been already been incorporated into sports equipment such as shoes, rackets, footballs, mouth guards, shoulder pads and headgears, etc. Various studies have been reported that the incorporation of nanocellulose into such products increases their durability and strength. This property of nanocellulose has reduced the degree of injuries that could occur to an athlete or a sportsman in the field. In the future, the utilization of nanocellulose in sports products will result in a low-cost process, ultra-strong products, and light-weight equipment. Many different studies and research works are yet to be conducted for the application of nanocellulose in sports. As an abundant and sustainable material, nanocellulose membranes are expected to expand their usage in the sports industry.

Table 8.2 Sports products incorporating nanocellulose exhibiting different characteristics.

Sports product	Properties	Sources	Treatment	Fabrication process	Instrument	Result	Advantages	Reference
Textile	Tensile strength	Waste viscose rayon fibers	Chemical	Samples were dried and conditioned at 65 ± 2% relative humidity (RH) and 27 ± 2 °C temperature	–	Improvement in the load-bearing capacity of the fiber	No effect on the rigidity of the material	[81]
Textile	Antimicrobial and antifungal	Cellulose manufactured by the My Baby Company in Iran	Acid treatment	Allicin-conjugated nanocellulose attached to cellulose textile by conjugation method and covalent and	XRD, FTIR, SEM, AATCC[a]	Antibacterial activity against *Staphylococcus aureus* obtained	In XRD crystalline phase size is 87.2517 Å. in FTIR difference observed at 3360 and 3387.98 cm-1.	[101]
Textile	Crease recovery angle	Waste viscose rayon fibers	Chemical	The test specimen was folded and compressed under the controlled condition of defined force	–	Crease recovery angle of fabric was improved with an increase in the concentration of nanoparticle	No effect on the rigidity of the material	[81]
Textile	Thermal stability	Cotton	Acid hydrolysis	Two reactions for 60 min and 75 min	X-ray diffraction (XRD), thermogravimetric, scanning electron microscopy (SEM)	Six CNCs suspension	Good crystallinity	[102]

(continued)

147

Table 8.2 (Cont'd)

Sports product	Properties	Sources	Treatment	Fabrication process	Instrument	Result	Advantages	Reference
Textile	Thermal property	Waste viscose rayon fibers	Chemical	2 cm × 8 cm fabric samples were tested at 100 mm/min	Differential scanning calorimetric measurements	An increase in temperature and the endothermic peak of nanocellulose-treated polyester occurred at 253.18 °C	Amorphous content increases from pure polyester to polyester/nanocellulose material	[81]
Textile	Water Absorbency	Waste viscose rayon fibers	Chemical	Polyester fabric treated with 1 g/L concentration of nanocellulose	Drop test and wicking height	Improvement in hydrophilicity of polyester treatment	Inherent hydrophilicity of the cellulose	[81]
Textile	Barrier property	Cotton	Sodium sulfate and sodium carbonate	Preparation of NFC and NFC-dye hydrogels, deposition of NFC-dye hydrogels on cotton fabric, testing	SEM, UV–vis spectroscopy	Variety of colors is achieved	Minimizing water pollution	[96]
Textile	Air permeability	Waste viscose rayon fibers	Chemical		Metefem air permeability tester	The air permeability of polyester fabric treated with nanocellulose was reduced	Occurrence of resistance	[81]

Textile	Barrier property	Cotton linters	Acidic hydrolysis	–	Electron microscopy, XRD, thermal analysis, zeta potential	Nanocrystals have an aspect ratio of 19, crystallinity of 91%	High hydrophilicity	[103]
Helmet	Thermal stability	*Urena lobata*	NaOH treatment	The fiber was compressed in a compression molding machine at 150–180 °C and 100 MPa pressure	Thermogravimetric analysis	More thermal stability (445 °C)	Great stability and thermal property	[83]
bicycles	Tensile strength	–	Chemical or thermal	Fibers of nanocellulose are 1 μm in length and thickness is around 4 and 20 nm so it is passed through a 1.1 mm hole and dried	–	The addition of acid leads to the cancellation of negative charges which in turn allows the fibrils to come close to each other and create a strong bond	Obtained fibers are strong and stable	[87]

[a]XRD, X-Ray Diffraction; FTIR, Fourier Transform Infrared Spectroscopy; SEM, scanning electron microscopy; AATCC, American Association of Textile Chemists and Colorists.

References

[1] P. Phanthong, P. Reubroycharoen, X. Hao, G. Xu, A. Abudula, G. Guan, Nanocellulose: extraction and application, Carbon Resour. Convers. 1 (2018) 32–43. https://doi.org/10.1016/j.crcon.2018.05.004.

[2] D. Rawtani, P.K. Rao, C.M. Hussain, Recent advances in analytical, bioanalytical and miscellaneous applications of green nanomaterial, TrAC Trends Anal. Chem. 133 (2020) 116109. https://doi.org/10.1016/j.trac.2020.116109.

[3] T. Zhong, R. Dhandapani, D. Liang, J. Wang, M.P. Wolcott, D. Van Fossen, H. Liu, Nanocellulose from recycled indigo-dyed denim fabric and its application in composite films, Carbohydr. Polym. 240 (2020) 116283. https://doi.org/10.1016/j.carbpol.2020.116283.

[4] K. Hua, M. Strømme, A. Mihranyan, N. Ferraz, Nanocellulose from green algae modulates the in vitro inflammatory response of monocytes/macrophages, Cellulose 22 (2015) 3673–3688. https://doi.org/10.1007/s10570-015-0772-2.

[5] O.G. Paniz, C.M.P. Pereira, B.S. Pacheco, S.I. Wolke, G.K. Maron, A. Mansilla, P. Colepicolo, M.O. Orlandi, A.G. Osorio, N.L.V. Carreño, Cellulosic material obtained from Antarctic algae biomass, Cellulose 27 (2020) 113–126. https://doi.org/10.1007/s10570-019-02794-2.

[6] A. Parnsubsakul, U. Ngoensawat, T. Wutikhun, T. Sukmanee, C. Sapcharoenkun, P. Pienpinijtham, S. Ekgasit, Silver nanoparticle/bacterial nanocellulose paper composites for paste-and-read SERS detection of pesticides on fruit surfaces, Carbohydr. Polym. 235 (2020) 115956. https://doi.org/10.1016/j.carbpol.2020.115956.

[7] Skočaj, M., 2019. Bacterial nanocellulose in papermaking. Cellulose. 26,6477–6488. https://doi.org/10.1007/s10570-019-02566-y.

[8] J.D.P. de Amorim, K.C. de Souza, C.R. Duarte, I. da Silva Duarte, F. de Assis Sales Ribeiro, G.S. Silva, P.M.A. de Farias, A. Stingl, A.F.S. Costa, G.M. Vinhas, L.A. Sarubbo, Plant and bacterial nanocellulose: production, properties and applications in medicine, food, cosmetics, electronics and engineering. a review, Environ. Chem. Lett. 18 (2020) 851–869. https://doi.org/10.1007/s10311-020-00989-9.

[9] Z. Kassab, S. Mansouri, Y. Tamraoui, H. Sehaqui, H. Hannache, A.E.K. Qaiss, M. El Achaby, Identifying Juncus plant as viable source for the production of micro- and nano-cellulose fibers: application for PVA composite materials development, Ind. Crops Prod. 144 (2020) 112035. https://doi.org/10.1016/j.indcrop.2019.112035.

[10] T. Galia, V. Ruiz-Villanueva, R. Tichavský, K. Šilhán, M. Horáček, M. Stoffel, Characteristics and abundance of large and small instream wood in a Carpathian mixed-forest headwater basin, For. Ecol. Manage. 424 (2018) 468–482. https://doi.org/10.1016/j.foreco.2018.05.031.

[11] A. Winter, B. Arminger, S. Veigel, C. Gusenbauer, W. Fischer, M. Mayr, W. Bauer, W. Gindl-Altmutter, 2020. Nanocellulose from fractionated sulfite wood pulp. Cellulose 27, 9325–9336. https://doi.org/10.1007/s10570-020-03428-8.

[12] H. Kargarzadeh, M. Mariano, J. Huang, N. Lin, I. Ahmad, A. Dufresne, S. Thomas, Recent developments on nanocellulose reinforced polymer nanocomposites: a review, Polymer (Guildf) 132 (2017) 368–393. https://doi.org/10.1016/j.polymer.2017.09.043.

[13] H.J. Hong, H. Yu, S. Hong, J.Y. Hwang, S.M. Kim, M.S. Park, H.S. Jeong, Modified tunicate nanocellulose liquid crystalline fiber as closed loop for recycling platinum-group metals, Carbohydr. Polym. 228 (2020) 115424. https://doi.org/10.1016/j.carbpol.2019.115424.

[14] S.S. Rana, M.K. Gupta, Isolation of nanocellulose from hemp (*Cannabis sativa*) fibers by chemo-mechanical method and its characterization, Polym. Compos. 41 (12) (2020) 5257–5268. https://doi.org/10.1002/pc.25791.

[15] Z. Kassab, I. Kassem, H. Hannache, R. Bouhfid, A.E.K. Qaiss, M. El Achaby, Tomato plant residue as new renewable source for cellulose production: extraction of cellulose nanocrystals with different surface functionalities, Cellulose 27 (2020) 4287–4303. https://doi.org/10.1007/s10570-020-03097-7.

[16] R.S.A. Ribeiro, B.C. Pohlmann, V. Calado, N. Bojorge, N. Pereira, Production of nanocellulose by enzymatic hydrolysis: trends and challenges, Eng. Life Sci. 19 (2019) 279–291. https://doi.org/10.1002/elsc.201800158.

[17] P. Phanthong, S. Karnjanakom, P. Reubroycharoen, X. Hao, A. Abudula, G. Guan, A facile one-step way for extraction of nanocellulose with high yield by ball milling with ionic liquid, Cellulose 24 (2017) 2083–2093. https://doi.org/10.1007/s10570-017-1238-5.

[18] P. Phanthong, G. Guan, Y. Ma, X. Hao, A. Abudula, Effect of ball milling on the production of nanocellulose using mild acid hydrolysis method, J. Taiwan Inst. Chem. Eng. 60 (2016) 617–622. https://doi.org/10.1016/j.jtice.2015.11.001.

[19] K. Lefatshe, C.M. Muiva, L.P. Kebaabetswe, Extraction of nanocellulose and in-situ casting of ZnO/cellulose nanocomposite with enhanced photocatalytic and antibacterial activity, Carbohydr. Polym. 164 (2017) 301–308. https://doi.org/10.1016/j.carbpol.2017.02.020.

[20] B.M. Cherian, A.L. Leão, S.F. de Souza, S. Thomas, L.A. Pothan, M. Kottaisamy, Isolation of nanocellulose from pineapple leaf fibres by steam explosion, Carbohydr. Polym. 81 (2010) 720–725. https://doi.org/10.1016/j.carbpol.2010.03.046.

[21] N. Srinivasababu, K.P. Kumar, Synthesis of nanocellulose fibrils/particles from cellulose fibres through sporadic homogenization. In: Lecture Notes in Mechanical Engineering, Springer Science and Business Media Deutschland GmbH, 2021, pp. 893–902. https://doi.org/10.1007/978-981-15-5463-6_79.

[22] M.A. Dahlem, C. Borsoi, B. Hansen, A.L. Catto, Evaluation of different methods for extraction of nanocellulose from yerba mate residues, Carbohydr. Polym. 218 (2019) 78–86. https://doi.org/10.1016/j.carbpol.2019.04.064.

[23] P.A. Penttilä, A. Várnai, J. Pere, T. Tammelin, L. Salmén, M. Siika-aho, L. Viikari, R. Serimaa, Xylan as limiting factor in enzymatic hydrolysis of nanocellulose, Bioresour. Technol. 129 (2013) 135–141. https://doi.org/10.1016/j.biortech.2012.11.017.

[24] Z. Karim, S. Afrin, Q. Husain, R. Danish, Necessity of enzymatic hydrolysis for production and functionalization of nanocelluloses, Crit. Rev. Biotechnol 37 (3) (2017) 355–370. https://doi.org/10.3109/07388551.2016.1163322.

[25] Y.T. Xiao, W.L. Chin, S.B. Abd Hamid, Facile preparation of highly crystalline nanocellulose by using ionic liquid, Adv. Mater. Res. 1087 (2015) 106–110. https://doi.org/10.4028/www.scientific.net/amr.1087.106.

[26] J.H. Jordan, M.W. Easson, B.D. Condon, Cellulose hydrolysis using ionic liquids and inorganic acids under dilute conditions: morphological comparison of nanocellulose, RSC Adv 10 (2020) 39413–39424. https://doi.org/10.1039/D0RA05976E.

[27] W.T. Wulandari, A. Rochliadi, I.M. Arcana, Nanocellulose prepared by acid hydrolysis of isolated cellulose from sugarcane bagasse, IOP Conf. Ser. Mater. Sci. Eng. 107 (2016) 012045. https://doi.org/10.1088/1757-899X/107/1/012045.

[28] Y. Guo, Y. Zhang, D. Zheng, M. Li, J. Yue, Isolation and characterization of nanocellulose crystals via acid hydrolysis from agricultural waste-tea stalk, Int. J. Biol. Macromol. 163 (2020) 927–933. https://doi.org/10.1016/j.ijbiomac.2020.07.009.

[29] V H.S. Onkarappa, G.K. Prakash, G.H. Pujar, C.R. Rajith Kumar, R. Betageri, Facile synthesis and characterization of nanocellulose from *Zea mays* husk, Polym. Compos. 41 (2020) 3153–3159. https://doi.org/10.1002/pc.25606.

[30] I. Hongrattanavichit, D. Aht-Ong, Nanofibrillation and characterization of sugarcane bagasse agrowaste using water-based steam explosion and high-pressure homogenization, J. Clean. Prod. 277 (2020) 123471. https://doi.org/10.1016/j.jclepro.2020.123471.

[31] K.K. Singh, T.K. Goswami, Design of a cryogenic grinding system for spices, J. Food Eng. 39 (1999) 359–368. https://doi.org/10.1016/S0260-8774(98)00172-1.

[32] I. Hamawand, S. Seneweera, P. Kumarasinghe, J. Bundschuh, Nanoparticle technology for separation of cellulose, hemicellulose and lignin nanoparticles from lignocellulose biomass: a short review, Nano-Struct. Nano-Objects 24 (2020) 100601. https://doi.org/10.1016/j.nanoso.2020.100601.

[33] Y. Wang, X. Wei, J. Li, F. Wang, Q. Wang, J. Chen, L. Kong, Study on nanocellulose by high pressure homogenization in homogeneous isolation, Fibers Polym. 16 (2015) 572–578. https://doi.org/10.1007/s12221-015-0572-1.

[34] M. Mahardika, H. Abral, A. Kasim, S. Arief, M. Asrofi, 2018. Production of nanocellulose from pineapple leaf fibers via high-shear homogenization and ultrasonication. fibers 6 (2), 28. https://doi.org/10.3390/fib6020028.

[35] S. Yang, Q. Xie, X. Liu, M. Wu, S. Wang, X. Song, Acetylation improves thermal stability and transmittance in FOLED substrates based on nanocellulose films, RSC Adv 8 (2018) 3619–3625. https://doi.org/10.1039/c7ra11134g.

[36] A. Isogai, Emerging Nanocellulose Technologies: Recent Developments, Adv. Mater. 32 (2020) 1–10, 2000630. https://doi.org/10.1002/adma.202000630.

[37] D. Haldar, M.K. Purkait, Micro and nanocrystalline cellulose derivatives of lignocellulosic biomass: a review on synthesis, applications and advancements, Carbohydr. Polym. 250 (2020) 116937. https://doi.org/10.1016/j.carbpol.2020.116937.

[38] I.W. Arnata, S. Suprihatin, F. Fahma, N. Richana, T.C. Sunarti, Cationic modification of nanocrystalline cellulose from sago fronds, Cellulose 27 (2020) 3121–3141. https://doi.org/10.1007/s10570-019-02955-3.

[39] K. Jin, Y. Tang, X. Zhu, Y. Zhou, Polylactic acid based biocomposite films reinforced with silanized nanocrystalline cellulose, Int. J. Biol. Macromol. 162 (2020) 1109–1117. https://doi.org/10.1016/j.ijbiomac.2020.06.201.

[40] K. Missoum, M. Belgacem, J. Bras, Nanofibrillated cellulose surface modification: a review, Materials (Basel) 6 (2013) 1745–1766. https://doi.org/10.3390/ma6051745.

[41] T. Keplinger, X. Wang, I. Burgert, Nanofibrillated cellulose composites and wood derived scaffolds for functional materials, J. Mater. Chem. A. 7 (2019) 2981–2992. https://doi.org/10.1039/C8TA10711D.

[42] A.J. Brown, XLIII.—On an acetic ferment which forms cellulose, J. Chem. Soc. Trans. 49 (1886) 432–439.

[43] P. Jacek, F. Dourado, M. Gama, S. Bielecki, Molecular aspects of bacterial nanocellulose biosynthesis, Microb. Biotechnol. 12 (2019) 633–649. https://doi.org/10.1111/1751-7915.13386.

[44] S. Sämfors, K. Karlsson, J. Sundberg, K. Markstedt, P. Gatenholm, 2019. Biofabrication of bacterial nanocellulose scaffolds with complex vascular structure. Biofabrication 11 045010. https://doi.org/10.1088/1758-5090/ab2b4f.

[45] P.A. Nizam, D. Gopakumar, Y.B. Pottathara, D. Pasquini, A. Nzihou, T. Sabu, Chapter 2 - Nanocellulose-based composites: fundamentals and applications in electronics. In: Nanocellulose-based composites for electronics, Elsevier, 2021, pp. 15–29.

[46] M.A. Akhlaghi, R. Bagherpour, H. Kalhori, Application of bacterial nanocellulose fibers as reinforcement in cement composites, Constr. Build. Mater. 241 (2020) 118061. https://doi.org/10.1016/j.conbuildmat.2020.118061.

[47] W. Czaja, E. Shwarz, D. Inselman, Bacterial derived nanocellulose textile material, US Patent Application 20200325600, 2020.

[48] D.A. Gopakumar, S. Thomas, O. F.A.T, S. Thomas, A. Nzihou, S. Rizal, H.P.S. Abdul Khalil, Nanocellulose based aerogels for varying engineering applications, in: S. Hashmi, I.A. Choudhury (Eds.), Encyclopedia of Renewable and Sustainable Materials, Elsevier, Amsterdam, 2020, pp. 155–165. https://doi.org/10.1016/b978-0-12-803581-8.10549-1.

[49] L.P. da Costa, Engineered nanomaterials in the sports industry, in: C.M. Hussain (Ed.), Handbook of Nanomaterials for Manufacturing Applications, Elsevier, Amsterdam, 2020, pp. 309–320. https://doi.org/10.1016/b978-0-12-821381-0.00014-4.

[50] R. Kamel, N.A. El-Wakil, A. Dufresne, N.A. Elkasabgy, Nanocellulose: from an agricultural waste to a valuable pharmaceutical ingredient, Int. J. Biol. Macromol. 163 (2020) 1579–1590. https://doi.org/10.1016/j.ijbiomac.2020.07.242.

[51] T. Ma, X. Hu, S. Lu, X. Liao, Y. Song, X. Hu, Nanocellulose: a promising green treasure from food wastes to available food materials, Crit. Rev. Food Sci. Nutr. (2020). https://doi.org/10.1080/10408398.2020.1832440.

[52] C. Sharma, N.K. Bhardwaj, Fabrication of natural-origin antibacterial nanocellulose films using bio-extracts for potential use in biomedical industry, Int. J. Biol. Macromol. 145 (2020) 914–925. https://doi.org/10.1016/j.ijbiomac.2019.09.182.

[53] M. Michelin, D.G. Gomes, A. Romaní, M.L.T.M. Polizeli, J.A. Teixeira, Nanocellulose production: exploring the enzymatic route and residues of pulp and paper industry, Molecules 25 (2020) 3411. https://doi.org/10.3390/molecules25153411.

[54] R. Saremi, N. Borodinov, A.M. Laradji, S. Sharma, I. Luzinov, S. Minko, Adhesion and stability of nanocellulose coatings on flat polymer films and textiles, Molecules 25 (2020) 3238. https://doi.org/10.3390/molecules25143238.

[55] Z. Wenjuan, S. Zhihua, The preparation of nano cellulose whiskers/polylactic acid composites, Proc. 2011 International Conference on Future Computer Science and Education, ICFCSE 2011 123–125. https://doi.org/10.1109/ICFCSE.2011.38.

[56] W.J. Zhen, Study on nanocellulose/starch composites, Adv. Mater. Res., 187 (2011) 544–547. https://doi.org/10.4028/www.scientific.net/AMR.187.544.

[57] S. Patil, S. Bhattacharya, M. Meenakshi, S.K. Patel, A critical influence of shapes and sizes on flow properties of different grades of microcrystalline cellulose, Innov. Int. J. Med. Pharm. Sci. 5 (2020) 1–8.

[58] V. Klar, H. Orelma, H. Rautkoski, P. Kuosmanen, A. Harlin, Spinning approach for cellulose fiber yarn using a deep eutectic solvent and an inclined Channel, ACS Omega 3 (2018) 10918–10926. https://doi.org/10.1021/acsomega.8b01458.

[59] T. Ghosh, P. Dhar, V. Katiyar, 2 Nanocellulose: extraction and fabrication methodologies. In: Cellulose Nanocrystals: An Emerging Nanocellulose for Numerous Chemical processes, De Gruyter, Berlin, Boston, 2020, pp. 23–48. https://doi.org/10.1515/9783110648010-002.

[60] M. Pereda, G. Amica, I. Rácz, N.E. Marcovich, Structure and properties of nanocomposite films based on sodium caseinate and nanocellulose fibers, J. Food Eng. 103 (2011) 76–83. https://doi.org/10.1016/j.jfoodeng.2010.10.001.

[61] I. Usov, G. Nyström, J. Adamcik, S. Handschin, C. Schütz, A. Fall, L. Bergström, R. Mezzenga, Understanding nanocellulose chirality and structure-properties relationship at the single fibril level, Nat. Commun. 6 (2015) 7564. https://doi.org/10.1038/ncomms8564.

[62] C. Salas, T. Nypelö, C. Rodriguez-Abreu, C. Carrillo, O.J. Rojas, Nanocellulose properties and applications in colloids and interfaces, Curr. Opin. Colloid Interface Sci. 19 (2014) 383–396. https://doi.org/10.1016/j.cocis.2014.10.003.

[63] S. Mondal, Preparation, properties and applications of nanocellulosic materials, Carbohydr. Polym. 163 (2017) 301–316. https://doi.org/10.1016/j.carbpol.2016.12.050.

[64] A. Dufresne, Nanocellulose processing properties and potential applications, Curr. For. Rep. 5 (2019) 76–89. https://doi.org/10.1007/s40725-019-00088-1.

[65] Y. Chu, Y. Sun, W. Wu, H. Xiao, Dispersion properties of nanocellulose: a review, Carbohydr. Polym. 250 (2020) 116892. https://doi.org/10.1016/j.carbpol.2020.116892.

[66] A. Dufresne, Nanocellulose: a new ageless bionanomaterial, Mater. Today. 16 (6) (2013) 220–227. https://doi.org/10.1016/j.mattod.2013.06.004.

[67] S.M. Salah, Application of nano-cellulose in textile, J. Text. Sci. Eng. (2013) 142. doi:10.4172/2165-8064.1000142.

[68] T.S. Roy, S.U.D. Shamim, M.K. Rahman, F. Ahmed, M.A. Gafur, The development of ZnO nanoparticle coated cotton fabrics for antifungal and antibacterial applications, Mater. Sci. Appl. 11 (2020) 601–610. https://doi.org/10.4236/msa.2020.119040.

[69] M.A. Corrales, D. Gierczycka, J. Barker, D. Bruneau, M.C. Bustamante, D.S. Cronin, Validation of a football helmet finite element model and quantification of impact energy distribution, Ann. Biomed. Eng. 48 (2020) 121–132. https://doi.org/10.1007/s10439-019-02359-1.

[70] M.L. Bland, C. McNally, J.B. Cicchino, D.S. Zuby, B.C. Mueller, M.L. McCarthy, C.D. Newgard, P.E. Kulie, B.N. Arnold, S. Rowson, Laboratory Reconstructions of bicycle helmet damage: investigation of cyclist head impacts using oblique impacts and computed tomography, Ann. Biomed. Eng. 48 (2020) 2783–2795. https://doi.org/10.1007/s10439-020-02620-y.

[71] E.P. Shore, A. O'connell, An Investigation into Player Compliance and Level of Protection Afforded by Mouthguards worn by Children Playing Sport in Ireland. (Ph.D. thesis) Trinity College, University of Dublin, 2020. http://www.tara.tcd.ie/bitstream/handle/2262/93992/SHORE_07706278_thesis_Final.pdf?sequence=1&isAllowed=y. (Accessed 21/12/2020).

[72] K. Anirudh, S.N. Yeole, A contemporary review on knee injuries and protective pads, in: Lecture Notes in Mechanical Engineering, Springer Science and Business Media Deutschland GmbH (2021) 493–499. https://doi.org/10.1007/978-981-15-6619-6_54.

[73] E. Casey, Sports and the environment: a life cycle assessment of children's football and hockey equipment, DePaul Discov. 9 (1) (2020) Article 4.

[74] R.J. Boergers, T.G. Bowman, M.R. Lininger, The ability to provide quality chest compressions over lacrosse shoulder pads, J. Athl. Train. 53 (2018) 122–127. https://doi.org/10.4085/1062-6050-346-16.

[75] A. Zare, A. Choobineh, H. Mokarami, M. Jahangiri, The Medical Gloves Assessment Tool (MGAT): developing and validating a quantitative tool for assessing the safety and ergonomic features related to medical gloves, J. Nurs. Manag. 29 (3) (2020) 591–60113188. https://doi.org/10.1111/jonm.13188.

[76] F. Tatàno, N. Acerbi, C. Monterubbiano, S. Pretelli, L. Tombari, F. Mangani, Shoe manufacturing wastes: characterisation of properties and recovery options, Resour. Conserv. Recycl. 66 (2012) 66–75. https://doi.org/10.1016/j.resconrec.2012.06.007.

[77] N.C. Jadhav, A.C. Jadhav, Waste and 3R's in Footwear and Leather Sectors, Springer, Singapore, 2020, pp. 261–293. https://doi.org/10.1007/978-981-15-6296-9_10.

[78] E.B. Heggset, B.L. Strand, K.W. Sundby, S. Simon, G. Chinga-Carrasco, K. Syverud, Viscoelastic properties of nanocellulose based inks for 3D printing and mechanical properties of CNF/alginate biocomposite gels, Cellulose 26 (2019) 581–595. https://doi.org/10.1007/s10570-018-2142-3.

[79] B. Wu, S. Wang, J. Tang, N. Lin, Nanocellulose in High-Value Applications for Reported Trial and Commercial Products, Springer, Singapore, 2019, pp. 389–409. https://doi.org/10.1007/978-981-15-0913-1_11.

[80] W. Hao, M. Wang, F. Zhou, H. Luo, X. Xie, F. Luo, R. Cha, A review on nanocellulose as a lightweight filler of polyolefin composites, Carbohydr. Polym. (243) (2020) 116466. https://doi.org/10.1016/j.carbpol.2020.116466.

[81] D.P. Chattopadhyay, B.H. Patel, Synthesis, Characterization and Application of Nano Cellulose for Enhanced Performance of Textiles, J. Text. Sci. Eng. 6 (248) (2016). doi:10.4172/2165-8064.1000248.

[82] L. Di Landro, G. Sala, D. Olivieri, Deformation mechanisms and energy absorption of polystyrene foams for protective helmets, Polym. Test. 21 (2002) 217–228. https://doi.org/10.1016/S0142-9418(01)00073-3.

[83] C.E. Njoku, J.A. Omotoyinbo, K.K. Alaneme, M.O. Daramola, Characterization of *Urena lobata* fibers after alkaline treatment for use in polymer composites, J. Nat. Fibers. (2020). https://doi.org/10.1080/15440478.2020.1745127.

[84] T. Ganapathy, R. Sathiskumar, P. Senthamaraikannan, S.S. Saravanakumar, A. Khan, Characterization of raw and alkali treated new natural cellulosic fibres extracted from the aerial roots of banyan tree, Int. J. Biol. Macromol. 138 (2019) 573–581. https://doi.org/10.1016/j.ijbiomac.2019.07.136.

[85] L. Pil, F. Bensadoun, J. Pariset, I. Verpoest, Why are designers fascinated by flax and hemp fibre composites?, Compos. Part A Appl. Sci. Manuf. 83 (2016) 193–205. https://doi.org/10.1016/j.compositesa.2015.11.004.

[86] M. Nagalakshmaiah, S. Afrin, R.P. Malladi, S. Elkoun, M. Robert, M.A. Ansari, A. Svedberg, Z. Karim, Biocomposites: present trends and challenges for the future, in: G. Koronis, A. Silva (Eds.), Green Composites for Automotive Applications, Elsevier, Amsterdam, 2018, pp. 197–215. https://doi.org/10.1016/B978-0-08-102177-4.00009-4.

[87] D. Mathijsen, Beyond carbon fiber: what will be the fibers of choice for future composites?, Reinf. Plast. 60 (2016) 38–44. https://doi.org/10.1016/j.repl.2015.12.003.

[88] P.G. Gan, S.T. Sam, M.F. Abdullah, M.F. Omar, Thermal properties of nanocellulose-reinforced composites: a review, J. Appl. Polym. Sci. 137 (2020) 48544. https://doi.org/10.1002/app.48544.

[89] A. Hebeish, S. Farag, S. Sharaf, T.I. Shaheen, Advancement in conductive cotton fabrics through in situ polymerization of polypyrrole-nanocellulose composites, Carbohydr. Polym. 151 (2016) 96–102. https://doi.org/10.1016/j.carbpol.2016.05.054.

[90] S.P. Singaraj, K.P. Aaron, K. Kaliappa, K. Kattaiya, M. Ranganathan, Investigations on structural, mechanical and thermal properties of banana fabrics for use in leather goods application, J. Nat. Fibers. 9 (1) (2019) 37–50. https://doi.org/10.1080/15440478.2019.1697982.

[91] T.Y. Pang, A. Subic, M. Takla, Thermal comfort of cricket helmets: an experimental study of heat distribution, Procedia. Eng. 13, 2011 252–257. https://doi.org/10.1016/j.proeng.2011.05.081.

[92] K.S. Prado, D. Gonzales, M.A.S. Spinacé, Recycling of viscose yarn waste through one-step extraction of nanocellulose, Int. J. Biol. Macromol. 136 (2019) 729–737. https://doi.org/10.1016/j.ijbiomac.2019.06.124.

[93] S.S. Nair, J. Zhu, Y. Deng, A.J. Ragauskas, High performance green barriers based on nanocellulose, Sustain. Chem. Process. 2 (2014) 1–7. https://doi.org/10.1186/s40508-014-0023-0.

[94] C. García, M.A. Prieto, Bacterial cellulose as a potential bioleather substitute for the footwear industry, Microb. Biotechnol. 12 (2019) 582–585. https://doi.org/10.1111/1751-7915.13306.

[95] P. Venkatraman, A.M. Gohn, A.M. Rhoades, E.J. Foster, Developing high performance PA 11/cellulose nanocomposites for industrial-scale melt processing, Compos. Part B Eng. 174 (2019) 106988. https://doi.org/10.1016/j.compositesb.2019.106988.

[96] A. Liyanapathiranage, M.J. Penã, S. Sharma, S. Minko, Nanocellulose-based sustainable dyeing of cotton textiles with minimized water pollution, ACS Omega 5 (16) (2020) 9196–9203. https://doi.org/10.1021/acsomega.9b04498.

[97] K. Fu, Z. Yang, Y. Pei, Y. Wang, B. Xu, YuH Wang, B. Yang, L. Hu, Designing textile architectures for high energy-efficiency human body sweat- and cooling-management, Adv. Fiber Mater. 1 (2019) 61–70. https://doi.org/10.1007/s42765-019-0003-y.

[98] A. Jebali, S. Hekmatimoghaddam, A. Behzadi, I. Rezapor, et al., Antimicrobial activity of nanocellulose conjugated with allicin and lysozyme, Cellulose 20 (2013) 2907–2987. https://doi.org/10.1007/s10570-013-0084-3.

[99] A. Kocić, M. Bizjak, D. Popović, G.B. Poparić, S.B. Stanković, UV protection afforded by textile fabrics made of natural and regenerated cellulose fibres, J. Clean. Prod. 228 (2019) 1229–1237. https://doi.org/10.1016/j.jclepro.2019.04.355.

[100] A. Sharma, M. Thakur, M. Bhattacharya, T. Mandal, S. Goswami, Commercial application of cellulose nano-composites – a review, Biotechnol. Rep. 21 (2019) e00316. https://doi.org/10.1016/j.btre.2019.e00316.

[101] R. Jafary, M.K. Mehrizi, S.h. Hekmatimoghaddam, A. Jebali, Antibacterial property of cellulose fabric finished by allicin-conjugated nanocellulose, J. Text. Inst. 106 (7) (2014) 683–689. https://doi.org/10.1080/00405000.2014.954780.

[102] M.M.Á.D. Maciel, K.C.C. e Carvalho Benini, H.J.C. Voorwald, M.O.H.H. Cioffi, Obtainment and characterization of nanocellulose from an unwoven industrial textile cotton waste: Effect of acid hydrolysis conditions, Int. J. Biol. Macromol. 126 (2019) 496–506. https://doi.org/10.1016/j.ijbiomac.2018.12.202.

[103] J.P.S. Morais, M. de Freitas Rosa, Men de sá Moreira de Souza Filho, L.D. Nascimento, D.M. do Nascimento, A.R. Cassales, Extraction and characterization of nanocellulose structures from raw cotton linter, Carbohydr. Polym. 91 (1) (2013) 229–235. https://doi.org/10.1016/j.carbpol.2012.08.010.

CHAPTER 9

Uses of nanocellulose in the environment industry

Garvita Parikh, Bansari Parikh, Aarohi Natu, Deepak Rawtani
School of Pharmacy, National Forensic Sciences University (Ministry of Home affairs, GOI), Gandhinagar, Gujarat, India

9.1 Introduction

Nanocellulose is kind of polysaccharide with an individual cellulose nanofiber (CNF). Because of its abundance, it is extensively used [1]. The nanocellulose particles can be mainly of two types: CNFs and cellulose nanocrystals (CNCs) which are used in papermaking as it is abundantly available and eco-friendly. But nanocellulose particles also exist in other forms, such as bacterial cellulose (BC), microfibrillated cellulose (MFC), which are primary constituent of cellulose and they can be extracted from wood and nanocrystalline cellulose [2]. Of all these nanocellulose types, the NFC are highly dependent on the fibril length [3]. These can be extracted from Avicel, ramie, wood, cotton, mulberry bark, etc. and can be produced by acid hydrolysis of cellulose [4]. And other are the bacterial nanocelluloses extracted from alcohols or low molecular weight sugars having a diameter of 20–100 nm. Biological, physical as well as chemical treatments are used to extract nanocellulose from cellulose [5]. Lately, nanomaterials have presented ability in effective detection and removal of chemical contaminants, organic pollutants, greenhouse gases, and biological agents. Because of high surface area and high reactivity, it is preferred over other conventional methods. As evident, nanocellulose has displayed good performance in applications such as water and wastewater treatment, air purification, papermaking, energy production and biomedical engineering [6], antibiotic formation, agriculture [7] as depicted in Fig. 9.1. Nanocellulose can absorb many pollutants such as heavy metals, dyes, dissolved organic pollutants, oil, and undesirable effluents. The other benefit of nanocellulose is one can regenerate this adsorbent [8]. This chapter summarizes various applications of nanocellulose in environmental industry in a broad way.

9.2 Nanocellulose-based adsorbent

Nanocellulose–based adsorbent materials are evolve out as a substitute to the expensive technology and energy-intensive based on activated carbon–based adsorption [6]. Nanocellulose is identified as a best source for effluent treatment as it has property of

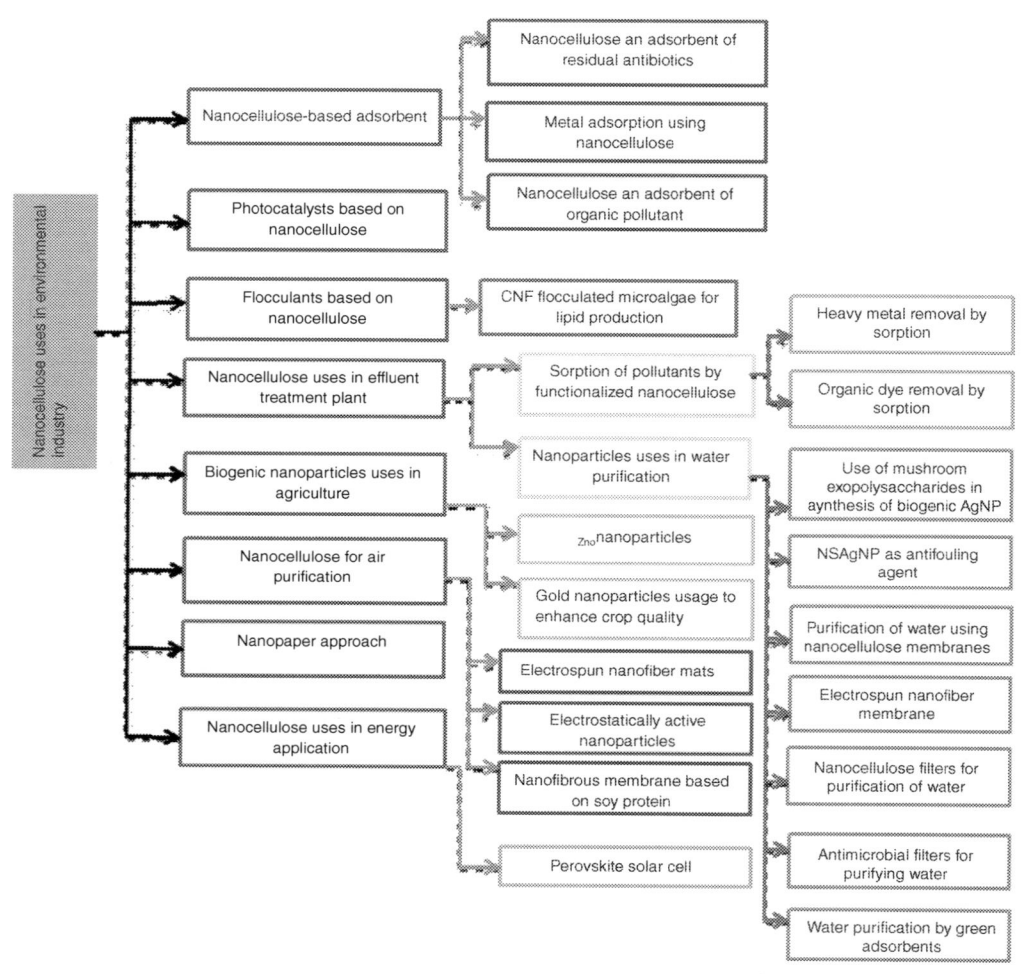

Fig. 9.1 Classification of nanocellulose usage in various industrial sectors.

naturally adsorbing the biomaterial. As nanometer-sized counterparts are smaller in size and also it possesses a wide surface area with improved porosity, which limits internal diffusion and effect of quantum size [9]. The improvement in adsorption efficiency of nanocellulose is due to the broad surface area, porosity, and crystalline nature [10] as depicted in Fig. 9.2. The insertion of amino group on MFC and nanocellulose surface improves the function of adsorption by removing arsenic [6] as depicted in Fig. 9.3.

9.2.1 Nanocellulose: an adsorbent of residual antibiotics

Nanocellulose is used as a surface-assimilative for the residual antibiotics that are often observed in aquaculture and industrial discharge with the antibiotic residues [6]. In this,

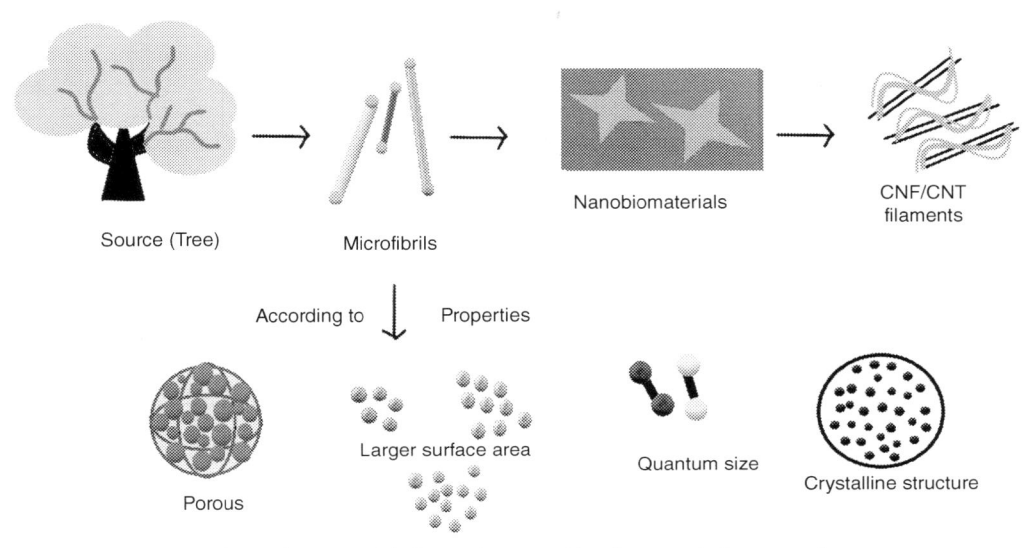

Fig. 9.2 Extraction of nanobiomaterials from plants and its characteristic properties.

cellulose is first turn into nanocellulose so that the surface plane rises at an exponential rate; besides its crystalline nature, the amount of surface hydroxy groups (–OH) also increases [34]. The surface –OH groups act as hydrophilic sites to accelerate nucleation and development of inorganic particles, consequently controlling the particle size, morphology, and crystallinity [6]. These –OH groups can be then be modified for selective and improved interactions by using variety of chemicals [11]. The hydrophobicity of nanocellulose substance can be increased by introducing hydrophilic groups, such as alcohols, carboxylic acids, etc., inside its polymeric chains as these will improve the adsorbance of antibiotics residue [9] as depicted in Fig. 9.4.

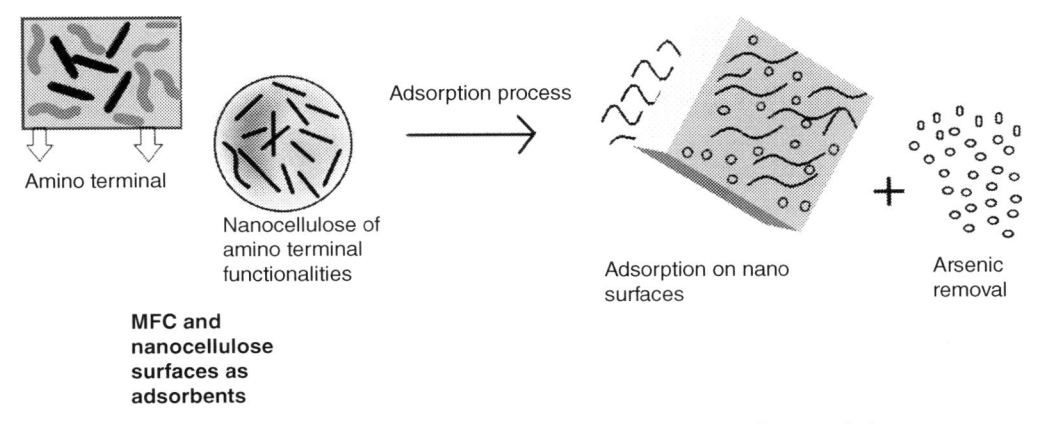

Fig. 9.3 Adsorption process by introduction of amino terminal on MFC and nanocellulose.

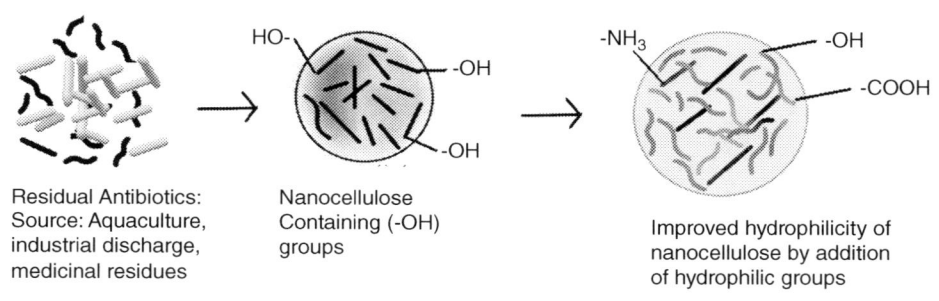

Fig. 9.4 Improved nanocellulose as surface-assimilative for superfluous antibiotics.

Besides, many antibiotics in pharma industries are used to treat the bacterial infections [12]. These antibiotics, when manufactured, causes environmental pollution affecting soil, aquatic plants, marine organisms when polluted water is released, affecting groundwater, surface water, leading to severe impacts on ecosystem and human health. One such antibiotic, amoxicillin, is also known to enter the water bodies. An effort by Omidvar et al. [13] to remove antibiotics from water bodies was made by preparing nanofilteration (NF) layers composed of novel asymmetric flat sheets by poly(ether sulfone)/1-methyl-2-pyrolidone/BrijS100 casting solutions. In comparison to others, BrijS100 resulted in detectable knock-back of amoxicillin of up to 99% [13].

9.2.2 Metal adsorption using nanocellulose

The functionalized nanocellulose material with carboxy groups can upgrade the adsorption process by carrying out the noxious metallic ions from the wastewater [6]. It is known that carboxy group has two lone electron pairs on oxygen, and hence it is necessary to form a chelate with a divalent metal [6]. In a study conducted by [34], authors synthesized a carboxylate-functionalized adsorbent by grafting maleic anhydride on primary −OH of CNC. This newly synthesized nanocellulose adsorbent is used for carrying out the cationic pigments from the liquid suspension by setting-up an arrangement of bidentate in between the pigment and the adsorbent's carboxy groups. Meanwhile, further nanocellulose adsorbent is modified with multiple carboxy functional groups for the surface assimilation of cobalt (II) and uranium (VI) [14]. The probability of the synthesized adsorbent can be increased by increasing surface carboxylic group concentration of the adsorbent as this will increase the possibility of removing several pigments or dyes such as methylthioninium chloride commonly called methylene blue and aniline green also called malachite green [6].

Other alternative is the utilization of amine, which adds chelating activity upon the pollutant anions, therefore increasing the adsorption capability of the nanocellulose materials [15]. Metal ions can easily adsorb upon the adsorbent material via ion exchange and electrostatic interaction till the amino groups are protonated easily under the acidic conditions [15]. The amino group possesses a single pair of electrons, which is combined to the nitrogen atoms and may form a covalent bond with metal ions [6]. The surface assimilation of metal ions is coordinated by hydroxy species and amino class is identified in the functionalized cellulose. Amino groups usually induce a high adsorption rate at pH below 7 and decrease at pH above 8. The amino groups modified with MFCs have different affinities for different metal ions and it follows the sequence of Cd (II) > Cu(II) > Ni(II) [16]. The modified nanocellulose adsorbent is capable to adsorb metal ions because of the presence of amino ($-NH_2$) on amino silane and/or $-OH$ groups on cellulose fiber and this adsorbent shows the capability to withdraw dyes [6].

9.2.3 Nanocellulose: an adsorbent of organic pollutant

Magnetic nanocellulose is utilized in the removal of organic impurities. Hokkanen et al. [11] reported that the calcium hydroxyapatite-microfibrillated cellulose was mainly utilized for the adsorbing the chromium Cr (VI)) from liquid suspension as depicted in Fig. 9.5. Calcium hydroxyapatite nanostructures can easily come in contact with the cellulose structures, which results in increase in nanocomposites properties due to their divalent character, strong hydrogen bond, and relatively small ionic radius [11].

In another study [14], authors chemically modified a composite polymer to produce a poly(itaconic acid)–poly(methacrylic acid)-grafted-nanocellulose/nano-bentonite composite as an efficacious adsorbent for taking away the cobalt (Co(II)) from liquid suspension [14]. This increases the approachability of the carboxy groups that are covalently connected to the inorganic matrix, and hence remove the organic pollutants from wastewater bodies [6].

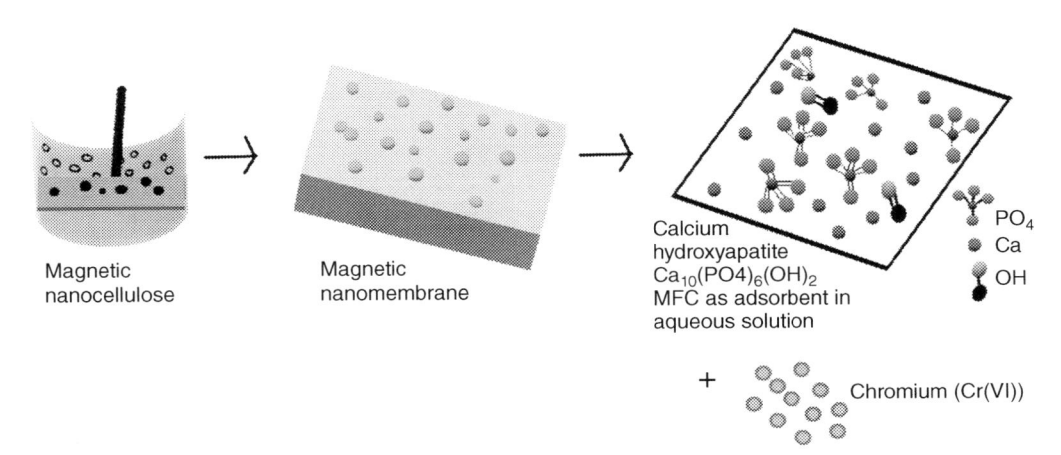

Fig. 9.5 Calcium hydroxyapatite microfibrillated magnetic nanocellulose as an adsorbent.

9.3 Photocatalysts based on nanocellulose

Nanostructured compounds based on the hybrid, organic along with inorganic, materials are evolved with the new properties having thermal, electrical, and mechanical features [6].

The cellulose-based substances are an eco-friendly biological material, thus they make more clean and sustainable environment [17]. Cellulose-derived materials are identified as the universally accessible inexhaustible organic reserves on the earth and provide biocompatible support because of its exceptional properties of hydrophilicity, chirality, and chemical variability [18]. Other than this, cellulose is considered as a reliable host substance for NPs because cellulose can make stability better and control the generation of NP [19].

For treating wastewater beneath UV or visible light, nanocellulose is utilized as a photocatalyst. Several nanocellulose materials such as nanocellulose–metal oxide (TiO_2, graphene oxide, ZnO, and Fe_2O_3) composites are utilized as photocatalysts to increase the degradation rate of organic contaminants compared to individual materials [6].

The nanofibrillated cellulose/titanium dioxide/magnetite/nanocomposites had a greater generation rate of photocatalytic hydrogen in comparison to the nanofibrillated cellulose/titanium dioxide sample [20]. When we incorporate the magnetic field and UV light, then it inhibits the photon degradation of the cellulosic structure and this hybrid structure shows high catalyst recyclability utilizing an outermost electromagnetic field [20] as illustrated in Fig. 9.6. Nanoparticles aggregate are difficult to reuse and recycle [21]. Therefore, a cellulose matrix is used for its 3D sieve-like composition that will behave as a catalyst strength and stops the agglomeration and development of NPs [6].

The implementation of a nanocellulose-based fine membrane and film shows better potential in effluent treatment because of its recycling purpose and ease of recollection [6].

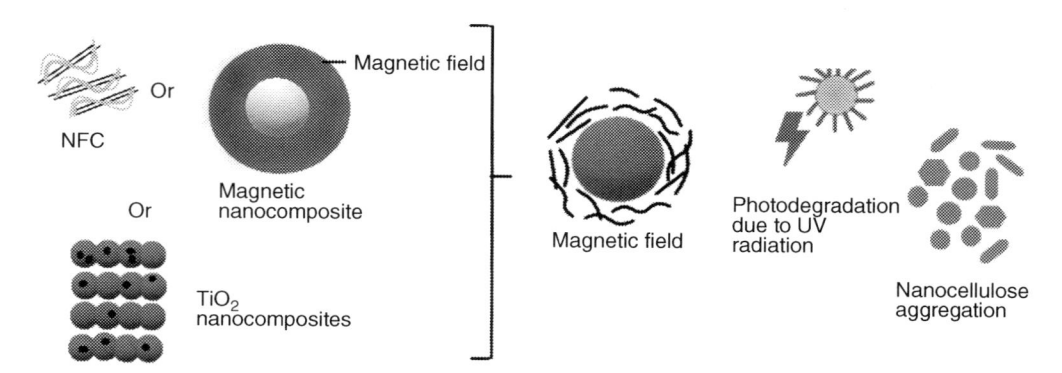

Fig. 9.6 Photocatalysts based on nanocellulose.

9.4 Flocculants based on nanocellulose

Generally, flocculation is usually carried out chemically by utilizing inorganic, multivalent cationic coagulator such as aluminum, iron salts, or artificial polyelectrolytes such as artificial polyacrylamides [6]. These synthetic materials cause environmental issues and are hazardous, nonbiodegradable, and toxic causing secondary pollution because of their derivatives, monomers, or intermediate products [22]. Because of that, there should be such a natural bio-based and sustainable flocculant that is suitable for wastewater purification as these chemical flocculants cause major secondary pollution to the environment.

Nowadays, biopolymer-based flocculants such as tannins, chitosan, cellulose, and alginate are used widely in many studies. Bio-based flocculants have proportions in the nanoscale and a great specific surface area [20]. Usually, there are two classifications of natural polymer–based flocculants [23]. One of these classification is obtained by modifying the natural polymers to construct flocculants [24] and the alternative class takes account of transplanting the natural polymers to manufacture seminatural flocculants [22].

Currently, nanocellulose materials are made effective by introducing cationic, anionic, or hydrophobic functional groups to the nanocellulose surfaces through a neutralization mechanism to generate effective flocculants [6]. The —OH group on the facet of cellulose enhances the functionality and produces highly effective flocculants [25]. The interconnection of cellulose with other compounds could be increased by introducing new classes in series to increase the hydrophilicity and surface polarity [6].

Other important mechanism is to chemically modify nanocellulose to produce highly charged material by using (2,2,6,6-tetramethylpiperidin-1-yl)oxyl (TEMPO)-mediated oxidation under mild and aqueous conditions [6]. This method is mainly useful for the conversion of surface hydroxymethyl groups to their respective carboxylic forms [26]. The anions available on the upper surface of the nanocellulose can prevent aggregation of nanocellulose through electrostatic repulsion. This electrosterically stabilized nanocrystal cellulose (NCC) may be produced by a three-step reaction: periodate, chlorite, and TEMPO oxidation [27]. It may be minimized by using less periodate oxidation and more TEMPO-mediated oxidation [27].

Various studies have been carried out to enhance the physicochemical structure of nanocellulose. For instance, Sun et al. [28] studied the efficacy of implementing the CNCs to flocculate Gram-negative bacteria (*Pseudomonas aeruginosa* PAO1). The author noticed that the efficiency of flocculating bacteria relies on the appearance of the cellulose colloidal particles because of this; the particles deplete less, the large colloids than the rod-shaped particles (usually observed in CNC) [28].

9.4.1 CNF-flocculated microalgae for lipid production

Without the cationic modification, CNF was utilized as a microalgal flocculant and alteration in metabolism was observed due to flocculated microalgae whose desulfurization was implemented but did not affected the efficiency of flocculation of CNF [29].

CNC was not treated by sulfur acid but tunicate CNF-flocculated microalgae were processed with sulfur acid. This can head to an advantage of increasing lipid production with a low cost of harvest [29].

9.5 Nanocellulose uses in effluent treatment plant

Currently, the chemicals utilized to process wastewater are based on artificially made organic or inorganic composite. Polyelectrolytes derived by oils are utilized for the removal of colloidal particles from effluent by coagulation and flocculation such as utilizing activated charcoal as surface-assimilative to withdraw impurities such as heavy metals and organic load from effluent [30]. These chemicals are related with negative health impacts and are expensive also. Moreover, nowadays there is a high requirement for ecofriendly and environmentally friendly chemicals to purify water. These green chemicals extracted from cellulose, that is nanocelluloses, are bio-based chemicals utilized for effluent processing and are based on nanoscale particles (nanofibrils) [30]. The two common anionic nanocelluloses (dicarboxylic acid and sulfonated anionic sulfonated cellulose (ADAC)) were used and identified as flocculants in the coagulation-flocculation processing of municipal effluent [30] as illustrated in Fig. 9.7 and they showed better performance by lowering the turbidness and chemical oxygen demand in municipal effluent by using combined coagulation-flocculation treated with a ferric coagulant.

9.5.1 Sorption of pollutants by functionalized nanocellulose

Biosorption means the capacity of biological material to adsorb heavy metals from effluent, in which the sorbent is of biological origin, for example, biopolymers or

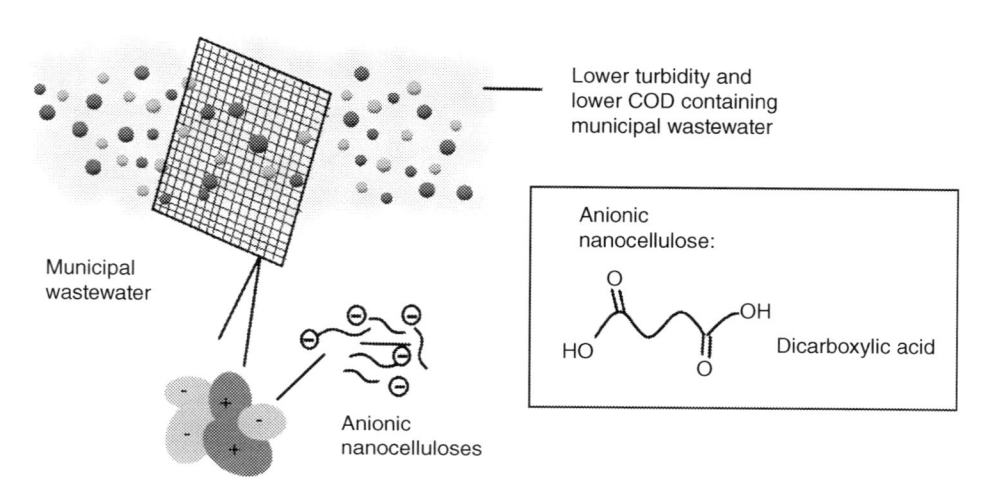

Fig. 9.7 Effluent treatment using coagulation- flocculation method.

microorganisms [31]. Various microbes are also utilized to uptake organic contaminants; biosorption is an eco-friendly solution for purifying water and soil remediation [31].

9.5.1.1 Heavy metal removal by sorption

Heavy metals are highly toxic contaminants in water that originates from either industrial or natural sources. They are present in various water forms and they form both cationic and anionic species [31]. For heavy metals uptake, a dominating species is required and the surface chemistry of the sorbent should be optimized so that it can maximize the ability, selectivity, and partitioning coefficient.

The sorption of negatively charged metallic species, such as chromates or arsenates, from the mixture can be obtained by functionalizing the nanocellulose with cationic species. Singh et al. [32] worked on the aminated CNC composed by sodium periodate oxidation can surface-assimilate 98% of dichromate ions in a 25 mg/mL mixture at pH of 2.5 [32]. Moreover the positively charged CNC may also be made using sulfated CNC reacting it with epoxypropyltrimethylammonium chloride (EPTMAC) [33]. The bisphosphonate nanocellulose acquired through consecutive oxidation with sodium periodate is then reacted with sodium alendronate and can withdraw vanadium in its metavanadate (VO_3^-) form with an utmost surface-assimilation ability of 194 mg/g at pH of 2 [22].

9.5.1.2 Organic dye removal by sorption

Organic contaminants in water represent a very vast diversity of compounds that include oils, dyes, pesticides, and pharmaceuticals products [31].

Organic dye pollutants generated by the fabric industry pose an environmental pollution mainly the water pollution. Organic dyes have aromatic structure and can display anionic, cationic, or nonionic properties. Their interactions with cellulose are important [31]. The CNCs composed by esterification of the surface hydroxyl groups with maleic anhydride [34] with carboxylic acid show an increased intake ability for numerous cationic dyes such as methylene blue, crystal violet, with highest crystal violet uptake of 244 mg/g at pH of 6 [34]. Another is the anionic dyes that commonly withdrew utilizing nanocellulose with functionality of cations, such as, cationic CNCs, amino-functionalized CNC which is formed by reacting it with ethylenediamine [1]. Moreover, CNC forming cross-linking microgels with polyvinyl amine can also be utilized, which shows high affinity for cationic along with anionic dyes, with highest surface-assimilation for Congo red, acid red and reactive light yellow of 896 mg/g, 1496 mg/g, and 1250 mg/g, respectively, as illustrated in Fig. 9.8 [31].

Other is the cationic CNFs that are adsorbents of anions in water, such as phosphate, fluoride, nitrate, and sulfate. In this, the surface-assimilation ability increases as the surface charge density of CNCs increases [31]. Cationic CNFs show highest specificity with regard to multivalent ions such as SO_4^{-2} and PO_4^{-3} than monovalent ions such as F^- and NO_3^- [35].

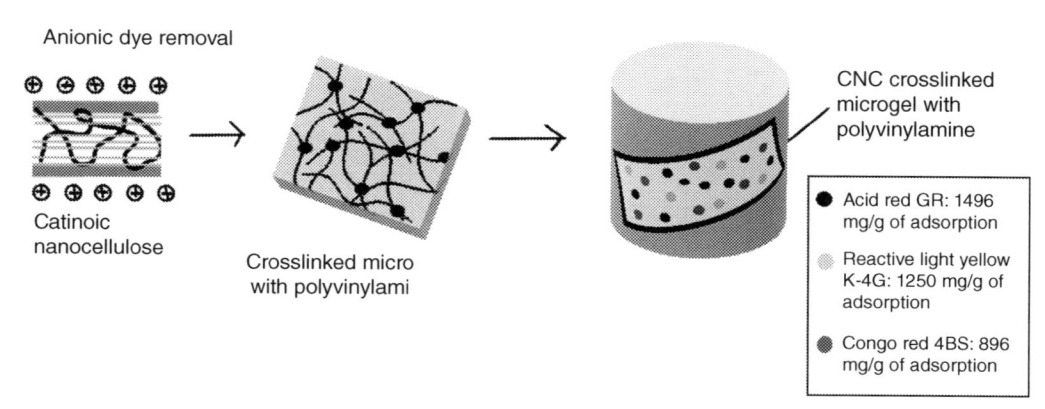

Fig. 9.8 Anionic dye removal.

9.5.2 NPs uses in water purification

9.5.2.1 Use of mushroom exopolysaccharides in synthesis of biogenic Silver NP (AgNP)

Mushroom exopolysaccharides, when used for synthesis of biogenic silver NPs, are collectively called myco-NPs [36]. Many organisms such as fungi, bacteria, yeast were tested for producing gold NPs, which gave a positive response displaying various sizes and shapes [37]. Adeeyo and Odiyo [36] showed that the reaction mixture was getting a stable color pattern after 24 h as *Lentinula edodes* reacted rapidly with $AgNO_3$ solution. This resulted in development of AgNPs in 24 h. The color change was observed from colorless to purple color. Thus, the end-point was found at purple color showing this approach as feasible, eco-friendly and the treated water was safe to drink [36].

9.5.2.2 NSAgNP as antifouling agent

When AgNP are fabricated on nanosilica surface, it is called NSAgNP which is utilized as an antifouling agent. At ambient temperature, silver ions containing protein are reduced, which results in the formation of NSAgNP which has excellent capacity of dye adsorption [24]. Also, antibacterial property was shown by NSAgNP against the biofilms of *Escherichia coli*, *P. aeruginosa*, and the planktonic cells [24]. This resulted in prohibiting the bacteria and also improved antifouling characteristics of the material.

9.5.2.3 Purification of water using nanocellulose membranes

To clean the wastewater, materials having the base of nanocellulose are chosen which can act as adsorbent, membrane, and filters [31]. These characteristics mainly rely on the operation mode and the dimension of pores.

There are various kinds of nanocellulose membranes as shown in Fig. 9.9, which are used for water purification such as functionalized nanocellulose loaded with electrospun

1. Impregnation of electrospun mats with nanocrystals

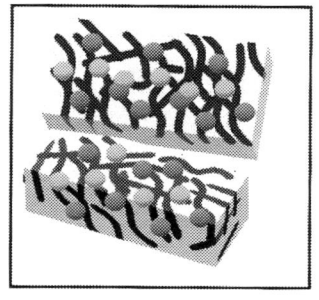

2. Impregnation of electrospun mats with functionalized NC

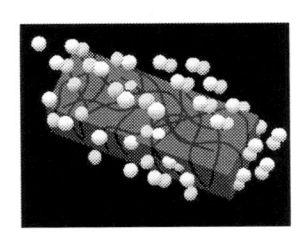

3. Electrospun nanofibrous membranes for bacteria and viruses

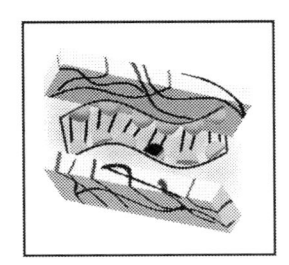

⊙ Nanocrystals on
● Electospun mats

Fig. 9.9 Nanocellulose membranes.

mats [38]. These mats with nanocrystals have isotropic or anisotropic structures varying in ratios. They have the capacity to charge the microporous membranes but are greatly relying on the electrospun layer [39].

Microorganisms such as bacteria and viruses are removed by these electrospun nanofibrous membranes (ENMs) . These membranes contain the cellulose nanocrystals (CNF) majorly and have capacity to remove bacteria such as *Brevundimonas diminuta* and *E. coli* [39].

Other nanocellulose membrane is made using cellulose acetate polymer-based electrospun mats as a support for CNCs, without any additional layer having a nonwoven support. This resulted in a cellulose acetate cellulose nanocrystal (CA-CNC) showing characteristic features such as having a membrane with microfiltration capacity and high porosity [40].

9.5.2.4 Electrospun nanofiber membrane for the removal of suspended solids and micron-sized particles

Due to porous structure, ENM has capacity to be used in liquid filtration techniques [41]. So, the uppermost layer of NF membrane has been altered using methacrylic acid and graft–copolymerization in which 150–200% increase on water flux [42] was found, resulting in energy-saving membranes. Also, Gopal et al. [43] developed an electrospun nanofibers for liquid separation, which lead to particulate removal. These electrospun layers were effective in removing and filtering 90% of microparticles from the solution mixture.

9.5.2.5 Nanocellulose filters for purification of water

For the fabrication of nanomaterials having cellulose to be layered, vacuum filtration is used as it is easy, comparatively quick, and easy. "Nanopaper approach" [44] is utilized to manufacture the nanocellulose by vacuum filtration as demonstrated in Fig. 9.10. In this process, a network of nanofibrils is formed giving a compact

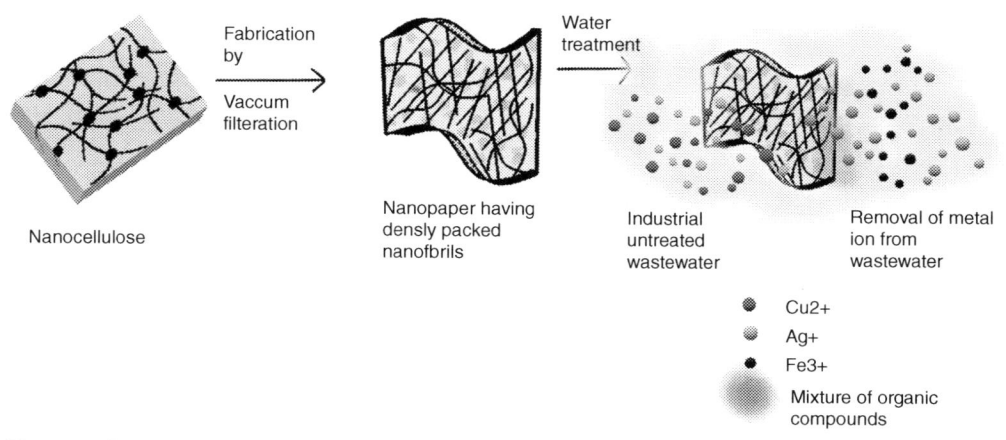

Fig. 9.10 "Nanopaper" approach.

appearance having pore dimension of 3–5 nm, which results in a thin film [31]. Metal ions such as Cu^{2+}, Ag^+, Fe^{3+}, fluorides, nitrates, phosphates, sulfates, humic acid, and organic compounds are also removed by Nanopaper approach, when the membranes are made [35].

The nanofiltration membrane prepared using nanopaper approach shows a diameter in the range of 5–30 nm, that resulted in small pore size that make the removal of molecules, with MW of 6–25 kDa, easy; nevertheless, the flux was low (100 L/m^2·h·bar) [44]. Similarly, membranes formed by the nanopaper approach can remove 6 kDa size of molecules through the size exclusion technique. These membranes, which are formed by nanopaper approach of transparent TEMPO cellulose nanofibril, have diameters below 5 nm [39]. The use of cellulose nanofibril (CNF) membranes with the group such as ammonium using a nanopaper technique is prepared and when mixed with nonfunctionalized CNF, these membranes have an ability to remove the nitrates from wastewater or other water streams [31]. The maximum capacity to produce flux in these layers is 150 L/m^2·h·bar [44].

When the nanocellulose and the CNF suspension are mixed, anisotropic membranes with multilayered structure are formed [45] then, they are dipped into the solution containing carboxyl surface groups having a dispersion medium of CNC [46]. These membranes have a smaller pore size. And if we treat this with acetone then the pore size will increase up to 194 Å, and the flux to 250 L/m^2·h·bar. These membranes help in removing maximum ions such as Fe^{+3}, Cu^{2+}, and Ag^+ [31].

9.5.2.6 Antimicrobial filters for purifying water

Nanocellulose filters have presented finer performance in removing virus from water. The results were observed when AgNP acting as an antimicrobial agent was used along

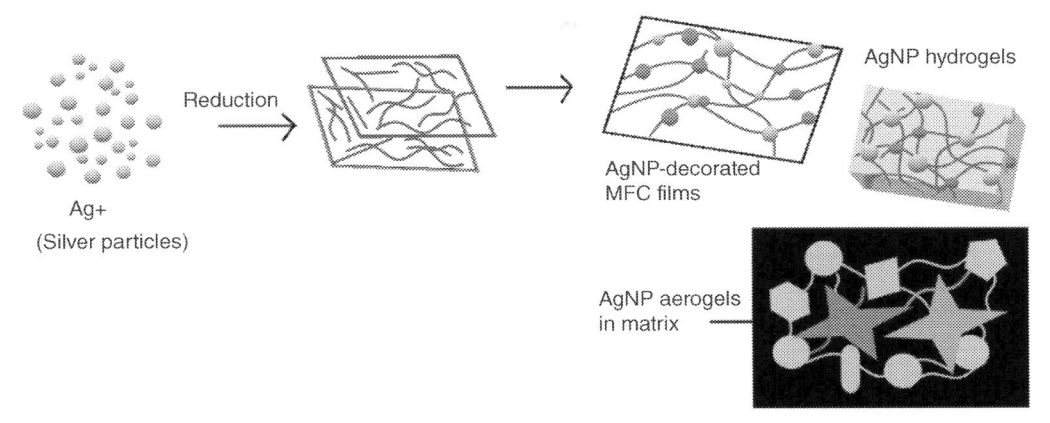

Fig. 9.11 AgNP preparations.

with the filter resulting in inhibition of growth of microorganisms, which increases the lifespan of the filter [2]. Antimicrobial material such as chitosan is used for purification of drinking water and its storage [47]. Due to considerable surface to volume ratios of AgNPs, these nanocelluloses possess greater ability to accommodate AgNPs, which provides them excellent antimicrobial properties [2].

On the nanocellulose surface silver-based NPs are amalgamated giving them the antimicrobial properties [48]. The AgNP/nanocellulose can be synthesized by reducing the Ag^+ in which the existence of nanocellulose is necessary. The techniques, such as UV irradiation and reducing chemicals introduction such as $NaBH_4$, are included in it. The AgNP aerogels, hydrogels, and decorated films can be prepared using reduction of silver particles by the reduction method [49]. According to a study, the mechanofluorochromism (MFC) gelation is facilitated by adding Ag^+ forming a hydrogel. Hydrogen bonding containing hydroxyl groups of MFC was responsible for rapid gelation, which resulted in the formation a complex structure between the carboxylate groups of the Ag^+ and MFCs [49]. Then, by using oven, the water is withdrawn from the hydrogel forming a transparent film. In comparison, an aerogel is formed giving a spongy appearance if the hydrogel is frozen and then water is removed from it [49] as shown in Fig. 9.11.

This nanocomposite can stop the growth of *Staphylococcus aureus* and *E. coli* as demonstrated in Fig. 9.12 [2].

9.5.2.6.1 Antibacterial filters for purifying water

NPs having stronger antimicrobial effects are used for purifying water [50]. The investigation revealed that alloy of Cu–Zn and Cu metal NPs damages the membrane and an increase in layer damage was observed at the dose of 20 μg/mL [51] but the effect on the other different NPs was not seen at all. This was noticed when a study was carried out by observing the interaction between cell membrane and Cu metal NPs, CuO, a binary

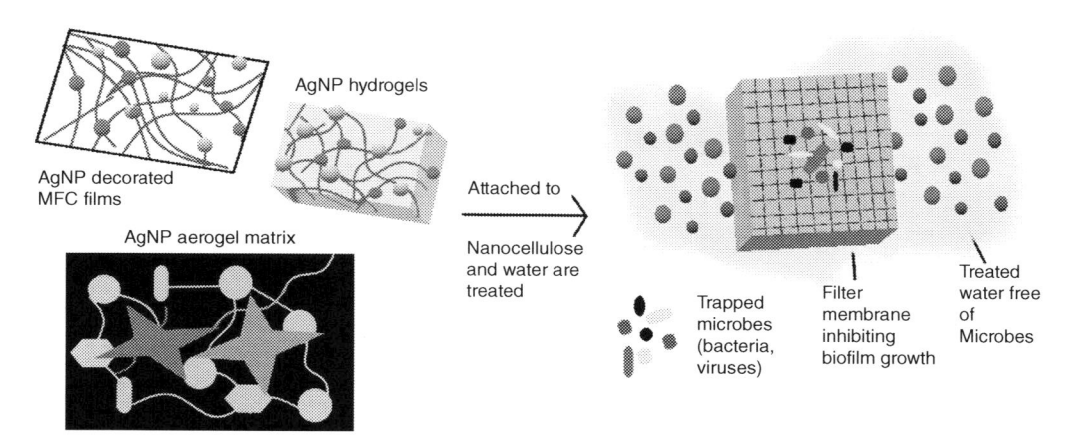

Fig. 9.12 AgNP-coated filter membranes utilized for removal of microbes.

of Cu–Zn alloy, and Cu metal particles of micron size. A study by [52] was conducted in two different steps: First, cupric acetate was incorporated to poly(vinyl alcohol) resulting in the formation of a fine film by moisture casting. In the second step, CuO particles were incorporated to polyurethane (PU) giving the composite nanofibers. The results showed that PU/CuO composite nanofibers had improved electrical conductivities leading to better results for using these nanofibers as carrier transport [52].

9.5.2.7 Water purification by green adsorbents

CNFs and CNCs were extracted from tunicates and hardwood pulp. These sources are used to recuperate platinum when it is modified with polyethylamine [53]. Another study showed that rice straw was treated physiochemically to modify the nanocellulose which was, in turn, used for removing toxic and harmful metals from wastewater. This offered an advantage as it was a green and environment-friendly technique having cost effectiveness which fitted well as an advantage for a wastewater industry [54]. Passionfruit peels acting as nanocellulose by hydrolyzing the natural component utilizing sulfuric acid suspension have a potential to act as an antibiotic drug carrier [55]. Tetracycline antibiotic was used to test the drug carrier by *in-vitro* release in phosphoric buffer medium.

9.6 Biogenic NPs uses in agriculture

9.6.1 ZnO NPs

To gain the high yield and good growth of food crops, an experiment was carried out on peanut seeds with varying concentrations on ZnONP [7]. A concentration of 1000 ppm was tested, which proved to be efficient and resulted in boosting the growth of roots and stem in peanuts.

9.6.2 Gold NPs usage to enhance crop quality

Gold and gold NPs are now extensively used to see their effects in the plants, which exhibit both the negative as well as beneficial effects. Also, the gold NPs increase the enzymatic activities, which plays a role of an antioxidant; and the radical scavenging has also been improved [56] leading to better crop yield, fruit yield, and the vegetative growth. An experiment was carried out by selecting five different concentrations (0, 10, 25, 50, and 100 ppm) of Gold NP and atomic absorption spectroscopy (AAS), and the presence of Gold NP in leaf tissue was confirmed [57]. After the experiment, the results revealed that the seed yield was increased at 10 ppm concentration, and also the content of reduced sugar and total sugar was raised up to 25 ppm [57]. Thus, the experiment proved to be successful as it increased the yield and growth of the crop.

9.7 Nanocellulose for air purification

Filtering air is the most often used methods for purification of air and sampling of aerosol [58]. Nowadays CNF filters are used widely for purification of air in many industries.

Using two techniques, different qualities of CNF are obtained. This can be done by manufacturing and modifying CNFs. These nanofibrils differ in quality due to the charges on surface of fibrils, one which has a negative charge and the other one has a neutral charge. These CNFs have the greater surface area and high strength, as shown in Fig. 9.13, which make them important for filtering the air [58].

In a method, TEMPO-mediated oxidation along with the oxidant called sodium hypochlorite is catalyzed to make the reaction rate faster with TEMPO (2,2,6,6-tetramethylpiperidine-1-oxy). This was another method suggested for filtration.

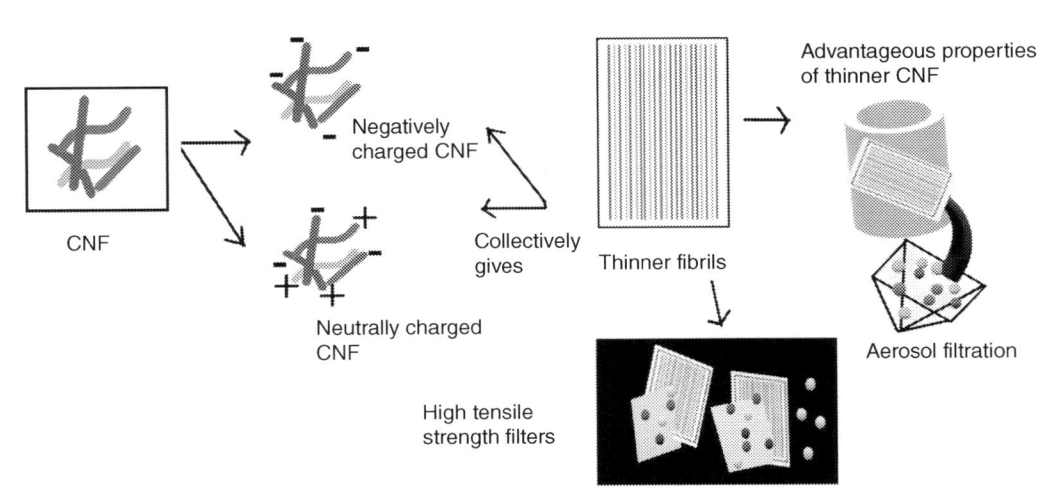

Fig. 9.13 Production of CNF as a filter for purification of air.

Also, on the surface of fibrils having cellulose, aldehyde and carboxyl groups are also attached. The nanofibrils separation is supported due to TEMPO-mediated oxidation resulting in the separation of each and every nanofibril, having widths of 3.5–20 nm and length between 0.2 and 2.2 μm [59]. This depends on homogenization conditions as well as oxidation [58].

Another technique, which results in development of a fresh nanofiber structure, is freezing and then drying. But this technique has a disadvantage of having a low manufacturing and can result in the formation of a complex structure. Here, different solid contents of the initial dispersion are used. A better preservative technique of the fibril structures is to preserve them in a solid content of 0.1% (C1 and T1) acting as a better preservative than the samples having a high content of solid (C2, C3, T2) [58]. All freeze-dried samples are of porous nature with diameter of 100 μm [60]. Nano-sized NaCl particles are utilized to estimate the ability of the collection of the particle. Payet's theoretical model [58] is used to know the particle penetration, but both the CNFs are used to make filters having a good performance [58].

9.7.1 Air purification by electrospun nanofiber mats

To check the factors related to the quality of the filter, Zhang et al. [61] studied and used electrospun nanofiber mats that showed a finer quality factor resulting in improved grade of the filter due to the use of various fine films of nanofiber mats giving the characteristics such as decrease in the diameter of the fibers.

9.7.2 Air purification by electrostatically active NPs

Polyacrylonitrile (PAN) nanofibers and TiO_2 nanofibers were utilized to observe the drop in the pressure and efficiency of air filtration by covering and fabricating them on nanofibrous filter media, which gave a hybridized form having greater sizes of pores and broader distribution of pore size [62]. Compared to the simple PAN fibers and TiO_2 fibers, the hybridized PAN nanofibers and TiO_2 nanofibers showed four times less pressure drop at 80 cm/s [62] of air velocity, which resulted in improved performance of the filters.

9.7.3 Antimicrobial air purification by using nanofibrous membrane based on soy protein

Soy proteins contain properties such as gelation, sorption, solubility, film formation. Many food processing industries and environmental industries have been inspired by the study carried out by Kinsella [63]. Further, Zhu et al. [64] created a nonpermeable compound layer using cellulose and concentrate of the soy protein, which possessed good physical properties. Thus, due to such multifunctionality properties, soy protein is gaining more attention. The soy protein isolate/polymide-6 (PA6)-silver nitrate system was electrospunned [65] followed by ultraviolet reduction, which proved successful and showed high performance against the simple air filters.

9.8 Nanopaper

Paper made up of CNFs displays excellent optical properties as the paper containing the nanofibrils has size lesser than the wavelength of visible light [66]. The immediate preparation of nanopaper is done by the sheet former which is semiautomatic [35]. In this study, the time for nanopaper preparation is 1 h for typical thickness of 60 μm preparation. In one more study, the pass number was increased by the homogenizer through chemical modifications resulting in nanopapers of increased tensile performance [67].

9.9 Nanocellulose uses in energy application

Nowadays nanocellulose-based nanocomposites are utilized in the manufacturing of fuel cell, Li-ion battery (LiB), and solar cell. Fuel cells are devices that can change chemical energy to electricity [2]. Oxygen is added to H_2 at the anode which is utilized as a fuel, whereas the oxygen is reduced to water at the cathode. An external circuit is connected and from there the electrons have a path to flow. BC has been found as best substratum to host a great number of nano-sized anode catalysts [2]. Precipitation of PdNPs on a BC membrane was obtained when a BC membrane was soaked in a 5 mM solution containing ammonium hexachloropalladate. After this, BC membrane was kept in between PdNP/BC nanocomposites then they are utilized as a membrane or layer electrode assembly in a fuel cell. Thus, this nanocomposite can be utilized in conversion of energy and its devices, as in the next step the maximum current was detected when H_2 was applied to PdNP/BC anode [2].

Electricity is generated with the help of solar cells using solar energy as a source. Nanocellulose paper (NCP) is also used to make solar cells as the NCP is smooth, transparent, and mechanically strong. This preparation of cells is based on the NCP having transparent property [66]. In this, indium tin oxide (ITO) film is coated on NCP by radio frequency magnetron sputtering of a target composed of 10% SnO_2 and 90% In_2O_3 [2]. After coating, the originally transparent paper became translucent. Then, the ITO-coated NCP is then treated with poly (3-hexylthiophene) (P3HT) and [6:6]-phenyl-C_{61}-butyric acid methyl ester. This acts as the layer absorbing the light [2]. The path length is then increased by the absorbing layer of this transparent paper, which results in increase in the solar light absorption. This NCP is used in photovoltaic applications and the solar cell which is made has a capacity to convert the power of 0.4 higher content (C2, C3, T2) [66].

Nowadays, LiBs are also based on nanocellulose. The advantage of using these batteries in the electronic products is that they have rechargeable characteristics. NCP has a great utility it gives a way to greater flexibility, shows high-performance, and is used in thin LIBs. NC is used as separator, electrode, and electrolyte in LiB cell [68] as shown in Fig. 9.14.

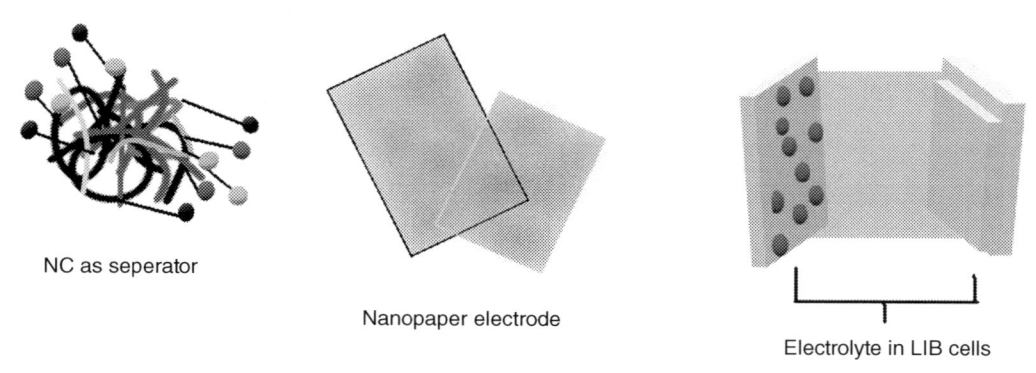

NC as seperator

Nanopaper electrode

Electrolyte in LIB cells

Fig. 9.14 Nanocellulose as separator, electrode, and electrolyte in the Li-ion batteries (LIB).

The graphite/MFC slurry is vacuum dried at 313 K for whole night, which results in production of graphite/MFC LIB anode [67]. Electrolyte is used in Li–ion batteries. The charging and discharging capacity of MFC/graphite anode shows similarity with graphite/poly(vinylidene fluoride) that is also called PVdF [2].

9.9.1 Perovskite solar cell as a source of energy

NCP having a layer of the resin called acrylic is used as a substrate to fabricate perovskite solar cells (PSCs) that are flexible [69] and can be modified by imparting properties such as transparency, biodegradability, substrates having low costs, and with a green approach toward environment has been created as an replacement to PSCs.

9.10 Conclusion

In the current chapter, the discussion is based on the use of nanomaterial focusing mainly on the use of nanocellulose in the environmental industry. Nanocellulose being nonabrasive, biodegradable, comprising lower densities has been widely used in the environmental industries nowadays. Such extraordinary characteristics and in the view of remarkable features along with the property of having abundant sources and cheap availability bring up nanocellulose into light by acting as an adsorbent and is used in various energy applications, preparation of photocatalysts based on nanocellulose, etc. Various methods of air purification and water purification in the industries with the help of nanocellulose have been discussed here. For the water purification, a method is discussed where; the rice straw was used to treat the toxic metals in the wastewater giving an environment-friendly approach for wastewater treatment. Another use of nanocellulose has been in the process of air purification where nanofibrous membrane based on the soy protein is been used. A technology related to nanopaper has also been discussed which summarizes various techniques used to prepare a nanopaper having size even smaller than the wavelength of a visible light. These are easy, cheap, straightforward, and convenient approaches of the use of nanocellulose in various environmental industries. Thus, conversion of cellulosic material into nanocellulose and using them in the environmental industries can be highly advantageous and can create a new way for using innovative technologies, providing much exposure to the country.

References

[1] S.F. Jin, Y. Chen, M. Liu, Nanocellulose applications in environmental protection, Adv. Mat. Res. 662 (2013) 198–201. https://doi.org/10.4028/www.scientific.net/AMR.662.198.

[2] H. Wei, K. Rodriguez, S. Renneckar, P.J. Vikesland, Environmental science and engineering applications of nanocellulose-based nanocomposites, Environ. Sci. Nano 1 (2014) 302–316. https://doi.org/10.1039/C4EN00059E.

[3] W. Wang, M.D. Mozuch, R.C. Sabo, P. Kersten, et al., Production of cellulose nanofibrils from bleached eucalyptus fibers by hyperthermostable endoglucanase treatment and subsequent microfluidization, Cellulose 22 (2015) 351–361. https://doi.org/10.1007/s10570-014-0465-2.

[4] D. Klemm, F. Kramer, S. Moritz, T. Lindström, et al., Nanocelluloses: a new family of naturebased materials, Angew. Chem. Int. Ed. 50 (24) (2011) 5438–5466. https://doi.org/10.1002/anie.201001273.

[5] V.N. Kumar, S. Rebello, E.M. Aneesh, R. Sindhu, et al., Nanocellulose in Paper Making, in: Amir Al-Ahmed, Inamuddin (Eds.), Advanced Applications of Polysaccharides and their Composites, 73, Mater. Res. Found, 2020, pp. 184–197.

[6] K.P.Y. Shak, Y.L. Pang, S.K. Mah, Nanocellulose: recent advances and its prospects in environmental remediation, Beilstein J. Nanotechnol. 9 (2018) 2479–2498. https://doi.org/10.3762/bjnano.9.232.

[7] S. Sabir, M. Arshad, S.K. Chaudhari, Zinc oxide nanoparticles for revolutionizing agriculture: synthesis and applications, Sci. World J. (2014) 1–8. https://doi.org/10.1155/2014/925494.

[8] N. Mahfoudhi, S. Boufi, Nanocellulose as a novel nanostructured adsorbent for environmental remediation: a review, Cellulose 24 (2017) 1171–1197. https://doi.org/10.1007/s10570-017-1194-0.

[9] T.S. Anirudhan, J.R. Deepa, S. Binusreejayan, Synthesis and characterization of multi-carboxyl-functionalized nanocellulose/nanobentonite composite for the adsorption of uranium (VI) from aqueous solutions: kinetic and equilibrium profiles, Chem. Eng. J. 273 (2015) 390–400. https://doi.org/10.1016/j.cej.2015.03.007.

[10] S. Olivera, H.B. Muralidhara, K. Venkatesh, V.K. Guna, K. Gopalakrishna, Y. Kumar K., Potential applications of cellulose and chitosan nanoparticles/composites in wastewater treatment: a review, Carbohydr. Polym. 153 (2016) 600–618. https://doi.org/10.1016/j.carbpol.2016.08.017.

[11] S. Hokkanen, A. Bhatnagar, E. Repo, S. Lou, M. Sillanpää, Calcium hydroxyapatite microfibrillated cellulose composite as a potential adsorbent for the removal of Cr(VI) from aqueous solution, Chem. Eng. J. 283 (2016) 445–452. https://doi.org/10.1016/j.cej.2015.07.035.

[12] J. Chaba, Synthesis of metal oxides coated carbon nanofibers and their application for removal of selected antibiotics in environmental matrices. (M.Sc. thesis) University of Johannesburg, 2018.

[13] M. Omidvar, S. Mousavi, M. Soltanieh, A.A. Safekordi, Preparation and characterization of poly (ethersulfone) nanofiltration membranes for amoxicillin removal from contaminated water, J. Environ. Health Sci Eng. 12 (2014) 18. https://doi.org/10.1186/2052-336X-12-18.

[14] T.S. Anirudhan, J.R. Deepa, J. Christa, Nanocellulose/nanobentonite composite anchored with multi-carboxyl functional groups as an adsorbent for the effective removal of Cobalt(II) from nuclear industry wastewater samples, J. Colloid Interface Sci. 467 (2016) 307–320. https://doi.org/10.1016/j.jcis.2016.01.023.

[15] X. Sun, L. Yang, Q. Li, J. Zhao, X. Li, X. Wang, H. Liu, Amino-functionalized magnetic cellulose nanocomposite as adsorbent for removal of Cr(VI): synthesis and adsorption studies, Chem. Eng. J. 241 (2014) 175–183. https://doi.org/10.1016/j.cej.2013.12.051.

[16] S. Hokkanen, E. Repo, T. Suopajärvi, H. Liimatainen, J. Niinimaa, M. Sillanpää, Adsorption of Ni(II), Cu(II) and Cd(II) from aqueous solutions by amino modified nanostructured microfibrillated cellulose, Cellulose 21 (2014) 1471–1487. https://doi.org/10.1007/s10570-014-0240-4.

[17] J. Yang, J. Yu, J. Fan, D. Sun, W. Tang, X. Yang, Biotemplated preparation of CdS nanoparticles/bacterial cellulose hybrid nanofibers for photocatalysis application, J. Hazard. Mater. 189 (2011) 377–383. https://doi.org/10.1016/j.jhazmat.2011.02.048.

[18] M. Gao, N. Li, W. Lu, W. Chen, Role of cellulose fibers in enhancing photosensitized oxidation of basic green 1 with massive dyeing auxiliaries, Appl. Catal. B 147 (2014) 805–812. https://doi.org/10.1016/j.apcatb.2013.10.015.

[19] J. Zeng, S. Liu, J. Cai, L. Zhang, TiO_2 immobilized in cellulose matrix for photocatalytic degradation of phenol under weak UV light irradiation, J. Phys. Chem. C 114 (2010) 7806–7811. https://doi.org/10.1021/jp1005617.

[20] X. An, D. Cheng, L. Dai, B. Wang, H.J. Ocampo, J. Nasrallah, X. Jia, J. Zou, Y. Long, Y. Ni, Synthesis of nano-fibrillated cellulose/magnetite/titanium dioxide ($NFC@Fe_3O_4@TNP$) nanocomposites and their application in the photocatalytic hydrogen generation, Appl. Catal. B 206 (2017) 53–64. https://doi.org/10.1016/j.apcatb.2017.01.021.

[21] K. Tu, Q. Wang, A. Lu, L. Zhang, Portable visible-light photocatalysts constructed from Cu_2O nanoparticles and graphene oxide in cellulose matrix, J. Phys. Chem. C 118 (2014) 7202–7210. https://doi.org/10.1021/jp412802h.

[22] T. Suopajärvi, J.A. Sirviö, H. Liimatainen, Cationic nanocelluloses in dewatering of municipal activated sludge, J. Environ. Chem. Eng. 8 (2017) 86–92. https://doi.org/10.1016/j.jece.2016.11.021.

[23] H. Zhu, Y. Zhang, X. Yang, H. Liu, L. Shao, X. Zhang, J. Yao, One-step green synthesis of nonhazardous dicarboxyl cellulose flocculant and its flocculation activity evaluation, J. Hazard. Mater. 296 (2015) 1–8. https://doi.org/10.1016/j.jhazmat.2015.04.029.

[24] S.K. Das, Md.M.R. Khan, T. Parandhaman, F. Laffir, A.K. Guha, G. Sekaran, A.B. Mandal, Nanosilica fabricated with silver nanoparticles: antifouling adsorbent for efficient dye removal, effective water disinfection and biofouling control, Nanoscale 5 (2013) 5549. https://doi.org/10.1039/c3nr00856h.

[25] S. Eyley, D. Vandamme, S. Lama, G. Van den Mooter, K. Muylaert, W. Thielemans, CO_2 controlled flocculation of microalgae using pH responsive cellulose nanocrystals, Nanoscale 7 (2015) 14413–14421. https://doi.org/10.1039/C5NR03853G.

[26] Y. Habibi, L.A. Lucia, O.J. Rojas, Cellulose nanocrystals: chemistry, self-assembly, and applications, Chem. Rev. 110 (2010) 3479–3500. https://doi.org/10.1021/cr900339w.

[27] D. Chen, T.G.M. van de Ven, Flocculation kinetics of precipitated calcium carbonate induced by electrosterically stabilized nanocrystalline cellulose, Colloids Surf. A Physicochem. Eng. Asp. 504 (2016) 11–17. https://doi.org/10.1016/j.colsurfa.2016.05.023.

[28] X. Sun, C. Danumah, Y. Liu, Y. Boluk, Flocculation of bacteria by depletion interactions due to rod-shaped cellulose nanocrystals, Chem. Eng. J. (2012) 476–481. https://doi.org/10.1016/j.cej.2012.05.114.

[29] S.I. Yu, S.K. Min, H.S. Shin, Nanocellulose size regulates microalgal flocculation and lipid metabolism, Sci. Rep. 6 (2016) 35684. https://doi.org/10.1038/srep35684.

[30] S. Terhi, Functionalized Nanocelluloses in Wastewater Treatment Applications, University of Oulu, Kuusamonsali (YB210), Linnanmaa, 2015.

[31] H. Voisin, L. Bergström, P. Liu, A. Mathew, Nanocellulose-based materials for water purification, Nanomaterials 7 (2017) 57. https://doi.org/10.3390/nano7030057.

[32] K. Singh, J.K. Arora, T.J.M. Sinha, S. Srivastava, Functionalization of nanocrystalline cellulose for decontamination of Cr(III) and Cr(VI) from aqueous system: computational modeling approach, Clean Technol. Environ. Policy 16 (2014) 1179–1191. https://doi.org/10.1007/s10098-014-0717-8.

[33] M. Hasani, E.D. Cranston, G. Westman, D.G. Gray, Cationic surface functionalization of cellulose nanocrystals, Soft Matter 4 (2008) 2238–2244. https://doi.org/10.1039/B806789A.

[34] H. Qiao, Y. Zhou, F. Yu, E. Wang, Y. Min, Q. Huang, L. Pang, T. Ma, Effective removal of cationic dyes using carboxylate-functionalized cellulose nanocrystals, Chemosphere 141 (2015) 297–303. https://doi.org/10.1016/j.chemosphere.2015.07.078.

[35] H. Sehaqui, A. Liu, Q. Zhou, L.A. Berglund, Fast preparation procedure for large, flat cellulose and cellulose/inorganic nanopaper structures, Biomacromolecules 11 (9) (2010) 2195–2198. https://doi.org/10.1021/bm100490s.

[36] A.O. Adeeyo, J.O. Odiyo, Biogenic synthesis of silver nanoparticle from mushroom exopolysaccharides and its potentials in water purification, CHEM 5 (2018) 64–75. https://doi.org/10.2174/1874842201805010064.

[37] M. Gericke, A. Pinches, Biological synthesis of metal nanoparticles, Hydrometallurgy 83 (2006) 132–140. https://doi.org/10.1016/j.hydromet.2006.03.019.

[38] H. Ma, C. Burger, H. Benjamin S, C. Benjamin, Ultra-fine cellulose nanofibers: new nano-scale materials for water purification, J. Mater. Chem. 21 (2011) 7507–7510. https://doi.org/10.1039/C0JM04308G.

[39] H. Ma, C. Burger, B.S. Hsiao, B. Chu, Ultra-fine cellulose nanofibers: new nano-scale materials for water purification, J. Mater. Chem. 21 (2011) 7507. https://doi.org/10.1039/c0jm04308g.

[40] L.A. Goetz, B. Jalvo, R. Rosal, A.P. Mathew, Superhydrophilic anti-fouling electrospun cellulose acetate membranes coated with chitin nanocrystals for water filtration, J. Membr. Sci. 510 (2016) 238–248. https://doi.org/10.1016/j.memsci.2016.02.069.

[41] S. Kaur, Z. Ma, R. Gopal, G. Singh, S. Ramakrishna, T. Matsuura, Plasma-induced graft copolymerization of poly(methacrylic acid) on electrospun poly(vinylidene fluoride) nanofiber membrane, Langmuir 23 (2007) 13085–13092. https://doi.org/10.1021/la701329r.

[42] V. Thavasi, G. Singh, S. Ramakrishna, Electrospun nanofibers in energy and environmental applications, Energy Environ. Sci. 1 (2008) 205. https://doi.org/10.1039/b809074m.

[43] R. Gopal, S. Kaur, Z. Ma, C. Chan, S. Ramakrishna, T. Matsuura, Electrospun nanofibrous filtration membrane, J. Membr. Sci. 281 (2006) 581–586. https://doi.org/10.1016/j.memsci.2006.04.026.

[44] A. Mautner, K.-Y. Lee, T. Tammelin, A.P. Mathew, A.J. Nedoma, K. Li, A. Bismarck, 2015. Cellulose nanopapers as tight aqueous ultra-filtration membranes. React. Funct. Polym. 86, 209–214. https://doi.org/10.1016/j.reactfunctpolym.2014.09.014.

[45] Z. Karim, A.P. Mathew, V. Kokol, J. Wei, M. Grahn, High-flux affinity membranes based on cellulose nanocomposites for removal of heavy metal ions from industrial effluents, RSC Adv. 6 (2016) 20644–20653. https://doi.org/10.1039/C5RA27059F.

[46] G. Metreveli, L. Wågberg, E. Emmoth, S. Belák, M. Strømme, A. Mihranyan, A size-exclusion nanocellulose filter paper for virus removal, Adv. Healthc. Mater. 3 (2014) 1546–1550. https://doi.org/10.1002/adhm.201300641.

[47] L. Zhang, X. Bai, H. Tian, L. Zhong, C. Ma, Y. Zhou, S. Chen, D. Li, Synthesis of antibacterial film CTS/PVP/TiO_2/Ag for drinking water system, Carbohydr. Polym. 89 (2012) 1060–1066. https://doi.org/10.1016/j.carbpol.2012.03.063.

[48] W. Xiao, J. Xu, X. Liu, Q. Hu, J. Huang, Antibacterial hybrid materials fabricated by nanocoating of microfibril bundles of cellulose substance with titania/chitosan/silver-nanoparticle composite films, J. Mater. Chem. B 1 (2013) 3477. https://doi.org/10.1039/c3tb20303d.

[49] H. Dong, J.F. Snyder, D.T. Tran, J.L. Leadore, Hydrogel, aerogel and film of cellulose nanofibrils functionalized with silver nanoparticles, Carbohydr. Polym. 95 (2013) 760–767. https://doi.org/10.1016/j.carbpol.2013.03.041.

[50] H. Palza, Antimicrobial polymers with metal nanoparticles, Int. J. Mol. Sci. 16 (2015) 2099–2116. https://doi.org/10.3390/ijms16012099.

[51] H.L. Karlsson, P. Cronholm, Y. Hedberg, M. Tornberg, L. De Battice, S. Svedhem, I.O. Wallinder, Cell membrane damage and protein interaction induced by copper containing nanoparticles—importance of the metal release process, Toxicology 313 (2013) 59–69. https://doi.org/10.1016/j.tox.2013.07.012.

[52] R. Nirmala, K.S. Jeon, B.H. Lim, R. Navamathavan, H.Y. Kim, Preparation and characterization of copper oxide particles incorporated polyurethane composite nanofibers by electrospinning, Ceramics Int. 39 (2013) 9651–9658. https://doi.org/10.1016/j.ceramint.2013.05.087.

[53] H.-J. Hong, H. Yu, M. Park, H.S. Jeong, Recovery of platinum from waste effluent using polyethyleneimine-modified nanocelluloses: effects of the cellulose source and type, Carbohydr. Polym. 210 (2019) 167–174. https://doi.org/10.1016/j.carbpol.2019.01.079.

[54] A. Kardam, K.R. Raj, S. Srivastava, M.M. Srivastava, Nanocellulose fibers for biosorption of cadmium, nickel, and lead ions from aqueous solution, Clean Technol. Environ. Policy 16 (2014) 385–393. https://doi.org/10.1007/s10098-013-0634-2.

[55] C.J. Wijaya, S.N. Saputra, F.E. Soetaredjo, J.N. Putro, C.X. Lin, A. Kurniawan, Y.-H. Ju, S. Ismadji, Cellulose nanocrystals from passion fruit peels waste as antibiotic drug carrier, Carbohydr. Polym. 175 (2017) 370–376. https://doi.org/10.1016/j.carbpol.2017.08.004.

[56] K.S. Siddiqi, A. Husen, Engineered gold nanoparticles and plant adaptation potential, Nanoscale Res. Lett. 11 (2016) 400. https://doi.org/10.1186/s11671-016-1607-2.

[57] S. Arora, P. Sharma, S. Kumar, R. Nayan, P.K. Khanna, M.G.H. Zaidi, Gold-nanoparticle induced enhancement in growth and seed yield of *Brassica juncea*, Plant Growth Regul. 66 (2012) 303–310. https://doi.org/10.1007/s10725-011-9649-z.

[58] L. Alexandrescu, K. Syverud, A. Nicosia, G. Santachiara, A. Fabrizi, F. Belosi, Airborne nanoparticles filtration by means of cellulose nanofibril based materials, J. Biomater. Nanobiotechnol. 07 (2016) 29–36. https://doi.org/10.4236/jbnb.2016.71004.

[59] G. Chinga-Carrasco, Cellulose fibres, nanofibrils and microfibrils: the morphological sequence of MFC components from a plant physiology and fibre technology point of view, Nanoscale Res. Lett. 6 (2011) 417. https://doi.org/10.1186/1556-276X-6-417.

[60] K. Syverud, H. Kirsebom, S. Hajizadeh, G. Chinga-Carrasco, Cross-linking cellulose nanofibrils for potential elastic cryo-structured gels, Nanoscale Res. Lett. 6 (626) (2011) 1–6. https://doi.org/10.1186/1556-276X-6-626.

[61] Q. Zhang, J. Welch, H. Park, C.-Y. Wu, W. Sigmund, J.C.M. Marijnissen, Improvement in nanofiber filtration by multiple thin layers of nanofiber mats, J. Aerosol Sci. 41 (2010) 230–236. https://doi.org/10.1016/j.jaerosci.2009.10.001.

[62] D. Cho, A. Naydich, M.W. Frey, Y.L. Joo, Further improvement of air filtration efficiency of cellulose filters coated with nanofibers via inclusion of electrostatically active nanoparticles, Polymer 54 (2013) 2364–2372. https://doi.org/10.1016/j.polymer.2013.02.034.

[63] J.E. Kinsella, Functional properties of soy proteins, J. Am. Oil Chem. Soc. 56 (1979) 242–258. https://doi.org/10.1007/BF02671468.

[64] Y. Zhu, E. Douglass, T. Theyson, R. Hogan, R. Kotek, Cellulose and soy proteins based membrane networks, Macromol. Symp. 329 (2013) 70–86. https://doi.org/10.1002/masy.201300032.

[65] Z. Jiang, H. Zhang, M. Zhu, D. Lv, J. Yao, R. Xiong, C. Huang, Electrospun soy-protein-based nanofibrous membranes for effective antimicrobial air filtration, J. Appl. Polym. Sci. 135 (2018) 45766. https://doi.org/10.1002/app.45766.

[66] L. Hu, G. Zheng, J. Yao, N. Liu, B. Weil, M. Eskilsson, E. Karabulut, Z. Ruan, S. Fan, J.T. Bloking, M.D. McGehee, L. Wågberg, Y. Cui, Transparent and conductive paper from nanocellulose fibers, Energy Environ. Sci. 6 (2013) 513–518. https://doi.org/10.1039/C2EE23635D.

[67] S.-J. Chun, S.-Y. Lee, G.-H. Doh, S. Lee, J.H. Kim, Preparation of ultrastrength nanopapers using cellulose nanofibrils, J. Ind. Eng. Chem. 17 (2011) 521–526. https://doi.org/10.1016/j.jiec.2010.10.022.

[68] L. Jabbour, R. Bongiovanni, D. Chaussy, C. Gerbaldi, D. Beneventi, Cellulose-based Li-ion batteries: a review, Cellulose 20 (2013) 1523–1545. https://doi.org/10.1007/s10570-013-9973-8.

[69] L. Gao, L. Chao, M. Hou, J. Liang, Y. Chen, H.-D. Yu, W. Huang, Flexible, transparent nanocellulose paper-based perovskite solar cells, NPJ Flex Electron. 3 (2019) 4. https://doi.org/10.1038/s41528-019-0048-2.

CHAPTER 10

Application of nanocellulose as nanotechnology in water purification

Birendra Bharti, Vibhanshu Kumar, Himanshu Kumar
Department of Water Engineering and Management, Central University of Jharkhand, Jharkhand, India

10.1 Introduction

Potable water is anticipated toward getting one of the most valuable wares on the earth around. World's population still lack daily access to sober drinking water, regardless of admittance to hygienic drinking water is a solitary significant basic liberties (UN General Assembly, Resolution 2010). Around 1 trillion folks do not actually have ingress to clean potable water [1,2] besides some 8,00,000 individuals dying every single year due to the ingestion of polluted drinking water, ineffective hand washing, and poor sanitation services [1]. In addition, the World Economic Forum assesses water shortage as the apex worldwide concern for environmental risks [3]. The predicted population of half the world will be in water-hassled regions by 2025 World Health Organization [4]. The widespread production of large-scale membrane filtration processes was fuelled by growing global water and wastewater treatment needs [5,6]. To establish instruments for treating these contaminants, research on new technologies and approaches to treatment is desperately needed. Availability of both drinking water and wastewater treatment are considered an essential component in the growth of human civilization [6,7].

Nanotechnology (NT) is the special branch of science which has wide range of specialty to controlling issue at the nuclear and subatomic level can possibly convey moderate and viable solutions for this issue in water disinfection/filtration, giving admittance to safe drinking water to a great many individuals [8] and thereby helping to eradicate hunger and meet the aim of sustainable growth. With the advent of NT, creation of cutting-edge nanomaterials [9] offers huge open doors for a broad scope of uses for discovery and remediation of ecological contaminants [10,11].

In the era of modern technology, proficient elimination and separation of toxic and pollutants materials from portable water always been an epic topic for discussion among common people and scientists. Some of contemporary researchers have successfully developed methodology and apparatus utilizing various natural occurring as well as synthetic polymeric contents for the proficient elimination and separation of toxic and pollutants materials from water [12,13]. However, its use is limited because of the low process capability, the insolubility, and the nonbiodegradability of a wide variety of solvents [14]. Late examinations have exhibited that film innovation can be effectively applied to

Nanocellulose Materials: Fabrication and Industrial Applications
DOI: https://doi.org/10.1016/B978-0-12-823963-6.00014-4

decontaminate wastewater delivered from the material, calfskin, food, electronic, journal enterprises, and urban wastewater [15,16]. In the widespread array of continuous effort, nanocellulose (NC) attracted these students of science and researchers for purging of water due to its special properties. To recapitulate the recent growth in the field of purging of portable water by the application of NC, we are incorporating few recent developments work in the technique of purging of water. Herein, we have also incorporated some of the pioneer work that has been carried out in the water purification by the application of NT.

10.2 NT-enabled multifunctional application of NC

NCs can clearly be divided into three major categories as cellulose nanofibers (CFs), cellulose nanocrystals (CNCs), and bacterial NC. Among various other viable practical nanomaterials, NC is attracting expanding interest for utilization in ecological remediation technologies because of its copious inimitable properties and functionalities. NC can be utilized to build up polymers, paper, and films as membranes, ensuring biodegradable and harmless ecosystem, bionanocomposites with wonderful improvement in material properties when contrasted with polymer networks of regular full-scale amalgamated materials [17]. These upgrades are credited to the amazing properties of cellulose strands, including high modulus, high rigidity, and great boundary properties, just as straightforwardness for compound surface change (Fig. 10.1). The NC-based

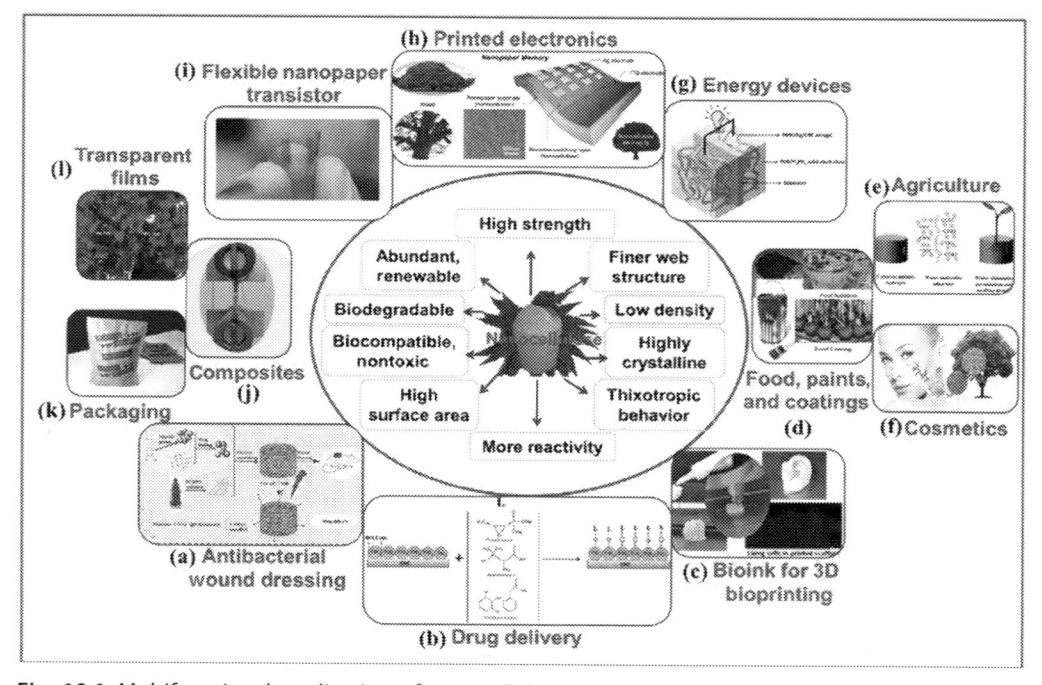

Fig. 10.1 Multifunctional application of nanocellulose reproduced after the permission (Ref: Yadav et al. [19], https://doi.org/10.1016/j.ijbiomac.2020.11.038).

mixture nanomaterials have tremendous expected applications nearby food bundling, biopharmaceuticals, biomedical, and beauty care products [18].

10.3 Effective role of NC as biosorbent

NC, which offers an amalgamation of absorption and adsorption with interesting cellulosic nature, has huge potential for another and green course to tackle the current substantial metal contamination issues. Fig. 10.2 shows the physicochemical properties and mechanism of NC as adsorbents for substantial metal particles from the points of view of cellulosic nature and its nano size. The high explicit territory of NC is relied upon to give countless dynamic locales on the outside of cellulose to immobilize metal particles. The particular surface region of cellulose nanofiber arranged to utilize a supercritical drying measure can be pretty much as high as 480 m^2/g [20].

The surface functional groups working as metal-restricting destinations on the biomass independent of miniature or nanoscale are viewed as liable for the immobilization of substantial metal particles. The presence of bountiful hydroxyl bunches on the outside of NC gives an interesting stage to huge surface adjustment to unite a horde of practical gatherings or atoms onto the cellulosic structure. The surface changes of C incorporate sulfonation, 2,2,6,6-tetra methyl piperdine-1-oxyl radical (TEMPO) intervened oxidation, phosphorylation, esterification, etherification, silylation, amidation, and so on meanwhile, numerous other physical and compound properties of NC can be custom-made by surface changes to meet the necessities of utilizations [21–24].

NC has great mechanical strength and unbending nature that can offer adsorbents with potential for use in high-pressure conditions in genuine water treatment applications.

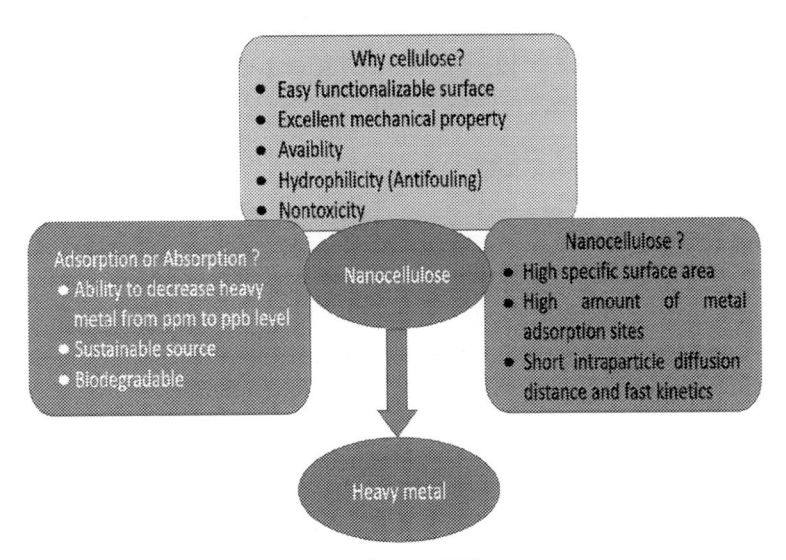

Fig. 10.2 Physicochemical properties of nanocellulose (NC).

Strength in the water climate, just as the hydrophilicity of NC, is additionally of benefit while utilizing in water treatment. The hydrophilicity is relied upon to lessen biofouling and natural fouling. Furthermore, NC, particularly for CNC, as a rule, has high crystallinity, which makes the adsorbents impervious to substance and organic consumption in a watery climate.

Feasibility of nanocellulose as water purifier.

In contemporary years, the application of NT in the field of water purification partakes in its peak of the era. There are numerous different nanomaterials and methodologies that have been discovered to purify the water. In line, NC permits the fitting of molecule surface science to encourage self-get together, controlled scattering inside a wide scope of grid polymers, which provides wide periphery for water purging. The first of two subcategories have immense potential as water purging. NC can be custom-made to the novel and altogether enhanced physical and synthetic properties. The fusion of high strength, synthetic latency, hydrophilic superficial science, and high surface territory makes NC as an extremely encouraging material for elite layers and channels, to specifically eliminate impurities from natural waters and industrial wastewater [25]. The high mechanical strength and inflexibility of NC are significant in high-pressure, water-treatment applications. NC have extraordinary notch of crystallinity properties that are artificially dormant in watery media [26] besides at high pH–values [27]. The inherent hydrophilicity of NC is required to lessen biofouling and natural fouling. These incredible possessions of NC have an immense expected possibility for wastewater treatment (shown in Fig. 10.3). In this context, Gopakumar et al. [68] discussed about NC-based cellulose pulp fibers having extraordinary properties. This NC comprises rehashing β-D-glucopyranose units covalently connected through acetal capacities between the hydroxyl gatherings of C4 and C1 carbon molecules that give it chirality and reactivity properties (Fig. 10.3A). Additionally, the bountiful accessibility of the surface hydroxyl bunches on the NC encourages different surface sciences that can be investigated for focusing on different pollutants in water. The presence of these hydroxyl bunches likewise makes a broad hydrogen holding network that makes cellulose insoluble in many fluid and natural solvents. Subsequently, on account of cellulose homogeneous responses, this hydrogen holding network should be disturbed to permit disintegration of the cellulose. The extraordinary facet ratio of the NC in an aqueous environment helps the creation of ultrafine network mechanisms explorable for the deletion and soaks in water of diverse contaminants. NC is hydrophilic in nature such that cellulose is used for raising the flux of the membranes with an antifouling hydrophilic covering.

Cellulose is an utmost bounteously discovered substance in the environment. It is a natural polysaccharide comprising β-d-glucopyranose entities amalgamated by

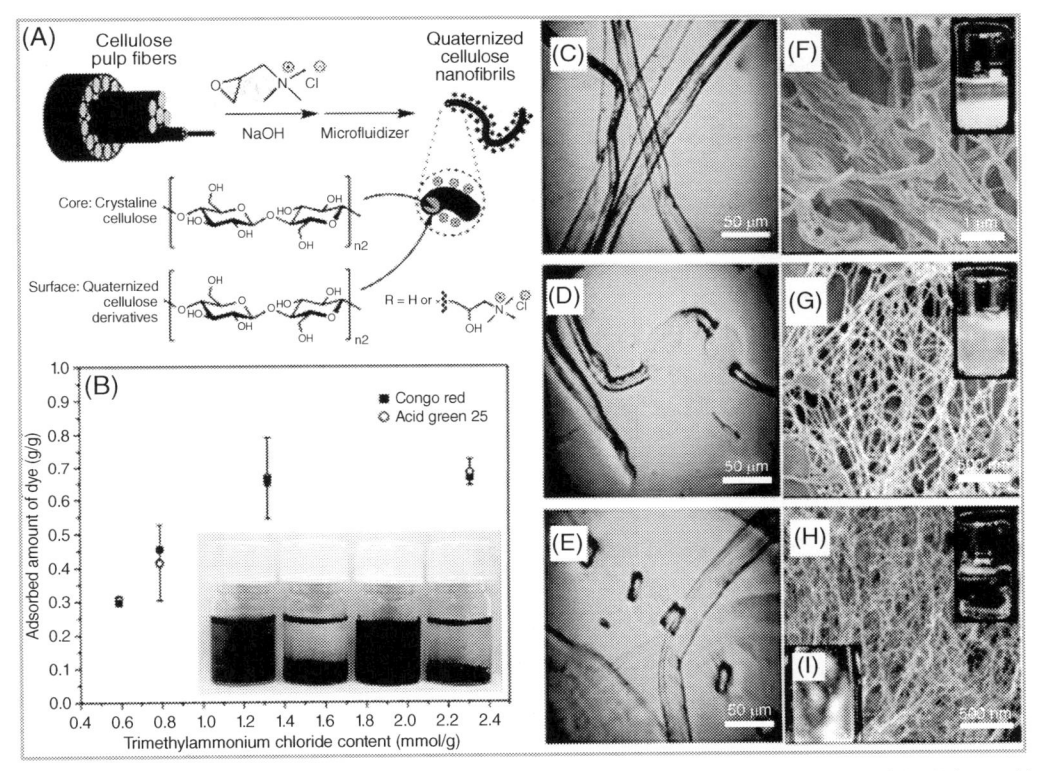

Fig. 10.3 Nanocellulose-based membrane for water purification (Ref: Gopakumar et al. [68], https://doi.org/10.1016/B978-0-12-813926-4.00004-5).

β-1,4-glycosidic yokes in a refined design. Cellulose manacles are in an arranged edifice, as there are no side manacles or kindling [28]. Typically happening cellulose-based nanomaterials have great auxiliary, mechanical, and ophthalmic possessions. Aside from utilizing cellulose and its subsidiaries in paper and bundling, car, individual consideration, development, and materials businesses, their use for wastewater remediation stays an extraordinary region of exploration because of genuine concerns. After sufficient synthetic adjustment on its surface, cellulose seems to show extraordinary adsorption capacity to contaminants, with the purpose of coordinating atoms containing essential gatherings, especially those wealthy in chemical elements (i.e., N, S, and O_2) [29]. Because of their natural abundance, ease of operation, and auxiliary assortments, CFs have been extensively deliberated [30,31]. As a functioning sorbent in pollutants and as a stabilizer for other dynamic particles, NC has created enthusiasm for this area.

The effectively functionalizable surface of NC takes into consideration the consolidation of various synthetic parities that may improve the authoritative and surface assimilation proficiency in contradiction of the contaminations in water, for example, colors, poisonous hefty metals, and so forth. In addition, CFs effectively functionalizable surface takes into consideration the fusion of synthetic parity which could expand the coupling productivity of contaminations to the CFs. CFs carboxylation by a long shot is the most read strategy for expanding their sorptive constraint shown by a graphical illustration in Fig. 10.4. The sorption of natural impurities has likewise been shown by means of synesthetic CF grids.

CFs can decrease innate hydrophilicity to increase the material sensitivity for hydrophobic compounds. The recent developments accomplished in nanoscience, in particular with nanocellulose, give a possible new form of material for use as nanoparticles (NPs) in water treatment. NC is effective retentive materials because of its high surface zone to volume proportion, minimal effort, profoundly normal cost, and intrinsic ecological idleness. The control of the superficial science of NC could be accomplished via the incorporation of conjointly natural and inorganic operational ties; consequently, this could go about as a water deterrent with the end goal eliminating oils remained in water system. Microfiltration (MF) films dependent on NC for water treatment have gotten significantly all the more intriguing in these days. The elements of NC besides the quality of the stuff can be ill-used in the manufacture of layers for water treatment. The expansion of NC to a polymer layer has for the most part been appeared to build

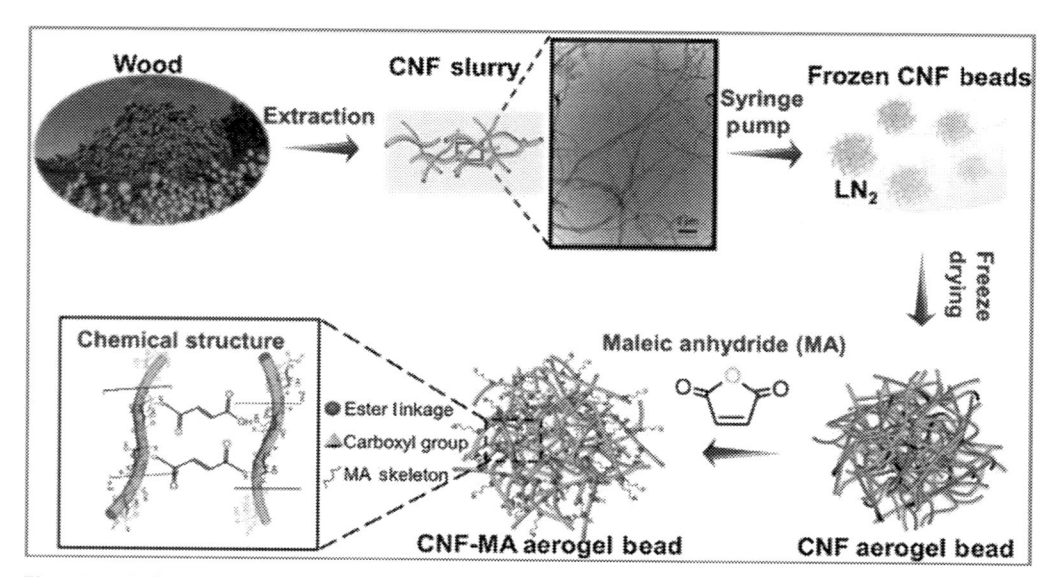

Fig. 10.4 Carboxylated cellulose nanofibers (CFs) for copper ion removal (Ref: Tang et al. [69], https://doi.org/10.1016/j.carbpol.2020.116397).

film hydrophilicity, in this way lessening the antifouling issues of the film. Each of these properties, for example, layer surface hydrophilicity, more prominent permeability, more noteworthy selectivity, and more prominent protection to make NC, has auspicious characteristics in lieu of wastewater recourse solicitations.

10.3.1 Application of cellulose CFs as water purifier

The characteristic of CFs and its quality can be utilized in the manufacture of films for recourse of water. As of late, numerous examinations have been accounted for on the CFs-based nanofibrous (NF) MF film for the recourse of water. Wang et al. equipped a new type CFs-based MF film that is accomplished in eradication of bacteria, microorganisms, and heavy metals. They exhibited an ultra-fine practical CF-enriched two-layer polyaniline/micro-Polyethylene Terephthalate (PET) fibrous pitch. They find that it has been made possible to fully extract bacteria (*Escherichia coli*) from the incredibly wide region of ultra-fine cellulose nanofibers by adding a significant number of loaded (carboxylate groups or adverse amine groups) membrane, achieving an log reduction value (LRV) of 4 for elimination of MS2 virus. The films proposed were exceptional in eliminating toxins with comparatively large pores and bacteria by scale exclusion (virus, heavy metals) of the infected water [32]. As a result of its unecofriendly possessions, hefty metals are viewed as the utmost natural toxins. Intended for instance, Cu^{2+}, Zn^{2+}, Cd^{2+}, and Pb^{2+} burned-through in high amounts can be cancer-causing to people. Because of high dissolvability and substantial metals in water bodies, they get collected in the environment. CFs have a unique role in fluid mechanics while filtration of water. The advancement of these cellulose nanofiber layers depended on the formation of a web-like framework with an exceptionally high charge density and a huge surface region per unit volume for the adsorption of impurity particles. Low-pressure films are a generally utilized drinking water treatment innovation to dismiss particles and microbiological pollutants. For the ecological and economical utilization of organic fragments for wastewater amendatory measures promotes cellulose nanomaterials (CNs) one of the dependable surface biosorbents in light of their enlarge surfacial zone-to-dimensions proportion, ease, usual enlarged plenitude, and organic inactivity [33]. Furthermore, poisonousness because of Cu^{2+}, Zn^{2+}, Cd^{2+}, and Pb^{2+} metal particles was effectively taken out from wastewater by utilizing adjusted cellulosic biopolymer. The adsorption of metal particles on adjusted cellulosic biopolymer estimated, as far as thermodynamic boundaries $\Delta H°$ and $\Delta G°$ uncovered that surface assimilation cycle was heat released and unconstrained [34].

Carboxylation of CNs is one which was examined widely strategically in the direction to upgrade their assimilation limit. It was discovered that presenting succinic corrosive onto CNCs brought about the expanded restricting effectiveness to Pb^{2+} and Cd^{2+} from watery arrangements [35]. The pace of sterilization of arrangement from these poisonous metal particles was additionally upgraded by the transformation of the corrosive carboxylic gatherings to sodiated carboxylates. Carboxylate-altered CFs exhibited 3−10% enlarged proficiencies to sorb Ni^{2+} and Cr^{3+}, notwithstanding Cd^{2+} and Pb^{2+} than unchanged CFs [36]. Different other examination bunches have announced that cellulose can viably adsorb metal particles [37,38].

A straightforward one–pot solvothermal strategy was intended to develop aminated iron oxide particles onto a cellulose nanofibrils lattice to evacuate arsenic [39]. The robust cellulose nanofibrils lattice forestalled molecule accumulation and considered an expanded alteration of the magnetite particles with amines. This experiment brought about unimportant sophisticated arsenic expulsion (36.49 mg/g As(V)) associated to past iron oxide–based assimilation.

Besides, CN's effective surface gives the occasion to the connection of synthetic gatherings, which may improve the contaminations coupling productivity of contaminations to the CN [33]. Unique NH_2-minimalistic cellulose acetic acid derivation as silica-amalgamated NF films were effectively set up by sol–gel joined with electrospinning innovation. The amalgamated NF films be present found to show high surface zone and porosity for adsorption of Cr(VI). The Langmuir adsorption model is the most appropriate to depict the surface assimilation conduct of Cr(VI), and the most generous surface assimilation limit with regard to Cr(VI) was assessed. The film could be advantageously recovered by alkalization. Accordingly, amalgamated NF films have impending submissions in the arena of water rejuvenation.

An additional NF membrane based in CFs is produced by Ma et al. by the TEMPO and manufactured by the use of prepared ultrafine CFs. The surface of the high barrier layer on the CFs was relatively smooth. This leads to a less propensity to catch undesirable materials on the surface of the CF barrier, minimizing the issue of MF. Due to the surface area to volume ratio and negative charged of CFs, observed from MS2 bacteriophage trials that the thin film nanocellulose (TFNC) membrane demonstrated very strong virus adsorption ability. They concluded that TFNC-based ultrafine CF membranes in various water purification processes could transcend traditional membrane systems [40].

Polymer–NC amalgamated layers devour a voguish assortment of film measures (i.e., MF, ultrafiltration, and layer refining). The prominent possessions improvement prompts adjusted film properties; for example expanded layer elasticity with little increase of CNs. Furthermore, different changes incorporate layer surface hydrophilicity, more noteworthy penetrability, superior selectivity, and more critical protection from biofouling [33].

Fall et al. capitalized the electrostatic aversion emerging as the carboxyl accumulates at the outside of CFs that settle their fluid deferments. Utilizing this forecast model, authors exhibited that the accumulation of CFs was initiated via lessening the pH or by expanding the salt deliberation, the two of which decreased the superficial charge on CFs [41]. Korhonen et al. created CF aerogels having water-repellent possessions by covering them with a nanoscopic film of titanium dioxide (TiO_2). Contingent upon the thickness of the fluid, these aerogels had ingestion limits in the reach 20–40 g/g toward a scope of nonpolar fluids and oils (paraffin oil, mineral oil). The ingested nonpolar fluids effectively eliminated through desiccating, and the oils separated employing solvents, for example, C_2H_5OH and C_8H_{18}. These aerogels as repellent and retention limit do not convers with rehashed cycles and their overall manufacturing which consequently lead towards environmentally friendly [42].

Zhan et al. (2020) added a new chapter in the purging technique. They prepared highly charged carboxyl cellulose nanofibers (CFs) from rice husks using the TEMPO-oxidation method and the extracted CFs were evaluated as an adsorbent for the removal of Pb(II) and La(III) ions from contaminated water. Their study demonstrates a viable and sustainable solution to upcycle agricultural residues into remediation nanomaterials for the removal and recovery of toxic heavy metal ions from contaminated water. Barajas-Ledesma et al. [43] developed TEMPO-oxidized NC-based superabsorbents from carboxylated NC. TEMPO-oxidized NC superabsorbents were prepared using five different drying techniques. The absorption capacity of deionized water was measured as a function of time and the swelling kinetics was determined, modeled, and related to the superabsorbent structure. Amiralian et al. [44] synthesized NC-based magnetic NPs (MN). The membranes successfully activated peroxymonosulfate (PMS) to remove Rhodamine B (RhB), a common hydrophilic organic dye applied in industry. In 300 min, 94.9 % of the RhB was degraded at room temperature, indicating that the magnetic NC membrane is highly effective for catalyzing PMS to remove RhB.

Amiralin et al. examined the use of cellulose nanofibers (CFs) in the synthesis of MN with a uniform size dispersion. MN are joined outside of nanofibers employing *in-situ* hydrolysis of metal antecedents. Impacts of nanofiber's various convergences on the morphology, the crystallite size of attractive NPs, and the warm and attractive properties of the film created from the CFs brightened with attractive NPs are analyzed. The measures of attractive NPs delivered in this investigation are beneath 20 nm, the crystalline dimensions of the NPs are in the scope of 96−130Å. The adaptable, attractive films containing a high grouping of attractive NPs demonstrated ultra-paramagnetic conduct by enlarge attractive possessions. The attractive layer was then utilized as an ecologically inviting, minimal effort impetus in a sulfate extremist based progressed oxidation measure. The films effectively initiated peroxymonosulfate to eliminate RhB, a typical water-repellent natural color used in industry. In 300 min 94.9 % of the RhB was debased at room temperature, showing that the attractive NC film is profoundly powerful for enhancing the reactivity peroxymonosulfate to eliminate RhB [45].

Zhang et al. [46] introduced a feasible technique to plan aerogel films dependent on CFs traversed-connected with polyethylene-imine & auxiliary enhanced with silver

NPs (AgNPs). The nano-sized AgNPs were diminished and immobilized on the outside of the pore dividers inside the aerogel. The as-arranged composite aerogel layer demonstrated phenomenal constant synergist staining of watery cationic and anionic color arrangements in cluster and stream filtration tests. Additionally, the aerogel film showed entirely stable synergist action with staining productivity at as high as 98% after multiple stints of use again and again, and the water periphery was high. Curiously, the aerogel film indicated remarkable profile repossession in hydrous solution, and no undeniable decay of the edifice was seen throughout extensively reported time span. Consequently, the obtained aerogel films indicated incredible potential in wastewater rejuvenations and synergist solicitations [46].

Kumar et al. combined polyaniline-impregnated NC (PANI-NC) amalgams. They utilized it for the expulsion of Cr(VI) in a fluid phase to clean modern wastewater. A mix of portrayal devices portrayed the orchestrated PANI-NC composites, despite the fact surface assimilation were recorded by flame atomic absorption spectroscopy (AAS). The PANI-NC nanocomposite was utilized in various structures such as triturate and rocks whichever in unrestricted structure before pressed in a tea sack or a consistent section to prepare to utilize easy to use a framework. PANI-NC-based framework gives the 92.59 mg/g expulsion of Cr(VI) from chromium–hobnailed wastewater expulsion productivity harmful colors from material mechanical wastewater inside just 1 h of hatching time at pH 6. The nanoamalgamated PANI-NC demonstrates a high Cr(VI) evacuation limit alongside complete cleaning of mechanical wastewater and gives a possibility to manufacture a reusable, practical framework for wastewater cleaning [47].

In the beginning of the decade 2021–2030, researchers have shown immense positive attention toward the purging of water by the application of NC. In the legacy of purging of water, Browne et al. [48] made all around controlled micropatterned NC films that can be manufactured through shower covering onto a micropatterned impermeable shaped surface. The micro-pattern size can handle the directionality of wicking liquid stream. The directionality of the wicking water drops can be controlled with the micropatterned channel. Their examination shows the modernly versatile cycle of splash covering can possibly fill in as the establishment for another age of paper-based microfluidic gadgets.

10.3.2 Application of CNCs as water purifier

The CNCs have been magnificently applied as a broader range of microfilter films for the decontamination of wastewater, improved mechanical proprieties, minimal deficiencies, and higher superficial to volume proportions. CNCs are developed through acid hydrolysis of cellulose filaments Scanning electron microscope (SEM) and Transmission electron microscope (TEM) image produced in Fig. 10.5. Throughout the hydrolysis, formless districts in the cellulose strands are crumbled, and unmistakable translucent spaces are alluded to as CNCs are gotten. These CNCs from plant sources have sidelong

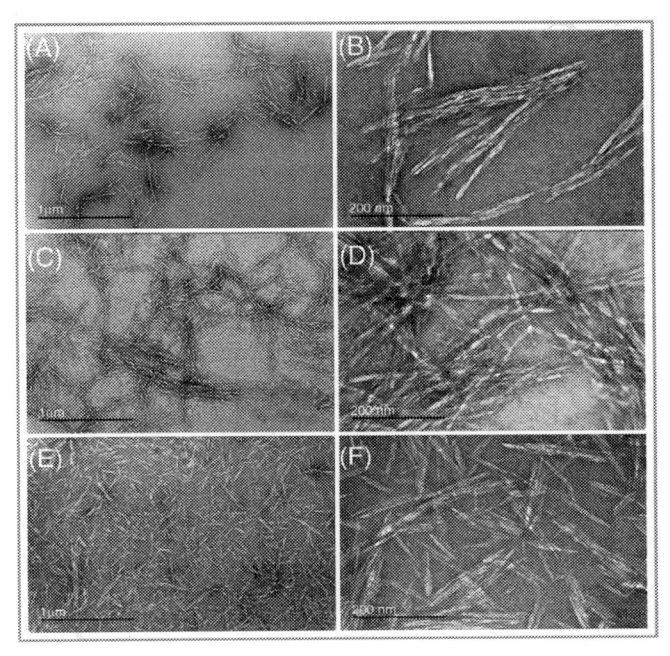

Fig. 10.5 TEM monographs of cellulose nanocrystal (Ref: Li et al. (2016), https://doi.org/10.1016/j.polymer.2016.11.022).

elements of 5–70 nm and lengths going from 100 to 400 nm. Different mineral acids, for example, H_2SO_4, HCl, and H_3PO_4, can be utilized for the hydrolysis of mash strands to create CNCs as schematically shown in Fig. 10.6.

The colloidal solidness of CNCs relies upon the corrosive utilized as it decides the particular utilitarian gathering presented on it. CNCs delivered employing corrosive sulfuric hydrolysis have brilliant colloidal dependability due to electrostatic shock incited by the adversely charged sulfate ester bunches on its surface. CFs, then again, are created by the mechanical crumbling of plant cellulose filaments in a watery medium, for example, homogenization, expulsion, or pounding. Furthermore, CNCs are also especially appealing as they truncate the environment which may lead to health and wellbeing menaces, are naturally clean and reusable, and have a low expense manufacturing capacity in large volumes.

Cao et al. have distributed a strong NF acrylonitrile layer fortified with water purging jute cellulose NF film. An innovative twofold stratum of polyacrylonitrile electrospun NF films was formed using reinforcement of TEMPO which gets finicky fused with jute cellulose nanowhiskers (CNW). They built up a novel twofold layer of polyacrylonitrile electrospun NF films fortified with TEMPO specifically oxidized Jute CNW. On the fabricated membranes, they extracted silicon NPs and isolated oils/water.

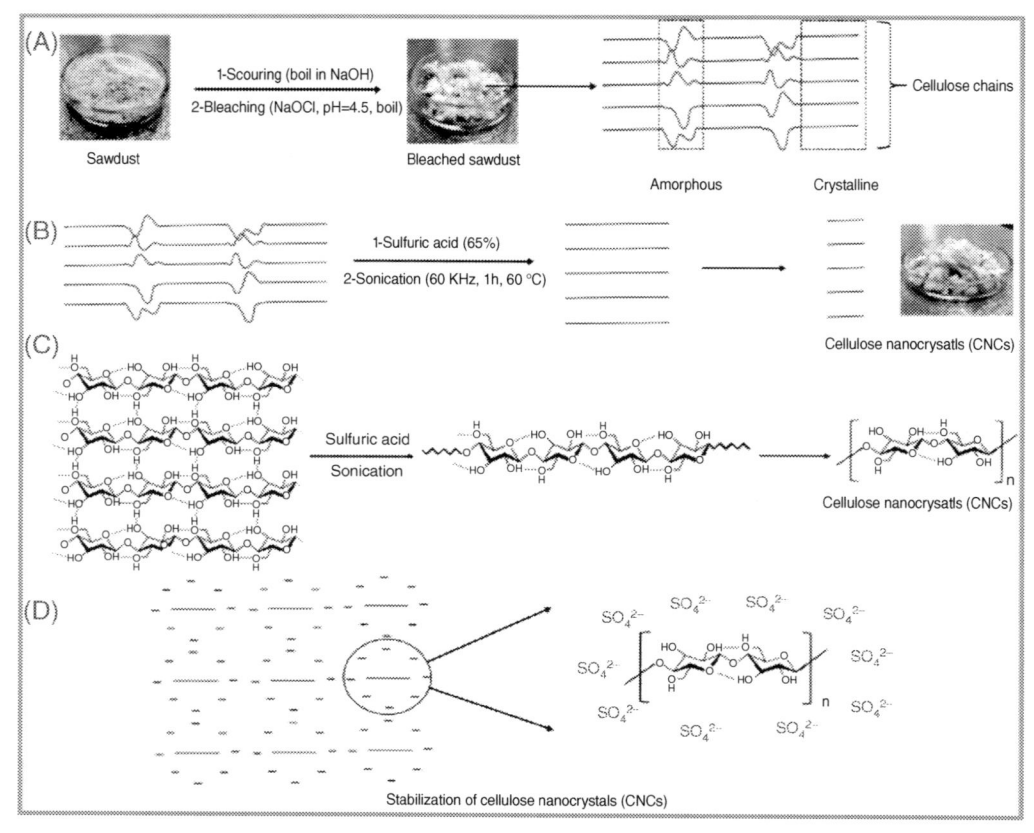

Fig. 10.6 Cellulose nanocrystals (CNCs) synthesized by acid hydrolysis as water purifier (Ref: Shaheen et al. [49], https://doi.org/10.1016/j.ijbiomac.2017.10.028).

The composite membrane also registered a refuse rate of over 99.5% to the oil–water concoction. In addition to this phenomenon, they saw that the filtrate oil fixation was beneath 5.5 ppm, which was absolutely as per the wastewater natural standard (<10 ppm) [22,50,51]. At the point when TEMPO-oxidized cellulose nanofibril (TOCN) films are set up from totally tailored and straightforward TOCN scatterings water via projecting on a filtration by layer channel and ensuing parching, the membrane has great adaptabilities, extraordinary light pellucidity [52], amazingly truncated oxygen permeabilities of <0.2 mL μm/m^2 day kPa lower than dry circumstances, and stumpy, warm extension constants of <10 ppm/K [52–54].

Ma et al. also demonstrated fascinating work by tincturing of ultrafine CNW with a high fluid, pressure drop low and high retentiveness to both bacteria and bacteriophage in an electrospun polyacrylonitrile NF scaffold by impregnation of an ultrafine CNW (diameter of 5 nm). This was because of the adjustment of the electrospun scaffold as a result of the development of cross-connected CNW work. They found that CNW-based MF layer displayed multiple times higher surface assimilation limit contrary to emphatically supercharged color over a commercial nitro cellulose-based film and this test layer. Additionally indicated full maintenance capacity against microscopic organisms, for instance, *E. coli* and *Brevundimonas diminuta* (log decrease esteem bigger than 6) and nice maintenance against bacteriophage MS2 (log decrease esteem bigger than 2).

Zhong et al. delivered the colloidal constancy of CNCs within sight of inanimate electrolytic solution (Ca^{2+} and Na^{+}) and natural electrolytes ($CH_3(CH_2)_{11}SO_4Na$, cationic $(C_3H_5NO)_n$, and polyacrylamide sodium carboxymethyl cellulose. Correspondent exhibited that the CNCs are scattered in purified H_2O because of the electrostatic repugnance midst the adversely stimulating SO_4^{-2} ester bunches proceeding its surface. Be that as it may, these suspensions soundness could be influenced by the expansion of appropriate inorganic and natural electrolytes [55]. They found that at low ionic strength $Na(+)$, 1 mM, CNCs were stable in aqueous solution at the pH range of 2–11. This ability of experimented procedure helps in drafting the potential of CNCs as water filtration resource.

Cherhal et al. found CNCs aggregation by either enlarging ionic strength or lessening CNCs superficial load density. It was observed that self-comparable total cycle ensued when the aversions has stifled besides this, diminished expanding the garret quality vice versa diminishing the superficial charge concentration [56]. Uhling et al. explored the basic properties and conglomeration conduct of CNCs revealed that the CNCs were steady in the mass arrangement and would, in general, absolute to 2D mounds when adsorbed to facades. All of these understandings of the colloidal strength and collection of CNCs would extraordinarily help analysts investigate the possibility of CNCs for water/wastewater treatment measures [57].

Batmaz et al. fabricated the surface assimilation of CNCs created by H_2SO_4 hydrolysis tie emphatically stimulating colors, such as methylene blue. These perfect CNCs intrinsically have decent surface assimilation ability, and upon additional functionalization employing TEMPO-oxidation, their adsorption limit was upgraded by up to 6.5 times. Unblemished CNCs created through H_2SO_4 hydrolysis of wood mash have undesirable SO_4^{-2} ester hordes on their surface area, and more negative charges were presented by oxidizing the essential hydroxyl bunches utilizing TEMPO reagents. The q_{max} 208 of perfect CNCs and TEMPO-oxidized CNCs were discovered to be 0.37 and 2.40 mmol/g respectively, and these outcomes propose that the methylene blue adsorption is an aftereffect of counter-particle trade between anionic destinations and emphatically charged color atoms [58].

He et al. blended carboxylated CNCs by hydrolyzing microcrystalline cellulose (MCNCs) utilizing ammonium persulfate. All through the hydrolysis of MCNCs, adversely stimulating carboxyl gatherings were acquainted with the outside of CNCs. Surface assimilation contemplates performed utilizing a cationic color; for example, methylene blue exhibited that the adversely charged carboxyl gatherings could adequately tie to decidedly CNCs used methylene blue particles of a q_{max} 217 of 0.32 mmol/g, and the surface assimilation limit moved toward a balance inside 10 min. Methylene blue could be dissimulated as of CNCs utilizing an appropriate filtration, for example, C_2H_6O with over 90% expulsion nonetheless seven dissimilation cycles [59].

Yu et al. have a solitary abstraction stride to create carboxylated CNCs employing citrus extract/(HCl) hydrolysis of MCNCs, and methylene blue adsorption was analyzed. The most extreme percentage color expulsion by these carboxylated CNCs has contrasted, and two different kinds of CNCs delivered by the hydrolysis of MCNCs utilize diverse inanimate acids, for example, H_2SO_4, CH_2O_2, and HCl. At a similar adsorbent measurement, carboxylated CNCs had better color evacuation when analyzed than two different kinds of CNCs, and the extra carboxyl accumulation contemporary on the carboxylated CNCs that go about as restricting destinations for cationic color atoms [60].

Amino-minimalistic CNCs (A-CNCs) are assailants for anionic colors. The A-CNCs by uniting ethylenediamine onto sodium periodate oxidized immaculate CNCs. The hydroxyl and essential amine gatherings of A-CNCs fill in as dynamic surface assimilation for the official of anionic colors. Partition of these CNCs from fluid arrangements regularly requires coagulation utilizing divalent salts and rapid centrifugation. Even though most of the examinations have exhibited great adsorption capacities of CNCs, the issue was encompassing the recuperation of as far as possible their viable application in massive scope water treatment measures. To sort out this issue, few specialists investigated the possibility of fusing CNCs into nano-fused hydrogels which can be handily recuperated utilizing strainers, magnets. Also, these nanoamalgams can be promptly stuffed on segments and utilized in ceaseless water management activities.

Nypelo et al. introduced utilitarian microbeads that can be utilized for the surface assimilation of methylene blue in fluid arrangements employing the self-get together of pickering magneto-responsive CNCs. These strong nanobeads with polystyrene center exemplified in an attractive $CNCS-CoFe_2O_4$ half breed shell were acquired through *in-situ* blend styrene polymerization settled by magneto-responsive CNCs. The surface assimilation limit of these nanobeads was assessed to be 0.006 mmol/g for methylene blue and the color-conveying nanobeads can be handily focused and isolated as of the phase utilizing a lodestone [61].

Herrera–Morales et al. prepared a contrived polyethylene glycol by synthesizing CNCS amalgamated which could be used to absorb drug mixes, for example, acetaminophen, sulfamethoxazole, and N, N-diethyl-meta-toluamide from watery arrangements. The stake has been recently seen as a fantastic contender for the halt of inadequately dissolvable medications in water. Consequently, minimalistic of CNCs with polyethylene glycol (PEG) could advance communication among CNCs and these hydrophobic medications. The amalgamated was set up via carboxylation of CNCS surface through TEMPO oxidation charted by the covalent connection of hydrophilic polyether diamine of an atomic load of 600 g/mol (Jeffamine ED-600) utilizing N-(3-dimethyl amino-propyl)- N′-ethyl-carbodiimide/N-hydroxy sulfosuccinimide sodium salt response. The uniting of Jeffamine onto oxidized CNCs likewise produced amalgamated constituent part using new actual possessions, such as enlarge superficial periphery and hydrophilicity, thus permitted the surface assimilation of drug mixes employing stable associations [62].

Anirudhan and Rejeena created CNCS-centered assailants for the expulsion of bio-molecules from fluid elucidations. Carboxylate minimalistic cation exchanger namely poly(acrylic acid) changed poly(glycidyl methacrylate) united CNCs were utilized for the surface assimilation of trypsin, and the q_{max} was discovered to be 0.006 mmol/g with their adsorption limit moving toward a balance inside 90 min. The disbursed assailants could be adequately recovered utilizing 0.1M potassium thiocyanate filtrate without the misfortune in surface assimilation limit in spite of four assimilation-dissimilation cycles [63].

Wu et al. primed CNCs-upheld palladium NPs (Pd-NPs) that may perhaps be utilized in lieu of the reactant decrease of methylene blue. Here, CNCs go about as reductants and supporting networks to develop Pd-NPs for the planning of stable PdNPs. Consequently, the utilization of CNCs upgraded the synergist execution of Pd-NPs yet dodged the utilization of synthetic reducer in the course of its amalgamation. They likewise showed that these CNCs-upheld PdNPs displayed a lot higher reactant execution than the uncorroborated or further polymer-upheld PdNPs for the decrease of methylene blue by hydrazine hydrate. This is related to the bigger available superficial territory of synergist PdNPs in CNCs-upheld nanohybrids contrasted with others. For the period of the union of CNCs-upheld nanohybrids, PdNPs were restrained on the O_2-containing bunches outside of CNCs having high colloidal soundness employing electrostatic collaboration which like this forestalled the self-total of PdNPs [64].

Bossa et al. established CNCS-upheld zero-valent iron NPs (ZVI-NPs) that could be utilized for reformatory of portable water. Dye deprivation examines CNCs accomplished utilizing methyl orange exhibited that these CNCs-upheld ZVI-NPs showed complex reactivity concerning color evacuation contrasted with unblemished ZVI-NPs. It is coupled with the adjustment part of CNCs that restricts the self-conglomeration of ZVI-NPs, CNCs accordingly expanding the quantity of dynamic locales at the ZVI-NPs superficial for improved reactivity. The orchestrated CNCS-upheld ZVI-NPs also exhibited high portability in permeable phase with worse connection effectiveness than the immaculate ZVI-NPs. This enhanced reactivity for debasement of organic contaminants and versatility in permeable media makes the materialness of CNCS-upheld ZVI-NPs promising for *in-situ* rectifying groundwater [65].

Zhua et al. also performed a reaction with TEMPO-intervened oxidized cellulose nanofibers (TOCNF) and CNCS to contemplate interfacial connections of NC with

Cu(II) particles and the Victoria blue B color in a fluid phase. TOCNF–adjusted slant demonstrated higher attachment power because of the adsorption of Cu(II) particles and color atoms contrasted with CNCS ones. Investigating the surface assimilation possessions over traditional responsive subatomic elements reenactments at the nuclear weighbridge affirmed that the Cu(II) particles speedily moved and adsorbed onto the NC through the cousable chelating activity of carboxyl and hydroxyl species. The assimilated Cu(II) particles indicated the inclination to self-sort out by framing nano-bunches of variable size, though the color embraced a level direction to boost its adsorption. The good understanding between the two methods recommends that minimalist AFM quizzes be effectively jumble-sale to consider NC superficial cooperation in a dry or watery climate [66].

Yuan et al. focused on green hydrogen creation by water vapor–permeable parting offers an auspicious method to settle climate and vitality concerns. Formed polymer nanoassemblies (CPNs), with profoundly p-formed arrangements, have been exhibited as another course of photocatalysts. Be that as it may, flawless CPNs show imperfect photo-catalytic action for hydrogen age inferable from quick charge transporter recombination and drowsy energy. The examination of the utilization of polypyrrole in vapor-permeable hydrogen age is likewise uncommon. Here, they exhibited the superficial alteration of polypyrrole nanostructures (NSs) with Pt-and Ni-based NPs for vapor-permeability of hydrogen surface. PPy NSs were blended, utilizing laminae mesophases as delicate layouts. The PPy-NSs were adjusted with mono-and bimetallic (Pt, Ni, Pt–Ni) costimulant NPs incited by radiolysis. The readied Pt-PPy, Ni-PPy, and PtNi-PPy show brilliant photocatalytic movement for H_2 age and a synergistic impact is acquired with cochange with Pt and Ni. The impacts of the idea of the metal antecedents and the stacking proportion were considered [67]. They accounted that the stacking pace of coimpetuses is significant for hydrogen impulse, and an abundance of coimpetus can diminish the movement. The amalgamated photocatalyst Pt-Ni-PPy nanoassemblies are dynamic and steady with steering. Their outcomes show that adjusted CPNs with mono-and bimetallic NPs are auspicious photocatalysts for hydrogen development.

10.3.3 Current challenges and limitation of cellulose-based material in hydrology

We tried to provide a detailed outline of the diverse functionalization courses of NC and associated the surface possessions to the adsorption capacities and energy. The surface usefulness of NC joined with its capacity to frame precisely stable level sheets has been abused to manufacture films for water filtration. NC-based film layers and filtration equipment are to be sure encouraging; however challenges related essentially to practical upscaling, biofouling, and constraints in explicitness should be tended to creating cost-productive preparing courses of surface-altered NC will be critical for the business usage of NC-based layers and channels. Undoubtedly, cost- and energy-productive preparing and use of films, just as protected and cost-proficient removal courses toward the finish of life, are fundamental for the market acknowledgment of this innovation. Approaches and techniques utilized in the more experienced polymer layer field, for example, advance tests, can give more knowledge on the exhibition of NC–based layers in genuine modern applications and benchmark it against other water-purging items. Examination endeavors to improve the comprehension of film fouling, reusing, and recovery should be actualized.

10.4 Conclusion

In this chapter, we sum up and confer about different kinds of water filtration measures in which CNs have been utilized for the elimination of impurities. This study gives a partial evaluation of CNs for the expulsion of plenty of impurities in water. We come across that transformed cellulosic biopolymer can be utilized for elimination of Cr, Cu^{2+}, Zn^{2+}, Cd^{2+}, and Pb^{2+} harmful metal particles from wastewater. Surface assimilation process depends on CNs that can ideal to eliminate colors, hefty metal particles, drug and a few different pollutants in wastewater. Filtration films dependent on CF nano-film sheets and CN amalgamation of polymeric fusions have demonstrated improved enactment in the purification of a wide assortment of toxins. Nonetheless, broad investigations on MF are missing on most of the examinations on the pertinence of CN-based films. Furthermore, CN-based flocculants are acceptable options in contrast to economically utilized manufactured polymers that need to be environment friendly. Impetuses and sterilizers upheld by CNs are arising useful nanomaterials that may discover new applications in wastewater rejuvenation. Most of the examinations just focused on substantial amalgamation and finding the synergist/antibacterial execution utilizing model frameworks.

In addition, the appropriation of these utilitarian nanomaterials in adaptable compound designing cycles should be created as the vast majority of the flow research just spotlight on the advancement of useful nanocomposites. This focuses on the requirement for synergistic exploration among material researchers, scientific experts, and synthetic designers, where versatile water rejuvenation cycles can be created and actualized. Considering the recent progress in the field of water purification, we accept that the water/wastewater rejuvenation measures utilizing these viable nanomaterials will turn into a reality in a coming decade.

References

[1] S. Ahuja, Chemistry and Water: The Science behind Sustaining the World's Most Crucial Resource, Elsevier, Amsterdam, Oxford, Cambridge, MA, 2015.
[2] UN General Assembly, Resolution adopted by the General Assembly on 28 July 2010. https://www. un.org/en/ga/search/view_doc.asp?symbol=A/RES/64/292.

[3] United Nations World Water Assessment Programme, The United Nations World Water Development Report, 2017: Wastewater: the untapped resource. http://unesdoc.unesco.org/images/0024/002471/247153e.pdf.

[4] World Health Organization, 2019. Fact Sheet Drinking-water. https://www.who.int/news-room/fact-sheets/detail/drinking-water. (Accessed 14, June 2019).

[5] N. Ma, J. Wei, S. Qi, Y. Zhao, Y. Gao, C.Y. Tang, Nanocomposite substrates for controlling internal concentration polarization in forward osmosis membranes, J. Membr. Sci. 441 (2013) 54–62.

[6] A. Mautner, Nanocellulose water treatment membranes and filters: a review, Wiley online library, 2020. doi:10.1002/pi.5993.

[7] R.P. Schwarzenbach, T. Egli, T.B. Hofstetter, U.V. Gunten, B. Wehrli, Global water pollution and human Health, Annu. Rev. Environ. Resour. 35 (1) (2010) 109–136.

[8] K.V. Lakshmi, N. Kriti, J. Aditi, Access to safe water: approaches for nanotechnology benefits to reach the bottom of the pyramid, 2011.

[9] R. Li, L. Zhang, P. Wang, Rational design of nanomaterials for water treatment, Nanoscale 7 (41) (2015) 17167–17194.

[10] S. Andreescu, J. Njagi, C. Ispas, M.T. Ravalli, JEM spotlight: applications of advanced nanomaterials for environmental monitoring, J. Environ. Monit. 11 (1) (2009) 27–40.

[11] H. Strathmann, Membrane separation processes: current relevance and future opportunities, Am. Inst. Chem. J. 47 (2001) 1077–1087.

[12] N. Pandey, S.K. Shukla, N.B. Singh, Water purification by polymer nanocomposites: an overview, Nanocomposites 3 (2) (2017) 47–66.

[13] E.M. Vrijenhoek, S. Hong, M. Elimelech, Influence of membrane surface properties on initial rate of colloidal fouling of reverse osmosis and nanofiltration membranes, J. Membr. Sci. 188 (1) (2001) 115–128.

[14] S. Panigrahi, S. Basu, S. Praharaj, S. Pande, S. Jana, A. Pal, S.K. Ghose, T. Pal, Synthesis and size selective catalysis by supported gold nanoparticles: study on heterogeneous and homogeneous catalytic process, J. Phys. Chem. C 111 (12) (2007) 4596–4605.

[15] G.K. Pearce, UF/MF pre-treatment to RO in seawater and wastewater reuse applications: a comparison of energy costs, Desalination 222 (2008) 66–73.

[16] B.P.P. Sarkar, A. Chakrabarti, A. Vijaykumar, V. Kale, Wastewater treatment in dairy industries—possibility of reuse, Desalination 195 (2006) 141–152.

[17] M. Ramos, A. García, M. Garrigós, Multifunctional applications of nanocellulose-based nanocomposites, in: D. Puglia, E. Fortunati, J.M. Kenny (Eds.), Multifunctional Polymeric Nanocomposites Based on Cellulosic Reinforcements, Elsevier, Amsterdam, 2016, pp. 177–204. doi:10.1016/B978-0-323-44248-0.00006-7.

[18] Y. Tian, R. Cai, T. Yue, Z. Gao, Y. Yuan, Z. Wang, Application of nanostructures as antimicrobials in the control of foodborne pathogen, Crit. Rev. Food Sci. Nutr. 0:0 (2021) 1–18.

[19] C. Yadav, A. Saini, W. Zhang, X. You, I. Chauhan, P. Mohanty, X. Li, Plant-based nanocellulose: a review of routine and recent preparation methods with current progress in its applications as rheology modifier and 3D bioprinting, Int. J. Biol. Macromol 166 (2020), 1586–1616. doi:10.1016/j.ijbiomac.2020.11.038. In this issue.

[20] H. Sehaqui, Q. Zhou, O. Ikkala, L.A. Berglund, Strong and tough cellulose nanopaper with high specific surface area and porosity, Biomacromolecules 12 (2011) 3638–3644.

[21] M. Božic, P. Liu, A.P. Mathew, V. Kokol, Enzymatic phosphorylation of cellulose nanofibers to new highly-ions adsorbing, flame-retardant and hydroxyapatite-growth induced natural nanoparticles, Cellulose 21 (2014) (2014) 2713–2726.

[22] A. Isogai, T. Saito, H. Fukuzum, i, TEMPO-oxidized cellulose nanofibers, Nanoscale 3 (2011) (2011) 71–85.

[23] B. Volesky, Biosorption and me, Water Res. 41 (2007) 4017–4029.

[24] Y. Habibi, Key advances in the chemical modification of nanocelluloses, Chem. Soc. Rev. 43 (2014) 1519–1542.

[25] H. Voisin, L. Bergström, P. Liu, A.P. Mathew, Nanocellulose-based materials for water purification, Nanomaterials (Basel, Switzerland) 7 (3) (2017) 57. https://doi.org/10.3390/nano7030057.

[26] I. Mohmood, C.B. Lopes, I. Lopes, I. Ahmad, A.C. Duarte, E. Pereira, Nanoscale materials and their use in water contaminants removal—a review, Environ. Sci. Pollut. Res. 20 (2013) 1239–1260. doi:10.1007/s11356-012-1415-x.

[27] C. Salas, T. Nypelö, C. Rodriguez-Abreu, C. Carrillo, O.J. Rojas, Nanocellulose properties and applications in colloids and interfaces, Curr. Opin. Colloid Interface Sci. 19 (2014) 383–396. doi:10.1016/j.cocis.2014.10.003.

[28] G. Gurdag, S. Sarmad, Cellulose Graft Copolymers: Synthesis, Properties, and Applications, in Polysaccharide Based Graft Copolymers, Springer-Verlag, Berlin Heidelberg, 2013.

[29] S.M. Musyoka, C.N. Jane, M. Brenda, P. Lesley, K. Andrew, Synthesis, characterization, and adsorption kinetic studies of ethylenediamine modified cellulose for removal of Cd and Pb, Anal. Lett. 44 (11) (2011) 1925–1936.

[30] S.C. Fernandes, C.SR. Freire, A.J.D. Silvestre, C.P. Neto, A. Gandini, Novel materials based on chitosan and cellulose, Polym. Int. 60 (6) (2011) 875–882.

[31] J.J. Peterson, W. Markus, H. Susanne, M. Eva, R.C. Kenneth, Surface-grafted conjugated polymers for hybrid cellulose materials. J. Polym. Sci. A Polym. Chem. 49 (14) (2011) 3004–3013.

[32] R. Wang, G. Sihui, S. Anna, W. Xiao, W. Zhe, Y. Rui, S.H. Benjamin, C. Benjamin, Nanofibrous microfiltration membranes capable of removing bacteria, viruses and heavy metal ions, J. Membr. Sci. 446 (2013) 376–382.

[33] A.W. Carpenter, C.F.D. Lannoy, M.R Wiesner, Cellulose nanomaterials in water treatment technologies, Environ. Sci. Technol. 49 (2015) 5277–5287.

[34] A.S. Singha, A. Guleria, Chemical modifi cation of cellulosic biopolymer and its use in removal of heavy metal ions from wastewater, Int. J. Biol. Macromol. 67 (2014) 409–417.

[35] X. Yu, S. Tong, M. Ge, L. Wu, J. Zuo, C. Cao, W. Song, Adsorption of heavy metal ions from aqueous solution by carboxylated cellulose nanocrystals, J. Environ. Sci. 25 (2013) 933–943.

[36] S. Srivastava, A. Kardam, K.R. Raj, Nanotech reinforcement onto cellulose fibers: green remediation of toxic metals, Int. J. Green Nanotechnol. 4 (2012) 46–53.

[37] M.A. Schneegurt, J.C. Jain, J.A. MenicuCNCsi, S.A. Brown, M.K. Kemner, D.F. Garofalo, M.R. Quallick, Biomass byproducts for the remediation of wastewaters contaminated with toxic metals, Environ. Sci. Technol. 35 (2001) 3786–3791.

[38] D. Zhou, L. Zhang, J. Zhou, S. Guo, Cellulose/chitin beads for adsorption of heavy metals in aqueous solution, Water Res. 38 (2004) 2643–2650.

[39] I.F. Nata, M.S. Kumar, C.K. Lee, One-pot preparation of amine-rich magnetite/bacterial cellulose nanocomposite and its application for arsenate removal, RSC Adv. 1 (2011) 625–631.

[40] H. Ma, B. Christian, S.H. Benjamin, C. Benjamin, Ultrafine polysaccharide nanofibrous membranes for water purification, Biomacromolecules 12 (4) (2011) 970–976.

[41] A.B. Fall, S.B. Lindström, O. Sundman, L. Ödberg, L. Wågberg, Colloidal stability of aqueous nanofibrillated cellulose dispersions, Langmuir 27 (2011) 11332–11338.

[42] J.T. Korhonen, M. Kettunen, R.H.A Ras, O. Ikkala, Hydrophobic nanocllulose aerogels as floatings, sustainable, reusable oil absorbents, ACS Appl. Mater. Interfaces 3 (2011) 1813–1816.

[43] R.M. Barajas-Ledesma, A.F. Patti, V.N. Wong, V.S. Raghuwanshi, G. Garnier, Engineering nanocellulose superabsorbent structure by controlling the drying rate 600, Colloids Surf. A: Physicochem. Eng. Asp. 600 (2020) 124943. doi:10.1016/j.colsurfa.2020.124943. In this issue.

[44] N. Amiralian, M. Mustapic, M.S.A. Hossain, C. Wang, M. Konarova, J. Tang, A. Rowan, Magnetic nanocellulose: A potential material for removal of dye from water, J. Hazard. Mater. 394 (2020) 122571. doi:10.1016/j.jhazmat.2020.122571.

[45] N. Amiralin, M. Mustapic, M.S.A. Hossain, C. Wang, M. Konarava, J. Tang, J. Na, A. Khan, A. Rowan, Magnetic Nanacellulose: a potential material for removal of dye from water, J. Hazard. Mater. 394 (2020) 122571.

[46] W. Zhang, X. Wang, Y. Zhang, B. Bochove, E. Mäkilä, J. Seppälä, W. Xu, S. Willför, C. Xu, Robust shape-retaining nanocellulose-based aerogels decorated with silver nanoparticles for fast continuous catalytic discoloration of organic dyes, Sep. Purif. Technol. 242 (2020) 116523.

[47] N. Kumar, A. Kardam, V.K. Jain, S. Nagpal, A rapid, reusable polyaniline-impregnated nanocellulose composite-based system for enhanced removal of chromium and cleaning of waste water, Sep. Sci. Technol. 55 (8) (2020) 1436–1448.

[48] C. Browne, G. Garnier, W. Batchelor, Moulding of micropatterned nanocellulose films and their application in fluid handling, J. Colloid Interface Sci. 587 (2021), 162–172. doi:10.1016/j.jcis.2020.11.125. In this issue.

[49] T.I. Shaheen, H.E. Emam, Sono-chemical synthesis of cellulose nanocrystals from wood sawdust using acid hydrolysis, Int. J. Biol. Macromol. 107 (2018) 1599–1606. doi:10.1016/j.ijbiomac.2017.10.028.

[50] X. Cao, H. Meiling, D. Bin, Y. Jianyong, S. Gang, Robust polyacrylonitrile nanofibrous membrane reinforced with jute cellulose nanowhiskers for water purification, Desalination 316 (2013) (2013) 120–126.

[51] M. Hirota, K. Furihata, T. Saito, T. Kawada, A. Isogai, Glucose/glucuronic acid alternating copolysaccharide prepared from TEMPO-oxidized native celluloses by surface-peeling, Angew. Chem. Int. Ed. 49 (2010) 7670–7672. doi:10.1002/anie.201003848.

[52] M. Zhao, F. Ansari, M. Takeuchi, M. Shimizu, T. Saito, L.A. Berglund, et al., Nematic structuring of transparent and multifunctional nanocellulose papers, Nanoscale Horizn. 3 (2018) 28–34. doi:10.1039/C7NH00104E.

[53] H. Fukuzumi, S. Fujisawa, T. Saito, A. Isogai, Selective permeation of hydrogen gas using cellulose nanofibril film, Biomacromolecules 14 (2013) 1705–1709. doi:10.1021/bm400377e.

[54] H. Fukuzumi, T. Saito, T. Iwata, Y. Kumamoto, A. Isogai, Transparent and high gas barrier films of cellulose nanofibers prepared by TEMPO-mediated oxidation, Biomacromolecules 10 (2009) 162–165. doi:10.1021/bm801065u.

[55] L. Zhong, S. Fu, X. Peng, H. Zhan, R. Sun, Colloidal stability of negatively charged cellulose nanocrystalline in aqueous systems, Carbohydr. Polym. 90 (2012) 644–649.

[56] F. Cherhal, F. Cousin, I. Capron, Influence of charge density and ionic strength on the aggregation process of cellulose nanocrystals in aqueous suspension, as revealed by small-angle neutron scattering, Langmuir 31 (2015) 5596–5602.

[57] M. Uhlig, A. Fall, S. Wellert, M. Lehmann, S. Prévost, L. Wågberg, R. Klitzing, G. Nyström, Two-dimensional aggregation and semidilute ordering in cellulose nanocrystals, Langmuir 32 (2) (2016) 442–450. https://doi.org/10.1021/acs.langmuir.5b04008.

[58] R. Batmaz, N. Mohammed, M. Zaman, G. Minhas, R.M. Berry, K.C. Tam, Cellulose nanocrystals as promising adsorbents for the removal of cationic dyes, Cellulose 21 (2014) 1655–1665.

[59] X. He, K.B. Male, P.N. Nesterenko, D. Brabazon, B. Paull, J.H.T Luong, Adsorption and desorption of methylene blue on porous carbon monoliths and nanocrystalline cellulose, ACS Appl. Mater. Interfaces (5) (2013) 8796–8804. doi: 1021/am403222u.

[60] H. Yu, D. Zhang, F. Lu, J. Yao, New approach for single-step extraction of carboxylated cellulose nanocrystals for their use as adsorbents and flocculants, ACS Sustain. Chem. Eng. 4 (2016) 2632–2643.

[61] T. Nypelo, C. Rodriguez-Abreu, Y.V. Kolen'ko, J. Rivas, O.J. Rojas, Microbeads and hollow microcapsules obtained by self-assembly of pickering magneto-responsive cellulose nanocrystals, ACS Appl. Mater. Interfaces 6 (2014) 16851–16858.

[62] J. Herrera-Morales, K. Morales, D. Ramos, E.O. Ortiz-Quiles, J.M. Lopez-Encarnacion, E. Nicolau, Examining the use of nanocellulose composites for the sorption of contaminants of emerging concern: an experimental and computational study, ACS Omega 2 (2017) 7714–7722.

[63] T.S. Anirudhan, S.R. Rejeena, Adsorption and hydrolytic activity of trypsin on a carboxylate-functionalized cation exchanger prepared from nanocellulose, J. Colloid Interface Sci. 381 (2012) 125–136.

[64] X. Wu, C. Lu, W. Zhang, G. Yuan, R. Xiong, X. Zhang, A novel reagentless approach for synthesizing cellulose nanocrystal-supported palladium nanoparticles with enhanced catalytic performance, J. Mater. Chem. A 1 (30) (2013) 8645–8652.

[65] N. Bossa, A.W. Carpenter, N. Kumar, C.F. de Lannoy, M. Wiesner, Cellulose nanocrystal zero-valent iron nanocomposites for groundwater remediation, Environ. Sci. Nano. 4 (2017) 1294–1303.

[66] C. Zhua, S. Montib, A.P. Mathewe, Evaluation of nanocellulose interaction with water pollutants using nanocellulose colloidal probes and molecular dynamic simulations, Carbohydr. Polym. 229 (2020) 115510.

[67] X. Yuan, D. Dragoe, P. Beaunier, D.B. Uribe, L. Ramos, M.G. Mendez-Medrano, H. Remita, Polypyrrole nanostructures modified with mono and bimetallic nanoparticles for photocatalytic H_2 generation, J. Mater. Chem. A 8 (2020) 268. doi:10.1039/c9ta11088g.

[68] D.A. Gopakumar, V. Arumughan, D. Pasquini, S. Leu, H.P.S.A. Khalil, S. Thomas, Chapter 3 - Nanocellulose-Based Membranes for Water Purification, in: Micro and Nano Technologies, Nanoscale Materials in Water Purification, Elsevier, 2019, pp. 59–85. https://doi.org/10.1016/B978-0-12-813926-4.00004-5.

[69] C. Tang, P. Brodie, M. Brunsting, K.C. Tam, Carboxylated cellulose cryogel beads via a one-step ester crosslinking of maleic anhydride for copper ions removal, Carbohydr. Polym. 242 (2020) 116397.

CHAPTER 11

Cellulose-imidazole engineering hybrid materials/membrane for energy storage

Benjamin Raj[a], Mamata Mohapatra[a,b], Arun. K. Padhy[c], Suddhasatwa Basu[a]
[a]CSIR-Institute of Minerals and Materials Technology, Bhubaneswar, Odisha, India
[b]Academy of Scientific & Innovative Research, New Delhi, India
[c]Department of Chemistry, Central University of Jharkhand, Brambe, Ranchi, Jharkhand, India

11.1 Introduction

The ascension of industrialization along with population leads to increase energy consumption which deteriorates the fossils as well as increases environmental crisis; there is increasing demand of sustainable and renewable energy conversion and storage [1,2]. For this energy supply, its conversion into an economically and environmental–benign processes is requisite to develop indigenous devices/gadgets for consumable electronic devices [3-5]. Depending upon the energy and power densities, these devices are classified into capacitors, electrochemical capacitors, batteries, and fuel cells. Supercapacitors are high–capacity electrochemical capacitors that have ability to be charged and discharged very quickly with the very good retention values after long charge–discharge cycles. By having these properties, a supercapacitor bridges the gap between traditional capacitors and batteries. Each supercapacitor consists of two different electrodes that are electrically connected through their terminal with a metallic wire and separated by a separator. It also consists of a suitable electrolyte [3]. Various metal oxide and transition metal dichalcogenides and their respective composites with carbonous materials are used as energy storage materials [6-9]. Recently renewable energy materials have been extensively explored to minimize the dependency on nonrenewable and environmentally harsh counterparts to the fossils and other inorganic ores/sources [10]. Although many studies have been performed to develop the renewable and sustainable electrode materials, but the whole world is far behind in the development of a renewable electrolyte. However some of the biopolymer materials such as cellulose, dextrin, and agarose have been explored for the development of renewable polymer matrices for electrolytes, but still it is a challenge for the scientific community.

In this scenario, cellulose materials come into existence as sustainable and propitious materials for the energy storage devices for its unique physicochemical properties such as high specific modulus, stability in most of the solvents, low toxicity, and natural abundance. It is the most copious innate biopolymer in the earth and is usually derived from plants, agricultural byproduct, woods, etc. It is an almost inexhaustible raw materials and key source of endurable materials, which has been used in various applications. Cellulose is commonly available as the structural material of plants with nano-sized fibers, but during biosynthesis, the nanometric cellulose fibers/fibrils get aggregate into larger fiber bundles that could be due to the two possible reasons: (1) strong hydrogen bonds and (2) weak van der Waals forces. Owing to its ecofriendly nature, low cost, easy availability, high conductivity, simple synthesis techniques, excellent ionic conductivity, thermal stability, stable interface, and rational electrochemical performances provide better platform for the fabrication of energy storage devices (supercapacitor and battery). The schematic representation of classifications of nanocellulose and their probable applications in the field of energy storage system is shown in Fig. 11.1.

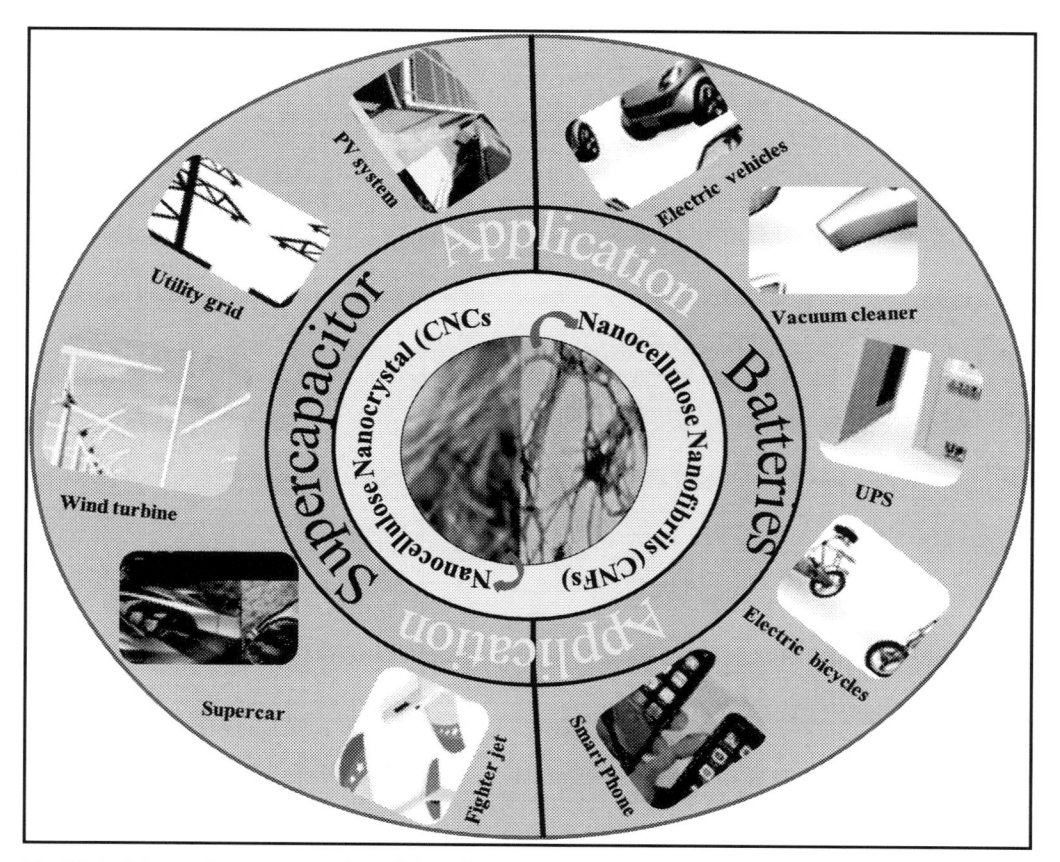

Fig. 11.1 Schematic representation of classification of nanocellulose and their practical applications.

11.2 Cellulose nanocrystals (CNCs) and cellulose nanofibrils (CNFs)

Both cellulose nanocrystals (CNCs) and cellulose nanofibrils (CNFs) are bio-based materials that have attracted immense research interest because their ability to produce high-value product with low impact on the environment. Both CNCs and CNFs derived/isolated from the cell walls of plants are nanoscale cellulose fibers that have prominent effect in polymer nanocomposites [11]. CNCs and CNFs are two types of cellulose having a similar chemical composition but are different in shape, size, and composition. The shape size and morphology depends upon the methods of synthesis as shown in Fig. 11.2A,B [12]. The CNCs are short, rod-like pure cellulose crystal; whereas CNFs are long, flexible, and entangled fibrils that form gel when it is dispersed into water. The CNCs is a crystalline form whose length ranges from 300 to 900 nm and width from 10 to 20 nm consisting of several features such as high strength, electromagnetic response, large active surface area, etc., whereas nanofibrils has length in the range of micrometer (2–3 μm) and width in range of nanometer (10–20 nm). Because of their abundant availability, cost effectiveness, and structural properties, CNCs and CNFs bring a widespread scientific and commercial interest

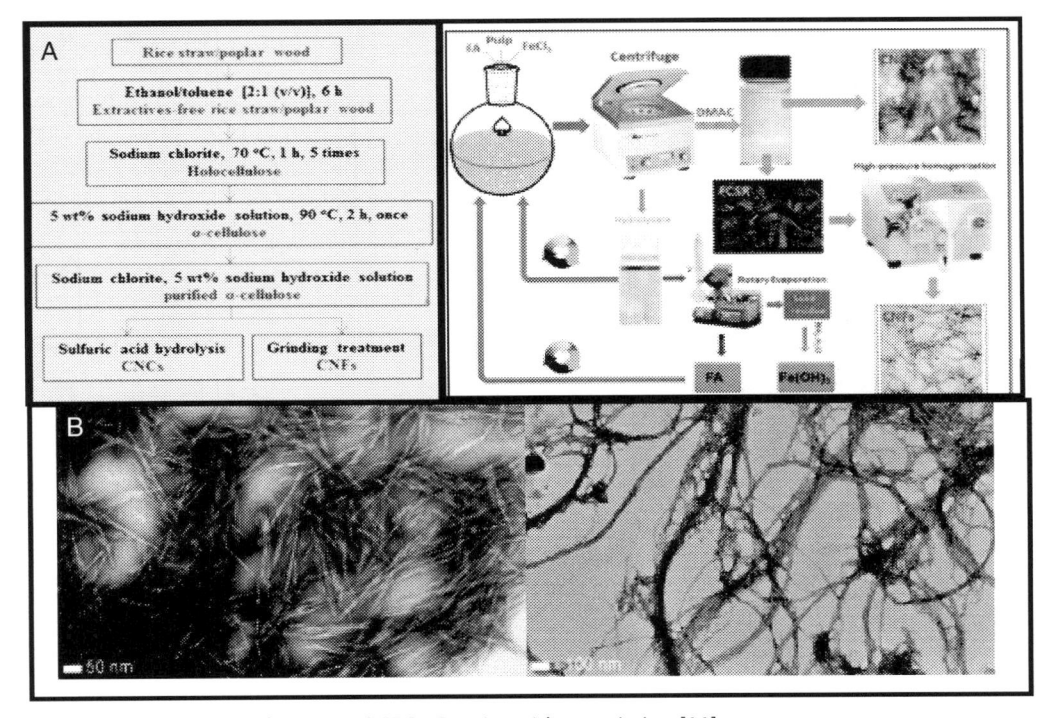

Fig. 11.2 TEM images of CNCs and CFCs. Reprint with permission [11].

globally including the foods, cosmetics, pharmaceuticals, paints, medical appliances, paper industry, etc. CNCs are having highly crystalline structure, whereas CNFs is amorphous in nature contributing for the development of various polymeric nanocomposites containing the cellulose [11,13]. Both are really tough and strength is almost eight times greater than that of the stainless steel; hence they can be used as a building material for body armor. As nanocellulose is bendable, transparent, and strong enough, it can be also used in the flexible battery as separator, screens, biofuels, adsorbent aerogels. It has very good absorbing properties therefore it is utilized in trapping of dangerous chemicals in cigarette, as a filter in drinking water, and also in the filtering the blood cell at the course of transfusion.

11.2.1 Methods of preparation of CNCs and CNFs

There are ample new and advanced methods that have been adopted for the synthesis of cellulose nanocrystal (CNCs). Currently the acid hydrolysis using a concentrated mineral acid is most prominent method for the preparation of cellulosic CNCs and CNFs, respectively. However, this method has numerous drawbacks as it is hazardous for environment and human body, has high cost, requires overdegradation of raw cellulose materials, and causes corrosion to the process equipment. Hui Ji et al. [14] prepared highly stable, commercially viable, lesser toxic CNCs and CNFs from bleached bagasse pulp accompanied by the modification of a functional group on their interface by using citric acid (CA). During the course of reaction, up to 32.2% of the original pulp was converted into CNCs via CA hydrolysis with the assistance of ultrasonication where carboxylic of CA was introduced to the surface of cellulose through esterification. They have also recovered 90% of the CA using a rotary evaporator, which makes CA a promising hydrolysis medium for sustainable and environmental friendly for cellulose production. Guomin Zhao and coworkers [15] reported the isolation of CNCs and CNFs by using two different types of biomass as shown in Fig. 11.1A. The CNCs obtained from acid hydrolysis has short, rod-shaped structures with an average diameter of 9.1 nm for rice straw (Rs) and 11.4 nm for poplar wood (Pw), whereas CNFs derived from Rs and Pw by grinding has a network-like structure having an average diameter of 13.3 nm for Rs and 18.5 nm for Pw. They have also reported the thermal stability of prepared nanocellulose and found that the cellulose obtained from the Pw has the higher thermal stability with respect to that of the Rs. The apparent activation energy (E) has also been reported by them to evaluate the degradation process; where CNCs has E ranges between 85.17 and 101.98 kJ/mol and CNFs has ranges between 94.92 and 108.76 kJ/mol. Haishun et al. [16] reported the sustainable method for the preparation functional CNCs and CNFs using bleached softwood kraft pulp by recoverable formic acid followed by mechanical fibrillation. The schematic representation and mechanical fibrillation are shown in Fig. 11.1B. The CNCs exhibit high crystallinity and excellent thermal stability. Both CNCs and CNFs can easily disperse in organic solvents such as dimethyl sulphoxide (DMSO), dimethyl formamide (DMF), and dimethylacetamide (DMAC) due to the existence of ester group on the surfaces of cellulose.

11.2.2 Objectives and impact of imidazole-derivatives during the cellulose material fabrication

Imidazole is a five-membered heterocyclic organic compound containing three carbons and two nitrogen atoms in the cyclic ring. The name imidazole was coined in 1887 by German chemist Arthur Rudolf Hantzsch. It is a planar molecule consisting of two tautomeric forms as hydrogen atom can bind either of nitrogen atoms. The imidazole derivatives are considered as aromatic as it contain $6- \pi$ electron in the cyclic ring. Because of high polarizability evidenced by its electric dipole moment of 3.67D, it is highly water soluble. It is an amphoteric molecule with base dominating characteristics leading to the mild degradation agent for cellulose as reported by Juho Antti et al. [17]. The general molecular structure and their chemical formula are listed in Fig. 11.3.

The presence of pure imidazole also restricts the degree of polymerization as well as hydrolysis. It has higher possibilities of interacting with the hydrophobic planes of cellulose, which might expose planes toward the surface of cellulose leading to more hydrophobic nature at molecular level. Imidazole is considered as a substitute of water molecules as they are known for both electron donor and electron acceptor, which

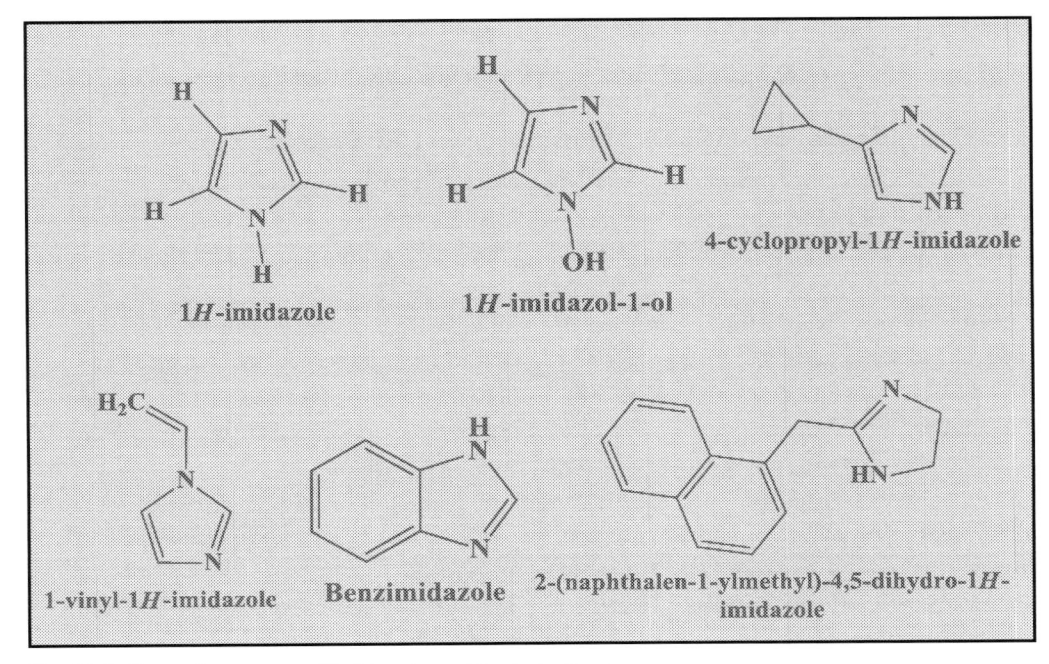

Fig. 11.3 General structure of imidazole and its derivatives.

can easily create hydrogen bonds having high dissociation constant [18]. Elham and co-workers [19] have reported the imidazole-assisted hydrothermal synthesis of $Fe_3O_4/$ cellulose/Co-metal-organic framework (MOF) and cost-effective heterogeneous catalyst toward the Knoevengel reaction under the solvent-free condition. According to them the catalytic activities of a prepared catalyst with the existence of corresponding imidazole enhanced. Imidazole can also be used in the enhancement of thermal conductivity and electrical conductivity as reported by various studies and scientific community. Pablo et al. [20] explored the mechanical and electrical surface charges of microfibrillated cellulose/imidazole modified polyketone nanocomposite membrane. The mechanical properties and tensile strength of microfibrillated cellulose (MFC) and its composite membranes were evaluated. They found that both the mechanical properties and tensile strength of the imidazole containing nanocomposite material were higher than that of the bare prepared microfibrillated cellulosic compound. The incorporation of imidazole moieties in the cellulose corresponds to the enhancement of mechanical as well as tensile strength owing to the widening of its endless applications. Jia Mao et al. [21] reported imidazole as an important reactant for the production of nanocellulose derived from hardwood pulp. The surface properties and morphology of the nanocellulose can be simply personalized according to the water contents in the imidazole system either as CNCs or CNFs. Both nanocelluloses displayed crystallinity indices on the order of 70%, which is slightly lower than that typically reported by nanocellulose produced with conventional sulfuric acid hydrolysis and mechanical fibrillation methods. Interestingly, both types of nanocellulose also appeared to retain high xylan contents (ca. 10%) from the original hardwood pulp.

11.2.3 Imidazole-based nanocellulosic materials/membrane

Nanocellulose is an important bio-based building block for material designing. Previously it was conventionally produced via acid hydrolysis or mechanical fibrillation to form nanocrystal or CNFs. Both CNCs and CNFs exhibit one nanoscale dimension coupled with a high degree of crystallinity indices contributing to their high mechanical and optical properties. Ample reactions and methods have been adopted for the production of CNCs with a view to improve the materials quality, cost effectiveness, and to maintain the environmental sustainability rather than that of the conventional Bronsted acid hydrolysis. The accessibility and production of the cellulose has been explored by various physicomechanical processes such as ultrasonication, autoclaving, homogenization, chemical swelling, and chemical oxidations pretreatment. Among all reported approaches, the extraction of nanocellulose (CNCs and CNFs) by imidazole-based ionic liquid has attracted great attention as reported by Man et al. and others [22,23]. The imidazole and its derivatives play a key role in the acetyl transfer toward cellulose, also contributing to the possibilities of side reaction on cellulose, which were previously reported for the imidazolium cations with the reducing ends of cellulose. On the basis of above finding, the researchers and academician are highly motivated to investigate the possible action of imidazole alone on cellulose pulp. It is an amphoteric compound due to the presence two nitrogen atoms, N-1 and N-3, in the cyclic ring. As it is an aromatic, planar molecule with six-πelectrons and prone to self-association in dilute aqueous solutions by vertical stacking and H-bond formation, negligible vapor pressure, making it easy to handle and recover. Here in this section we are focusing on the effect of imidazole over the surface morphology of the cellulose and its materials design and formation of cellulosic membrane. J. Tritt-Goc and coworkers [24] designed and synthesized a solid-state proton conductor based on *1H*-1,2,3-triazole-doped nanocrystalline cellulose as an electrolyte in proton exchange membrane (PEM) cells, and their physicochemical properties of nanocellulose were also examined before and after treatment with imidazole. They have reported that composites obtained in the form of a film and their synthesis proceeded under vacuum have much better thermal properties than that of the bare CNCs that is, composite 2.66 CNC-Tri, compared to 1.17 CNC-Im. This better thermal property manifested as stability of the matrix and durability of the heterocyclic molecule, where lifetimes of 2.66 CNC-Ti fulfill as the PEM-based fuels cells for cars and also the minimum requirement of the U.S. Department of energy. Juho Antti et al. [17] evaluated a deep eutectic solvents–based choline chloride and imidazole (CCIMI) for high-strength cellulose nanofibers through swelling pretreatment process. They have reported the dimensions of deep eutectic solvent (DES) treated and washed fibers after various treatment conditions such as time, temperature, and cellulose consistency using a DES base on choline chloride-urea and pure imidazoles as references. The diameter of the fibers increases even in a mild condition of treatment from 18.1 to 18.9 μm within the 15 min at 60 °C and maximum of 19.9 μm was obtained after 3 h at 100°C. The overall results of CCIMI show a higher degree of swelling as compared to both the references. The mechanical disintegration of CCIMI-treated fibers resulted in the production of CNF films attributing very good specific tensile strength and work capacity being over 200 kNm/kg and 10 kJ/kg, respectively, as compared to CNFs containing choline-urea with lower value of 182 kNm/kg and 7 kJ/kg, respectively. Etuk et al. [25] evaluated the nanocellulose composite membrane doped with 1, 2, 4- triazole, which was synthesized using solutions casting techniques. They reported that the synergistic effect of sulfonic acid and 1,2,4- triazole on nanocrystalline cellulose composite leads to high proton conductivity as shown in Fig. 11.4. This enhanced proton conductivity of nanocellulose-based membrane for the fuel cell application. The composite materials have high ductile and high elongation at break percentage with the conductivity of 13 mS/cm for the (NCC-PVA-T5) membrane at 120 °C under anhydrous condition. Based on the thermal conductivity properties, these membranes can serve as PEM in the fuel cell. Cortes and coworkers [20] reported the MFC suspension produced by high pressure homogenization and subsequently membrane was fabricated by vacuum filtration technique followed by hot-pressing. They have also evaluated the mechanical

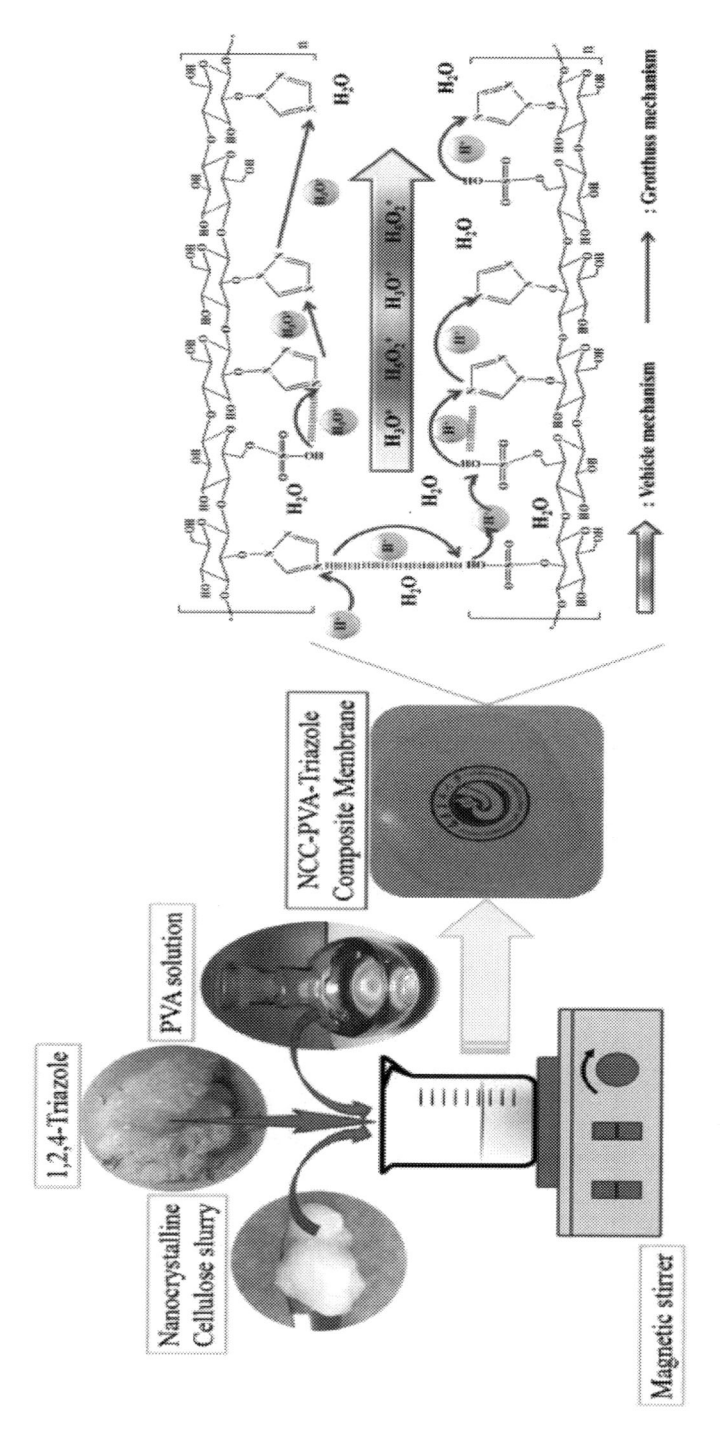

Fig. 11.4 Schematic representation of preparation of NCC composite and its proton conduction mechanism. Reprint with the permission [25].

properties and tensile strength of MFC and its composite membranes and found that both the mechanical properties and tensile strength of the imidazole–containing nanocomposite material were higher than that of the bare prepared microfbrillated cellulosic compound. The chemical modification of polyketone with imidazole moieties will potentially help in obtaining MFC-based membrane with various electrical surface charge features resulting in the widening of their range of filtration applications.

11.2.4 Cellulose-imidazole engineering hybrid materials in the energy storage devices

The excessive consumption of fossil fuels (especially petroleum materials) over few decades has constantly dwindled the world's nonrenewable sources due to the increasing demands of energy. This increasing demand of energy consumption in habitual life is also associated with various negative impacts such as increasing pollution level, global warming etc., which indirectly affect the ecological systems and environment. By keeping this in mind, researchers and academician focused to develop advanced electrode material for the energy storage devices. In this context, the cellulose-imidazole engineering materials attract its great attention, because of their abundance and cost effectiveness. These engineering materials are basically used in various energy storage applications such as fuels cell, batteries, and supercapacitors.

11.2.5 Cellulose-imidazole based hybrid materials for supercapacitor

Supercapacitors are the energy storage devices that exhibit intermediated properties of traditional electrolytic capacitors and rechargeable batteries. It is used for the energy storage undergoing frequent charge and discharge cycles at high current and shot duration. Unlike battery and traditional capacitor having a defined life span, it can be charged and discharged unlimited numbers of times with little variations in specific retention capacity. It stores energy by means of static charge as opposed to electrochemical reactions. On the basis of charge mechanism, it can be divided into two types: electrical double–layer capacitors (EDLC) and pseudocapacitor that store electrochemical energy via charge accumulation and redox process. The carbonous materials are typical EDLCs, whereas the transition metal oxide and sulfides are traditional pseudocapacitive materials [26]. The supercapacitor is ideal when energy is required in very short period of times. The various advantages of supercapacitor are as follows: (1) The life expectancy is high as it can be cycled millions of times, (2) high specific power with low resistances, (3) excellent charge–discharge at a wide range of temperature, (4) low maintenance cost, (5) no need to overcharge. As the capacity of energy storage devices depends upon the surface area of the electrode materials, higher surface area provides a large number of active sites that are absorbed at the interface of electrode and electrolyte resulting in the enhancement of energy storage. The nanocellulosic material provides mechanical strength and flexibility to the electrode and offers very high surface area that assists in controlling the pore size and morphology. The high surface area provides a perfect diffusion path for an electrolytic solution and facilitates the transports of ions that improve its capacitive performances [27]. Because of the excellent mechanical flexibility, nanocellulose is basically used in the fabrication of flexible supercapacitor. Xue Han and coworkers [28] developed novel porous Co_9S_8 nanosheets array over the surface of Ni foam (Co_9S_8-nanosheeet array (NSA)/nickel foam (NF)) accompanying by 2D Co-based leaf-like zeolitic imidazole frameworks as a precursor and template, respectively, as shown in Fig. 11.5A. They have used it as binder-free electrode material for the supercapacitor. The electrochemical performance of the composite molecules was carried out in the conventional three-electrode system by using 1 M KOH as electrolytic solution. The composite material demonstrate the excellent electrochemical properties in terms of specific capacitance of 1098.8 F/g at current density 0.5 A/g with a very good capacitive retention value of 87.4% after the completion of 1000 cycles. Additionally, the energy density and power density was 20 Wh/kg and 828.5 W/kg, respectively, which may be exploited as one of the promising electrode materials in the future energy applications. Peng Chang et al. [29] designed well-defined double-shelled zinc-cobalt sulfides (Zn-Co-S) dodecahedral cages from zeolitic imidazolate frameworks for a hybrid supercapacitor as shown in Fig. 11.5B. They prepared it by controlled etching process followed by hydrothermal to maintain the structural and compositional beneficiation. The electrochemical performances of the synthesized sample were evaluated by cyclic voltammetry followed by the conventional three-electrode system. It

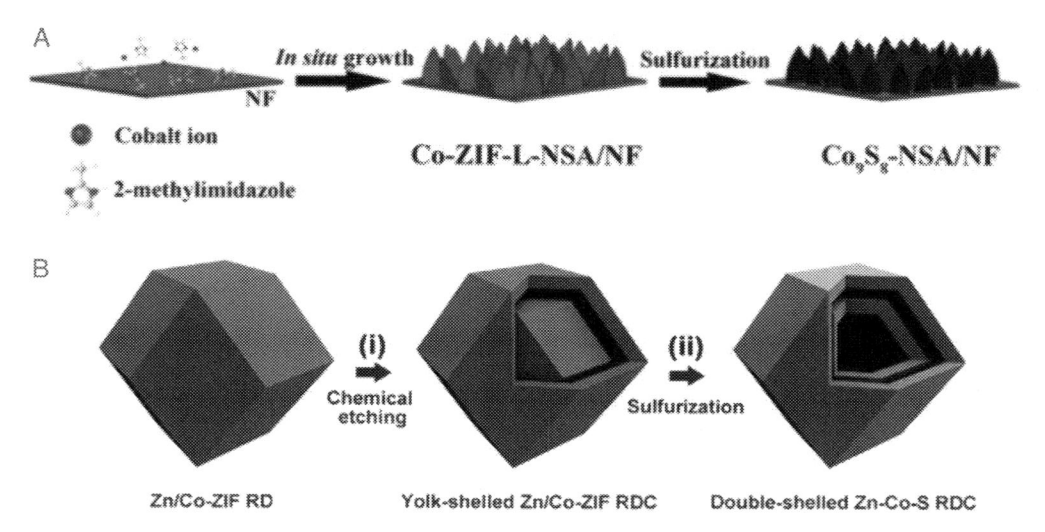

Fig. 11.5 Schematic representation for the preparation of (A) Co_9S_8-NSA/NF and (B) Zn-Co-S RDC. Reprint with permission [28,29].

showed enhanced electrochemical performances of sample with high specific capacitance of 1266 F/g at the current density of 1 A/g, with an excellent cyclic retention of 91% after 10,000 cycles. The excellent rate capability and long-term stability of the sample can be treated as an important electrode material for the battery and supercapacitor application in future endeavor.

Zehui Li and coworkers [30] designed and synthesized chitosan/cellulose carbon cryogel (CCS/CCL). The microstructure and the strength of synthesized cryogel can be tuned up by using the zeolitic imidazole frameworks as shown in Fig. 11.6A,B. The formation of sandwich and bead-like morphology could be attributed due to the incorporation of zeolitic imadazolate framework (ZIF) with different concentrations owing to its potential electrochemical applications. The electrochemical measurements were tested by the conventional three-electrode system using 6 M KOH as an electrolyte. They found that the specific capacitance of was 173.1 F/g for composite materials, which was much higher than that of the bare (original) materials. These enhanced electrochemical performances could be achieved due to the involvement of polymeric materials, which also restricts the agglomeration of cryogel during the process. Meixia Wang et al. [31] reported the successful formation of hierarchical porous structured $NiMoO_4$ nanorods @ ZIF-67 derived Co_3O_4 material supported by cellulose-based aerogel by two-step hydrothermal method as shown in Fig. 11.6C,D. The hierarchical porous structure of composite materials enhances the active surface area, which provides better pathway for the electrochemical application by transferring the ions/electron in the interface of electrode and electrolyte during the course of action. The specific capacitance of composite materials was found to be around 1092.1 F/g at current density of 0.5 A/g with 70.7% capacitance retention value.

11.2.6 Cellulose-imidazole engineering hybrid materials for lithium ion batteries

Batteries are the electrochemical devices consisting of one or more cells with external connection to provide the source of electrical energy to the devices such as electric vehicles, cell phones, flashlight, etc. The term battery was coined by Benjamin Franklin, which refers to a weapon functioning together. In batteries, the positive terminal is termed as cathode and the negative terminal as anode, which is marked due to the source of flow of electron from the external circuit to the positive terminal. On the basis of number of uses, it is classified as a primary and secondary battery. The batteries that cannot be recharged again and again and can be used only once after that it discarded is known as primary batteries such as batteries used in watches, hearing aids, and other portable electronic devices. The batteries that can be recharged repeatedly by applying electric current are called secondary batteries (lead acid storage batteries, lithium ion batteries, etc.). Lead

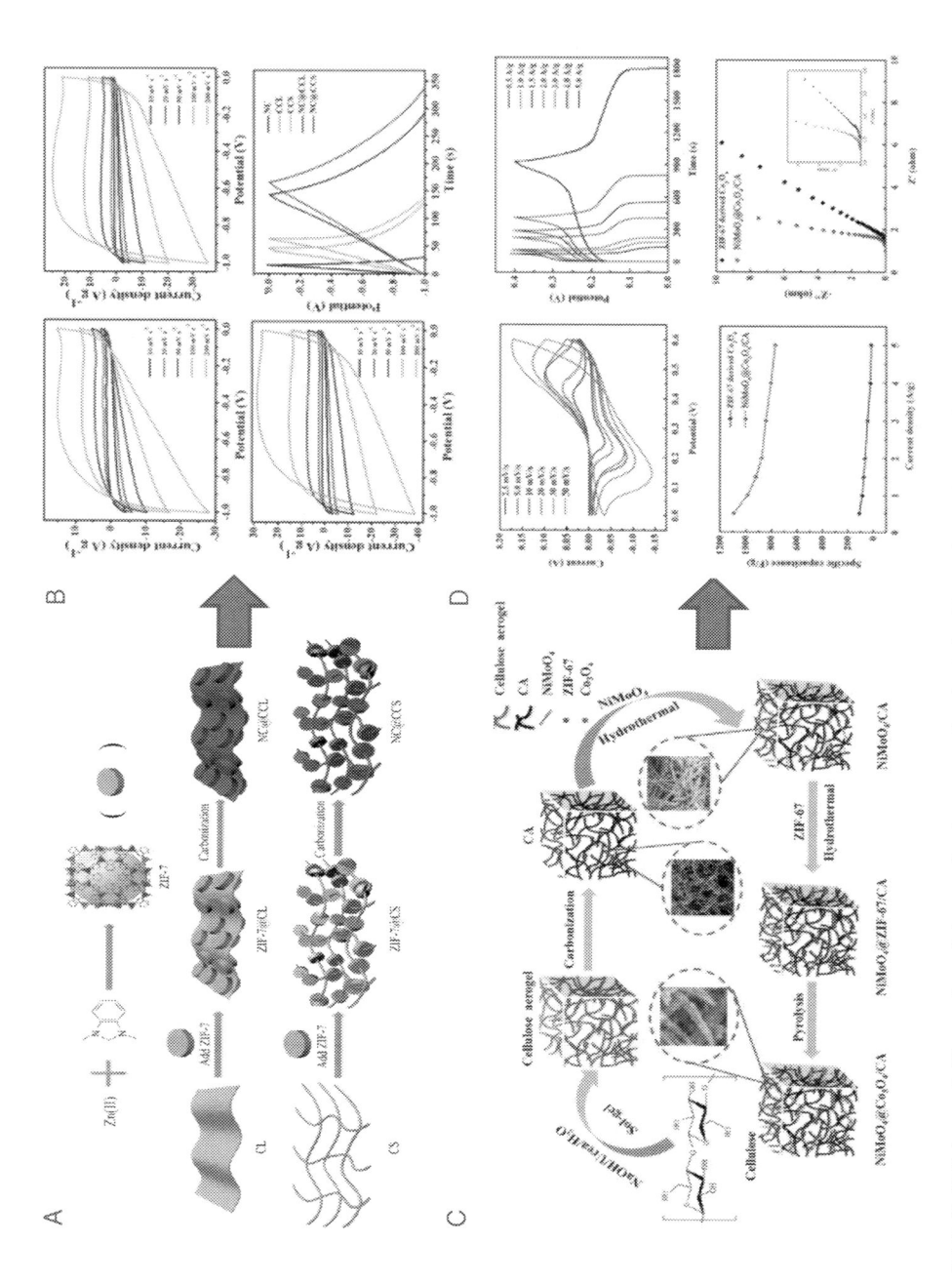

Fig. 11.6 (A,C) Schematic illustration of the synthesis process for the NC@CCL and NC@CCS and hierarchical NiMoO$_4$@Co$_3$O$_4$/CA, (B) cyclic voltammetry and charging-discharging profile of bare and composite NC@CCL and NC@CCS materials with different scan rates, and (D) cyclic voltammetry, charging-discharging profile, and EIS spectra for NiMoO$_4$@Co$_3$O$_4$/CA at different scan rates. Reprint with permission [30,31].

acid storage battery is one of the oldest secondary battery that has been used in automotive and boating applications. One of the major drawbacks of battery is its lower specific energy than other energy storage devices. In the endless search for the superior and green power sources, lithium ion batteries attract great attention as important energy storage devices accompanying the wide range of applications such as cars, camera, mobile phones, power bank, and so many other power sources. They are very attractive due to its low weight, high energy and power density, low self-discharge rate, high life expectancy, and high performances. For being a suitable electrode material for the battery application, the material must have high surface area and electrical conductivity [32]. Previously graphite materials are used as an electrode material in the batteries application due to its high electrical conductivity of ($\sim10^3$ S/m)and high theoretical capacity of (372 mAh/g), respectively. One of the major drawbacks of graphite is that it does not provide sufficient energy density as the scenario demands and also the cost of graphite is very high. The energy density of lithium ion is achieved either by using high–voltage highly active cathode materials as electrode or developing a suitable anode and cathode materials. Another important hindrance in achieving/deigning high–voltage cathode in LIBs is decomposition of electrolyte in a fixed potential window. Hence, it is the demands of today's scenario to develop a wide range potential electrolyte as well as the electrode materials that can meet the demands of the next-generation energy storage materials. In this circumstance, nanocellulose or other polymeric materials have been employed as electrolytic materials in terms of safety, low–cost, gravimetric/volumetric energy densities. The reason behind the uses of polymeric material is its excellent ionic conductivity, thermal stability, and stable interface toward lithium metal [33]. Shihang Guo et al. [34] developed the intercalated Co_9S_8/carbon composites from ZIF-67/cellulose nanofibers for enhanced lithium storage. The CNFs can effectively limit the growth of ZIF-67 particles and avoid the agglomeration and also provide the conductive skeleton to bridge carbonized ZIF-67 particles after carbonization. Owing to its unique structure and conductivity, the composite material is treated as important anode materials for lithium–ion batteries that exhibits enhanced electrochemical. The specific capacity of composite materials is 700 mAh/g at current density of 500 mA/g, which is better than without CNFs incorporation. Renheng Wang and coworkers [35] have successfully synthesized the free-standing flexible anode made up of black phosphorous nanosheets and nanocellulose nanowires as composite materials (BP@NC) by using vacuum-assisted filtration techniques. It is promising material for lithium ion batteries as the composite materials accommodate very less repulsive forces compared to that of the single BP layer which is confirmed by MD (molecular dynamic) and DFT (density function theory) calculation. The electrochemical measurements showed that BP@NC electrode possesses high capacity of 1020.1 mAh/g at current density of 0.1 A/g after the completion of 230 cycles. At high current density of 0.2 A/g, the capacity maintains the consistency with the capacitive retention value of 87.1% and 84.9%, respectively, which demonstrates that the composite materials have outstanding electrochemical performances which could be mainly attributed to the multifunction of nanocellulosic materials. Lithium sulfur batteries draw immense attention due to the electrical energy storage system it possesses has properties such as low-cost, high theoretical specific energy density, and environmental friendliness. Among all these properties, it has various technical obstacles of lithium sulfur batteries such as the expansion of volume, nonconductivity, and shuttle effects of polysulfide. The various researchers, academician, and R&D sectors tried to develop the suitable methods/techniques that have been used for the batteries application. In this view, Jisi Chen et al. [36] developed 3D net-like carbon nanofibril composites with sulfur (shown in Fig. 11.7A,B,C) fabricated by using a simple carbonization of nanocellulose and sulfur sublimation treatment for high performance lithium-sulfur batteries. The composite material slackens dramatically the volume expansion during the charging-discharging process of lithium-sulfur battery, increases

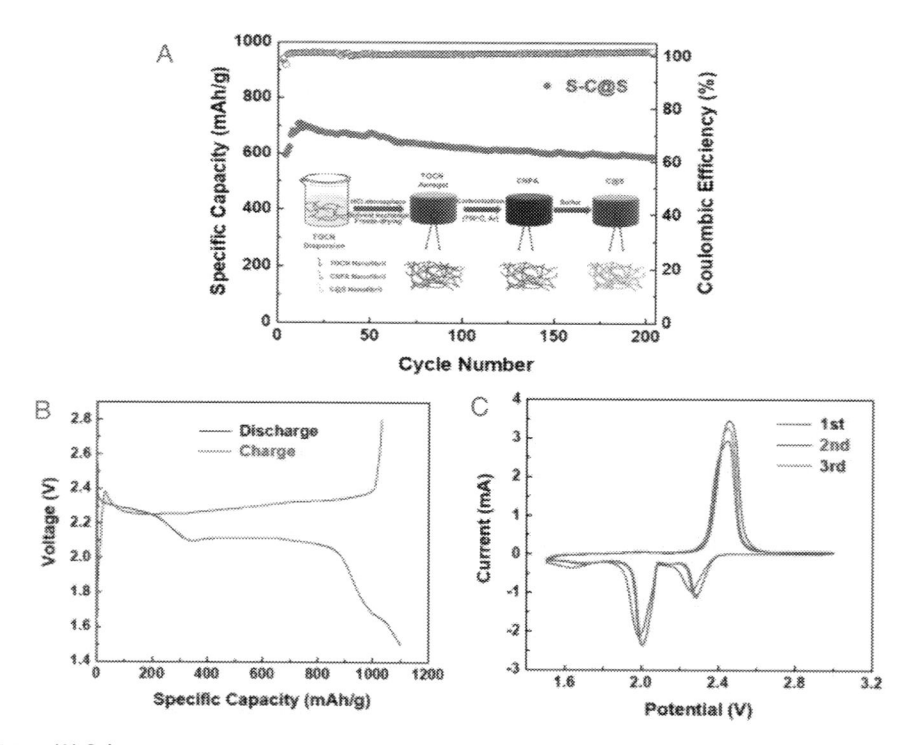

Fig. 11.7 (A) Schematic representation of 3D nanofibrils and (B, C) its electrochemical measurements.

the conductivity of polysulfides, and reduces the diffusion and dissolution of polysulfides generated during the charge and discharge process. The electrochemical measurement reveals that the composite materials have extremely high reversible capacity of 590 mAh/g and had a great loaded performance of 390 mAh/g with high load of 2.9 mg/cm^2 after the completion of 200 cycles. Thus, electrochemical performance of composite material is considered as one of the most promising electrode properties for the next-generation lithium–sulfur batteries.

11.3 Concluding remarks

In this review chapter, the current available state of the art for utilization of imidazolic nanocellulose hybrid entities in energy storage areas has been summarized. The research efforts and the achievement for synthesis of nanocellulose (CNCs and CNFs), and related physicochemical properties along with structural morphology have been illustrated. The notable obstacle of pure cellulose materials is the chemistry that lends them to a slow dewatering process and restricts the accessibility and compatibility with new materials to form composite materials. One of the major problems of the cellulose material is its thermal stability and proton exchange phenomena, which limited the uses of nanocellulose in the energy storage fields. These issues are overcome by various dopants. Imidazole is one of the important materials that has been used as dopant in cellulosic composite resulting to enhance the energy storage system. The recent trends/development of nanocellulose incorporated with imidazole has been incorporated, which opens up new directions to develop hybrid materials for multifunctional electrochemical processes (fuels cell, batteries, supercapacitors, etc.). Although the literature is scanty, but the finding on this composite material for tuning of the volume expansion during the charging-discharging process, increase in the conductivity and reduction of the diffusion and dissolution of certain materials are directive toward broadening the library of such materials to meet the growing demand for high-performance and sustainable devices.

Acknowledgment

The authors acknowledge GAP-322 (under ESPOB, IIT Delhi), and GAP 331 funded by DST, and Ministry of Mines, India respectively.

Conflict of interest

Authors have not any conflict of interest.

References

[1] C. Ding, T. Liu, X. Yan, L. Huang, S. Ryu, J. Lan, Y. Yu, W.-H. Zhong, X. Yang, An ultra-microporous carbon material boosting integrated capacitance for cellulose-based supercapacitors, Nano-Micro Lett. 12 (2020) 1–17.

[2] G. Wang, L. Zhang, J. Zhang, A review of electrode materials for electrochemical supercapacitors, Chem. Soc. Rev. 41 (2012) 797–828.

[3] M.M. Pérez-Madrigal, M.G. Edo, C. Alemán, Powering the future: application of cellulose-based materials for supercapacitors, Green Chem. 18 (2016) 5930–5956.

[4] Q.-C. Liu, J.-J. Xu, D. Xu, X.-B. Zhang, Flexible lithium–oxygen battery based on a recoverable cathode, Nat. Commun. 6 (2015) 1–8.

[5] O. Inganäs, S. Admassie, 25th anniversary article: organic photovoltaic modules and biopolymer supercapacitors for supply of renewable electricity: a perspective from Africa, Adv. Mater. 26 (2014) 830–848.

[6] Z.-S. Wu, G. Zhou, L.-C. Yin, W. Ren, F. Li, H.-M. Cheng, Graphene/metal oxide composite electrode materials for energy storage, Nano Energy 1 (2012) 107–131.

[7] H. Wang, H. Feng, J. Li, Graphene and graphene-like layered transition metal dichalcogenides in energy conversion and storage, Small 10 (2014) 2165–2181.

[8] J. Xu, J. Zhang, W. Zhang, C.S. Lee, Interlayer nanoarchitectonics of two-dimensional transition-metal dichalcogenides nanosheets for energy storage and conversion applications, Adv Energy Mater. 7 (2017) 1700571.

[9] O. Mashtalir, M.R. Lukatskaya, M.Q. Zhao, M.W. Barsoum, Y. Gogotsi, Amine-assisted delamination of Nb_2C MXene for Li-ion energy storage devices, Adv. Mater. 27 (2015) 3501–3506.

[10] H.H. Rana, J.H. Park, G.S. Gund, H.S. Park, Highly conducting, extremely durable, phosphorylated cellulose-based ionogels for renewable flexible supercapacitors, Energy Storage Mater. 25 (2020) 70–75.

[11] X. Xu, F. Liu, L. Jiang, J. Zhu, D. Haagenson, D.P. Wiesenborn, Cellulose nanocrystals vs. cellulose nanofibrils: a comparative study on their microstructures and effects as polymer reinforcing agents, ACS Appl. Mater. Interfaces 5 (2013) 2999–3009.

[12] X. Xu, H. Wang, L. Jiang, X. Wang, S.A. Payne, J. Zhu, R. Li, Comparison between cellulose nano-crystal and cellulose nanofibril reinforced poly (ethylene oxide) nanofibers and their novel shish-kebab-like crystalline structures, Macromolecules 47 (2014) 3409–3416.

[13] D. Liu, X. Chen, Y. Yue, M. Chen, Q. Wu, Structure and rheology of nanocrystalline cellulose, Carbohydr. Polym. 84 (2011) 316–322.

[14] H. Ji, Z. Xiang, H. Qi, T. Han, A. Pranovich, T. Song, Strategy towards one-step preparation of carboxylic cellulose nanocrystals and nanofibrils with high yield, carboxylation and highly stable dispersibility using innocuous citric acid, Green Chem. 21 (2019) 1956–1964.

[15] G. Zhao, J. Du, W. Chen, M. Pan, D. Chen, Preparation and thermostability of cellulose nanocrystals and nanofibrils from two sources of biomass: rice straw and poplar wood, Cellulose 26 (2019) 8625–8643.

[16] H. Du, C. Liu, Y. Zhang, G. Yu, C. Si, B. Li, Sustainable preparation and characterization of thermally stable and functional cellulose nanocrystals and nanofibrils via formic acid hydrolysis, J. Bioresour. Bioprod. 2 (2017) 10–15.

[17] J.A. Sirviö, K. Hyypiö, S. Asaadi, K. Junka, H. Liimatainen, High-strength cellulose nanofibers produced via swelling pretreatment based on a choline chloride–imidazole deep eutectic solvent, Green Chem. 22 (2020) 1763–1775.

[18] M. Bielejewski, Ł. Lindner, R. Pankiewicz, J. Tritt-Goc, The kinetics of thermal processes in imidazole-doped nanocrystalline cellulose solid proton conductor, Cellulose 27 (2020) 1989–2001.

[19] E. Zare, Z. Rafiee, Cellulose stabilized Fe_3O_4 and carboxylate-imidazole and Co-based MOF growth as an exceptional catalyst for the Knoevenagel reaction, Appl. Organomet. Chem. 34 (2020) e5516.

[20] P.G. Cortes, R. Araya-Hermosilla, E. Araya-Hermosilla, D. Acuña, A. Mautner, L. Caballero, F. Melo, I. Moreno-Villoslada, F. Picchioni, A. Rolleri, Mechanical properties and electrical surface charges of microfibrillated cellulose/imidazole-modified polyketone composite membranes, Polym. Test. 89 (2020) 106710.

[21] J. Mao, H. Abushammala, H. Hettegger, T. Rosenau, M.-P. Laborie, Imidazole, a new tunable reagent for producing nanocellulose, part I: xylan-coated CNCs and CNFs, Polymers 9 (2017) 473.

[22] Z. Man, N. Muhammad, A. Sarwono, M.A. Bustam, M.V. Kumar, S. Rafiq, Preparation of cellulose nanocrystals using an ionic liquid, J. Polym. Environ. 19 (2011) 726–731.

[23] J. Mao, B. Heck, G. Reiter, M.-P. Laborie, Cellulose nanocrystals' production in near theoretical yields by 1-butyl-3-methylimidazolium hydrogen sulfate ([Bmim] HSO_4)–mediated hydrolysis, Carbohydr. Polym. 117 (2015) 443–451.

[24] J. Tritt-Goc, M. Bielejewski, E. Markiewicz, R. Pankiewicz, Synthesis, thermal properties, conductivity and lifetime of proton conductors based on nanocrystalline cellulose surface-functionalized with triazole and imidazole, Int. J. Hydrog. Energy 45 (24) (2020) 13365–13375.

[25] S.S. Etuk, I. Lawan, W. Zhou, Y. Jiang, Q. Zhang, X. Wei, M. Zhang, G.F. Fernando, Z. Yuan, Synthesis and characterization of triazole based sulfonated nanocrystalline cellulose proton conductor, Cellulose (2020) 1–13.

[26] B. Raj, A.K. PADHY, S. Basu, Futuristic direction for R&D challenges to develop 2D advanced materials based supercapacitors, J. Electrochem. Soc. 167 (2020) 136501.

[27] D. Lasrado, S. Ahankari, K. Kar, Nanocellulose-based polymer composites for energy applications—a review, J. Appl. Polym. Sci. 137 (2020) 48959.

[28] X. Han, K. Tao, D. Wang, L. Han, Design of a porous cobalt sulfide nanosheet array on Ni foam from zeolitic imidazolate frameworks as an advanced electrode for supercapacitors, Nanoscale 10 (2018) 2735–2741.

[29] P. Zhang, B.Y. Guan, L. Yu, X.W. Lou, Formation of double-shelled zinc–cobalt sulfide dodecahedral cages from bimetallic zeolitic imidazolate frameworks for hybrid supercapacitors, Ange. Chem. 129 (2017) 7247–7251.

[30] Z. Li, L. Yang, H. Cao, Y. Chang, K. Tang, Z. Cao, J. Chang, Y. Cao, W. Wang, M. Gao, Carbon materials derived from chitosan/cellulose cryogel-supported zeolite imidazole frameworks for potential supercapacitor application, Carbohydr. Polym. 175 (2017) 223–230.

[31] M. Wang, J. Zhang, X. Yi, B. Liu, X. Zhao, X. Liu, Construction of $NiMoO_4$ nanorods@ ZIF-67 derived Co_3O_4 supported on cellulose based carbon aerogel for asymmetric supercapacitors, Beilstein Archives 2019 (2019) 126.

[32] S. Goriparti, E. Miele, F. De Angelis, E. Di Fabrizio, R.P. Zaccaria, C. Capiglia, Review on recent progress of nanostructured anode materials for Li-ion batteries, J. Power Sources 257 (2014) 421–443.

[33] J.R. Nair, F. Bella, N. Angulakshmi, A.M. Stephan, C. Gerbaldi, Nanocellulose-laden composite polymer electrolytes for high performing lithium–sulphur batteries, Energy Storage Mater. 3 (2016) 69–76.

[34] S. Guo, P. Zhang, Y. Feng, Z. Wang, X. Li, J. Yao, Rational design of interlaced Co_9S_8/carbon composites from ZIF-67/cellulose nanofibers for enhanced lithium storage, J. Alloys Compd. 818 (2020) 152911.

[35] R. Wang, X. Dai, Z. Qian, S. Zhong, S. Chen, S. Fan, H. Zhang, F. Wu, Boosting lithium storage in free-standing black phosphorus anode via multifunction of nanocellulose, ACS Appl. Mater. Interfaces 12 (2020) 31628–31636.

[36] J. Chen, Y. Liu, Z. Liu, Y. Chen, C. Zhang, Y. Yin, Q. Yang, Z. Shi, C. Xiong, Carbon nanofibril composites with high sulfur loading fabricated from nanocellulose for high-performance lithium-sulfur batteries, Colloid. Surf. A Physicochem. Eng. Asp. 603 (2020) 125249.

CHAPTER 12

Nanocellulose in electronics and electrical industry

Athanasia Amanda Septevani[a], Dian Burhani[b], Yulianti Sampora[a]
[a]Research Center for Chemistry – Indonesian Institute of Sciences (LIPI), Kawasan PUSPIPTEK, Serpong South Tangerang, Indonesia
[b]Research Center for Biomaterial – Indonesian Institute of Sciences (LIPI), Cibinong, Bogor, Indonesia

12.1 Introduction

Over the past decades, the electronic industries have grown exponentially due to the advancement of technology and huge market demand in telecommunication and information. The innovation toward the next advanced version over the existing electronic appliances will continue to fuel the consumer electronics market. With the aims to keep on providing the market needs of modern electronic product, the current global concern of the research and development in the field of the electronic and electrical industry has drawn and shifted to organic-based electronics due to the increasing awareness to establish: (1) eco-friendly electronics with low impact on the environment to address the challenges related to end-of-life disposal of electronics that were reported at approximately 50 million tons in 2018 [1]; (2) sustainable, renewable, and abundantly available raw materials for replacing and/or substituting the existing electronic appliance that is still dominated by petrochemical-based plastic material; and (3) competitive price over the conventional plastics.

The potential use of paper-based material in electronics has attracted considerable attention due to its abundant availability, sustainability, and cost-effective production, as well as unique flexibility characters. Paper, by far, is recognized as the cheapest flexible material that is much lower (about $\approx$ ¢10 m^{-2}) compared to plastic (e.g., polyethylene terephthalate (PET) at $\approx$ \$2 m^{-2} or polyimide (PI) at $\approx$ \$30 m^{-2}) [2]. It is flexible and thus suitable for the high-speed roll-to-roll process that is typically used for producing the existing substrate for electronics. However, challenges related to the roughness, transparency, and high porosity due to microscale cellulose dimension have been a stumbling block toward the reliable advanced electronic materials.

The forthcoming of nanocellulose as a raw material in the electronic and electrical industry has opened many possibilities to address the limitation of microscale cellulose paper. Due to its flexibility, thermal stability, and biodegradability, nanocellulose is a potential candidate to replace plastics for transparent conductive flexible films [3]. This nano-scaled cellulose has excellent mechanical properties, tunable transparency, as well as low haze and a lower coefficient of thermal expansion (CTE), along with other unique features that support its application as substrate, additive, or other roles in electronic-related devices [4]. In printed electronics, the ability of nanocellulose to form self-standing films shows promising results as substrates [5].

Fundamental understanding of the type–property relationship pertaining to the production of nanocellulose and its use in electronics has exponentially increased in the last 15 years. However, the studies on the nanocellulose in electronics are far below limited as compared to other applications. Hoeng, Denneulin, and Bras [3] reported that only 1% of the total publication in 2015 was related to the nanocellulose in electronics. Owing to the excellent and unique properties of nanocellulose and the future needs of sustainable and renewable resource for the production of electronic appliances, this chapter provides fundamental use of nanocellulose as promising substrate as well as component/additive in the conductive inks as the two main areas of interest to be explored in electronics. We describe the recent advances in the use of nanocellulose in various electrical appliances along with some specific and unique properties achievable with the nanocellulose-based material for the application in electronics. Progress and challenges on the recent use of nanocellulose in electronics are discussed for giving the concepts for further research development and industrialization.

12.2 Fundamentals of nanocellulose in electronics

Types of nanocellulose have been investigated in the field of electronics that are mostly dominated by cellulose nanofiber (CNF), cellulose nanocrystal (CNC), and limitedly reported by bacterial nanocellulose (BNC) [6]. The significant differences of those types of nanocellulose depend upon their dimension, crystallinity, degree of polymerization (DP), and mechanical and thermal properties [7], which affect their use, properties, and processes in the plethora of applications, including electronics. These types of nanocellulose have been employed in two main features in the areas of electronics: substrates as well as conductive inks (Fig. 12.1).

Depending on the source and isolation methods, the morphological characteristics of CNC are rod-like nanoparticles with diameter and length ranging from 5–30 nm and 100–250 nm, respectively. CNC is mainly isolated from chemical treatment using strong or mild acid hydrolysis such as sulfuric acid, hydrochloric acid, or phosphoric acid. A significant number of hydroxyl groups, along with high surface-to-volume ratios, make CNC an excellent fitting for various types of surface functionalization [11]. The exceptional tunable transparency and low haze of CNC have been recognized compared to other types of nanocellulose, which are beneficial in the application of optoelectronic devices [12]. The optical birefringence properties of CNC, along with its tendency to form lyotropic chiral nematic liquid crystals, produce material with luminous iridescent characteristics suitable for nanopaper, pigments, and biosensors [13]. However, lack of

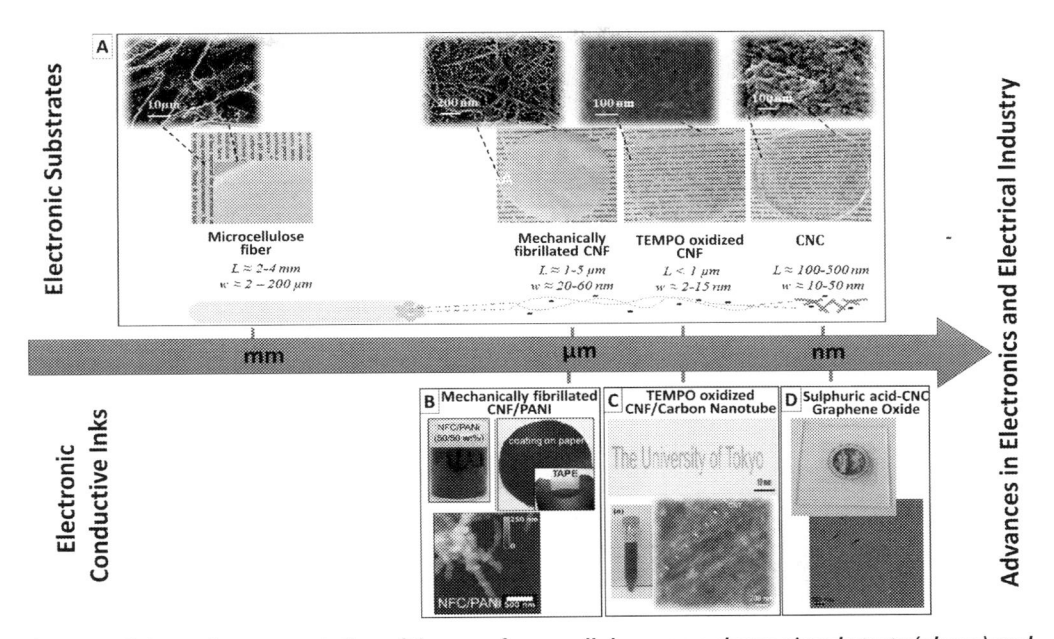

Fig. 12.1 *Schematic representation of the use of nanocellulose as an electronic substrate (above) and electronic conductive inks (below) concerning the types and dimensions of nanocellulose.* (A). Different types of nanocellulose as an electronic substrate adapted from [3] with permission from The Royal Society of Chemistry; (B). Coated conductive inks containing CNF and polyaniline on paper, adapted from [8] with permission from Elsevier; (C) Injected inks contain TEMPO-mediated CNF/CNT printed on copy paper, adapted from [9] with permission from American Chemical Society; (D). Coated CNC/graphene oxide on glass substrate, adapted from [10], with permission from Elsevier.

energy–dissipating amorphous phase or high crystallinity CNC results in a brittle film of a solely CNC. Therefore, their application in flexible electronics is limited and more favored for nanocomposite applications with polymers [14,15].

Meanwhile, CNF forms long and entangled fiber with a flexible network. Depending on the processing methods, the dimension of CNF ranges from 2 to 60 nm in diameter, and few microns in length [3]. The length of the obtained CNF can be reduced from hundreds of nanometer to a few microns by extending shear treatments [14]. CNF consists of both amorphous and crystalline parts. Generally, the crystallinity degree of CNF from softwoods and hardwoods ranges from 40% to 80% [6]. They can be produced by mechanical fibrillation, enzymatic treatments, and 2,2,6,6–tetramethylpiperidine–1–oxy (TEMPO) mediated oxidation. Low density, low thermal expansion coefficient, high mechanical strength, and tunable surface properties make CNF a promising building block for creating advanced materials [16]. For printed electronics, studies show that most of the electronic substrates are made by CNF rather than CNC.

Despite having the same chemical structure as CNC and CNF, BNC is derived from the secretion of the microbial organism, mostly by bacterial strain of *Gluconacetobacter xylinum*. The structural morphology of BNC is a ribbon shape nanofibers in the form of a web-like network to create a three-dimensional network [17]. Similar to CNF, BNC has a long entangled structure with a typical diameter in nanometer and length within the range of nano up to micron meter. Due to its high water retention, high mechanical strength, and biocompatibility, the studies on BNC is dominantly reported in biomedical [18]. However, several studies also reported the utilization of BNC as a substrate in organic light-emitting devices [19], flat panel display [20], and electrical insulating paper [21]. Nonetheless, the production cost of BNC should be considered due to expensive fermentation technology and time consumption. The cost of the CNF and CNC is expected to be lower than that of BNC due to the use of more abundant resources of lignocellulose biomass processed via mechanochemistry.

12.3 Cellulose nanopaper as a substrate in electronic devices

Most of the studies reported that the use nanocellulose as a potential substrate could substitute or replace the existing plastic and or paper-based matrixes. As such, the required properties, including mechanical strength, optical transparency and haze, porosity, and roughness, determine the parameter for electronic applications [22]. For electronic device purposes, electrical properties such as dielectric constant and resistivity are required. Meanwhile, optical properties such as transparency and haze are the essential properties in optoelectronic [23]. Paper made by nanocellulose, also known as cellulose nanopaper (CNP), has many advantages over plastic as well as paper substrate, as summarized in Table 12.1.

Compared to the conventional paper, CNP possesses very high transparency due to the high packing density of nanofiber as well as significant smaller porous cavities and thus lower light intensity of backward scattering toward the higher transmission. CNP has a low CTE similar to glass as well as the higher thermal durability and

Table 12.1 Comparison between cellulose nanopaper, conventional paper, and plastic [3,24,25].

Characteristics	Cellulose nanopaper	Conventional paper	Plastic (polyethylene terephthalate)
Surface roughness (nm)	0.2 – 5	5,000 – 10,000	0.5 – 5
Porosity (%)	20 – 40	50	–
Pore Size (nm)	10 – 50	3,000	–
Optical transparency at 550 nm (%)	90	20	90
Coefficient of thermal expansion (ppm/K)	<8.5	28 – 40	50–200
Young's modulus (GPa)	7.4 – 14	0.5	2 – 2.5
Printability	good	excellent	poor

dielectric constant than conventional plastics, due to its high mechanical properties of CNF [23]. In addition CNP is light in weight, highly foldable, and more flexible than glass or silicon wafer. Modified CNP can also endure higher processing temperatures than conventional plastic. The thermal degradation temperature of CNP was found to be 300 °C. Even after 30 min, no change in color and dimensional stability were observed at 200 °C. These properties have favorably opened the application in flexible electronics such as transparent electrodes, organic solar cells, transistors, antenna, and memory [26–31].

The common process for the fabrication of CNP typically uses nanocellulose with concentration less than 2% in water dispersion followed by the removal of the water content to obtain dried CNP using (1) vacuum filtration followed by oven drying [25,32] or hot pressing [33] of the wet nanocellulose cake; and (2) casting followed by drying to evaporate the water using either oven drying [27,34] or room temperature drying [35,36]. Fig. 12.2 shows the methods for preparing CNP with regard to their pros and cons of the process and their obtained properties of CNP.

These typical methods of the preparation of CNP have been reported to result in the distinctive properties of the final paper. Yang et al. [38] reported that CNP produced from casting possessed a smoother surface morphology compared to that of CNP from vacuum–filtration. After the filtration process, the separation procedure is conducted to separate the nanocellulose gel from the membrane (filter paper) followed by the drying process. The separation and transfer process are postulated to attribute the rougher surface of the obtained CNP from filtration compared to that from the casting process. Meanwhile, the CNP produced from casting could be easily peeled off from the smooth surface of Petri dish after evaporation without the transfer process leading to smooth surface morphology. As a result of the smoother surface characteristic, CNP obtained from film casting exhibits greater transparency in comparison to the ultrafiltration, due to negligible light scattering. Similar observation on the surface roughness and transparency between these two common methods was also reported by Sehaqui et al. [36].

It is also worth noting that the vacuum filtration process requires a membrane with a pore size diameter of up to a nanometer, which is typically expensive, leading to an increase in production cost. Nevertheless, CNP fabrication via casting is time-consuming and susceptible to wrinkling problems leading to poor mechanical properties. Under constrained clamping applied in a vacuum filtration setup, wrinkling can be avoided. The rapid vacuum process also helps in removing the moisture content effectively, thus reduces the moisture concentration gradient in the surface of nanocellulose gel and helps in reducing the wrinkling problem. As a result, the improved mechanical properties are achieved by the CNP obtained from vacuum filtration compared to that of CNP from casting [36].

Casting Method

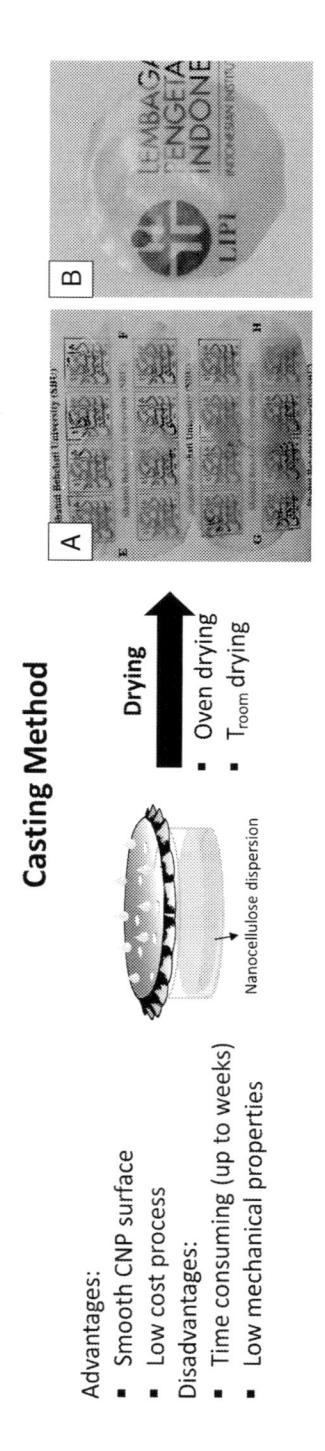

Drying

- Oven drying
- T_{room} drying

Nanocellulose dispersion

Advantages:

- Smooth CNP surface
- Low cost process

Disadvantages:

- Time consuming (up to weeks)
- Low mechanical properties

Vacuum-filtration method

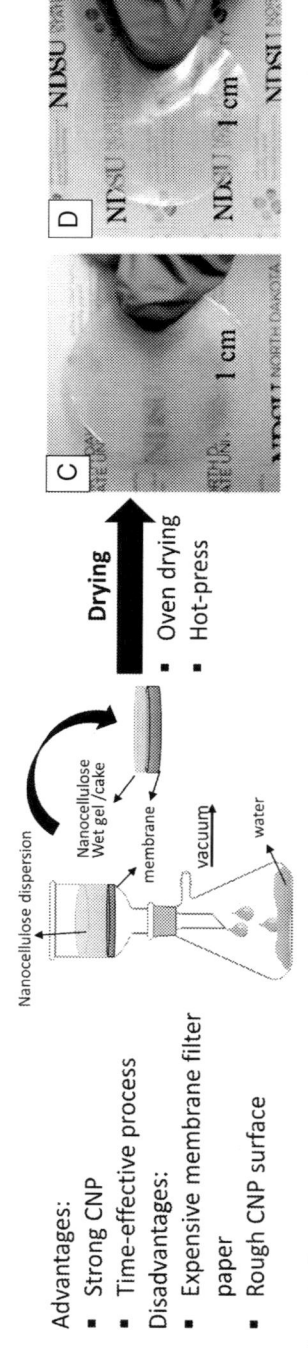

Nanocellulose dispersion

Nanocellulose Wet gel/cake

membrane

vacuum

water

Drying

- Oven drying
- Hot-press

Advantages:

- Strong CNP
- Time-effective process

Disadvantages:

- Expensive membrane filter paper
- Rough CNP surface

Fig. 12.2 Common routes to prepare CNP from (A) mechanical fibrillated CNF reproduced from [37] with permission from Elsevier; (B) sulfuric acid-CNC, unpublished image from the author's collection; (C), (D) hybrid CNP containing CNF-only and CNC/CNF-83/17, respectively, reproduced from [31] with permission from The Royal Society of Chemistry.

12.3.1 Promising characteristics and challenges of nanocellulose as electronic substrates

12.3.1.1 Transparency and low haze

Traditional paper is opaque, but when the dimension of the fiber is reduced to nanoscale, the paper becomes transparent due to the denser structure [39,40]. The improved transparency is caused by the reduced interstice dimension between the nanofibers. The nanofibers are so densely packed that light scattering inside the film or at the surface is prevented by the absence of cavities. The CNP shows a more homogeneous diffraction index, which further minimizes backscattering [41,42]. Haze is described as the ratio between the transmitted light through CNP (deviated from the incident light beam of an angle above 2.5°) and the total transmitted light intensity [42]. Generally, optical transparency has the opposite value with haze [22]. CNP with low haze is bright transparent, while CNP with a high haze is hazy transparent or translucent [12]. Studies reported that the maximum transparency of glass, plastic, and CNP is about 90%, with a very low haze of around <20% [22]. Clear and transparent CNP with low haze is a proper fitting as a substrate for the optoelectronic devices such as transparent electrodes or transistor substrates, flexible display, and touch screen [12,31]. Meanwhile, high haze CNP is a promising material for optical diffusers in LED lighting [12]. Due to its high diffuse transmittance, CNP with high haze has been developed in other electronic-related applications such as skylights, indoor lighting, and solar cells [31].

Apart from the effect of preparation method of CNP on the optical transparency as discussed earlier, types of nanocellulose significantly influence the transparency. CNF produced from mechanical methods such as high-pressure homogenizer [37] typically exhibits more opaque and lower transparency compared to the CNF produced from chemically TEMPO-mediated oxidation [38]. The highest transparency is achieved from CNF-based TEMPO-mediated oxidation, which exhibits optical transparency of up to 90% at the wavelength of 600 nm, which is comparable to a plastic substrate [15].

Controlling the degree of fibrillation can also result in tunable transparency. Wakabayashi et al. [43] used membrane filtration with TEMPO-oxidation cellulose. They found that a low degree of fibrillation of TEMPO-oxidation cellulose/water dispersion, which consists of nanofibrils, unfibrillated fibers, and fiber bundles, took a shorter time than the higher degree of fibrillation, in terms of film preparation using membrane filtration. The higher degree fibrillated nanofiber achieved the most increased transparency. The presence of CNCs with lower aspect ratio can also help in improving transparency. Xu et al. [31] demonstrated the influence of increasing CNC ratio in CNC/CNF composite on the transparency of CNP using vacuum-assisted filtration. Increasing the CNC ratio showed a decrease in optical haze and an increase in transparency. This result was attributed to the stable individual CNC in the suspension due to electrostatic repulsion, while CNF tended to agglomerate in a low concentration suspension. Further, CNC also acted as a rigid nanofiller to modify the surface morphology, roughness, and porosity of CNP.

Several reports on the application of BNC as transparent CNP could also be achieved due to its high light scattering at the interface of nanofibers and air [15]. Wu et al. [44] also produced BNC suspension via TEMPO-mediated oxidation followed by mild mechanical treatment as the nanocellulose material for CNP. The BNC-based CNP was prepared by cast-drying method and achieved light transmittance of 83% at 550 nm. Legnani et al. [45] successfully prepared transparent BNC with transparency of up to 88%, which was compatible to be used as a substrate for organic light-emitting diodes (OLED).

12.3.1.2 Porosity and roughness

Surface roughness and porosity are significant parameters to be controlled in the fabrication of CNP as substrates in electronic-related devices because they affect the transparency, haze, and mechanical strength [14,23,31]. Surface smoothness of CNP is mostly in the range of 2–5 nm, with a porosity of 20–40% [29] and a pore size < 50 nm. The smoothness allows the homogeneous deposition of coating thin films on the surface of CNP while hindering the possibilities of the film to crack and deform during the printing process. Chinga–Carrasco et al. [46] reported that smoothness of surface combined with sufficient surface energy resulted in better print resolution and homogenous coating ink in inkjet (IJ) printing.

The characteristic of fibrillated fiber also affects the active layer morphology, which is a significant property for a solution-processed device. Technically, a pore size of 3–5 μm causes the printed line to be wavy contour, which leads to inaccurate printing and affects the production cost [42]. Chinga–Carasco et al. [46] studied how the roughness of CNP was affected by the residual fibers that were not fibrillated into nanofibrils during the homogenization process. They found that the residual fibers have a significant effect on the surface roughness at wavelengths > 80 μm, which corresponded to the lateral dimensions of cellulose fibers.

As discussed earlier that the fabrication process of CNP has been reported to affect the smoothness and thus the porosity. Several techniques have been used to improve the porosity, including the use of supercritical drying [29], solvent exchange, flocculation before the film formation [14], as well as the dispersion of nanofibrils in organic solvents before the production of CNP [47]. Solvent exchange is favored and considered a simpler method to control the pore size of CNP [48]. CNP modified by swelling in different types of solvent, enhance their ability to form hydrogen bonds and thus exhibiting a significant improvement in pore size [48].

Porosity also significantly affects the mechanical properties of CNP. Henriksson et al. [49] found that nanofibrillar network structure and high mechanical nanofibril performance were related to porosity. CNP with porosity imparted by sufficient network could form a denser structure consisting of more CNF to commence load transfer, and thus resulting in a higher elastic modulus, tensile strength, and toughness. Meanwhile, at a higher porous network, there is a decrease in hydrogen bond density, which lowers the fibril–fibril interfacial strength [50]. Generally, porosity has an inverse relation with Young's modulus, yield strength, and tensile strength [14]. A more detailed discussion about the mechanical properties of CNP is discussed in the following section.

12.3.1.3 Mechanical properties

CNP made either from CNC, CNF, or BNC typically exhibits high mechanical properties with the value of Young's modulus ranging from 79 to 220 GPa [51-53], tensile strength of 100–1000 MPa, strain to failure of 3–10%, and toughness of 3–25 MJ/m^3 [50]. The mechanical properties bestowed by nanofibrils significantly rely on the understanding of the structure formation and interfibrillary interactions, as well as the mechanism of deformation in bulk [54]. Several factors influencing the mechanical properties of CNP include DP, orientation, moisture, counterions, and pH, as well as porosity, as discussed earlier.

Degree of polymerization (DP)

Extensive experiments [49,55] have demonstrated that CNP made from the same material but with higher DP shows a consistent increase of tensile strength and strain at break. Fang et al. [56] studied the role of DP on the mechanical properties of CNP and successfully fabricated superstrong yet lightweight anisotropic nanocellulose films. The obtained CNP showed excellent tensile strength, modulus, and toughness of $1{,}014.7 \pm 69.1$ MPa, 60.2 ± 0.8 GPa, and 15.2 ± 1.9 MJ/m^3, respectively. Interestingly, the DP of cellulose plays a vital role in increasing the mechanical properties of the CNP when the lignin content is lower than or equal to 3.3%. The higher DP of cellulose, the better the mechanical properties of the CNP could be achieved. Therefore, to get high toughness and tensile strength, it is essential to maintain high DP of the cellulose [14].

Orientation

Generally, nanocellulose possesses excellent mechanical properties; however, its potential could not fully be achieved due to random-in-plane or random-in-space orientation and distribution of nanocellulose [80]. Ansari et al. [57] reported that higher mechanical properties of CNP were produced in the film with highly oriented nanocellulose. Hence, several studies have been devoted to the production of the unidirectional orientation of CNP to enhance the mechanical properties. The hypothesis is to adjust the direction of fiber with the same direction of tensile strength and thus enabling to achieve high mechanical resistance by anisotropic alignment of the crystalline phase. The alignment of nanocellulose is usually conducted by fully hydrating it to allow enough interfibrillar motion for disentanglement and rearrangement of the CNF's orientation. At the beginning of the alignment, porosity removes and fixes cohesive fibrillar network into higher rigidity and lower elongation [14]. As CNF is longer and more flexible than CNC, they are more challenging to be aligned. Several methods could be used to produce oriented CNP, including dynamic sheet former [58], stretching with mechanical forces [59], and cold drawing [60].

Counter ions and pH

TEMPO-oxidation CNF (TOCNF) with its carboxylic functional groups allows ion exchange with various counterions. This event enables the modification and the colloidal characteristic of CNF. Numerous experiments have proved that selected multivalent counterions increase the mechanical properties of CNP. However, at ambient humidity, their effect is small and therefore, it is best to observe them in a hydrated state, where ionic interaction is more imminent [14]. The effect of humidity or moisture on the CNP's properties is strongly related to the presence of abundantly available hydroxyl groups on the surface of nanocellulose. A further detailed discussion on the effect of humidity on the mechanical properties will be elaborated in the following section. Further, CNP from charged TOCNF shows better mechanical properties when prepared at high pH. The dispersion state of CNF is associated with this result; at high pH (>9), COO^-Na^+ at the surface of the fibrils allows high electrostatic repulsion leading to a better dispersion. Meanwhile, at low pH (<3), COOH dominates the surface resulting in a decrease in repulsion and aggregates formation. The presence of these aggregates affects the organization of CNF in CNP, reducing the mechanical properties [57].

12.3.1.4 Hydrophilicity

Nanocellulose is highly hydrophilic due to its natural advantages of abundant hydroxyl groups on its surface. The presence of hemicellulose and lignin in the nanocellulose derived from lignocellulosic material influences the hydrophilic characteristics. While cellulose contains linear hydrophilic chemistry, hemicellulose has a branched hydrophilic polymer chain, which hinders the formation of strong hydrogen bonds between inter- and intramolecules of individual nanofiber and consequently reduces the mechanical properties of CNP [14]. The presence of aromatic and cross-linked structure of lignin poses a more hydrophobic structure that can cause yellowish and brittle CNP over the period under the UV light exposure [2]. As highly transparent and flexible CNP is a fundamental requirement in the application of electronics, the removal of hemicellulose and lignin through the pretreatment process would thus is highly required in the case of nanocellulose derived from lignocellulose resources. Li et al. [61] investigated the role of lignin and hemicellulose in CNP, showing that the presence of hemicellulose and lignin could improve the hydrophobicity of CNP due to the presence of hydrophobic xylan and lignin, which hindered the formation of hydrogen bonds between cellulose and water.

The inter—intra connection of individual nanocellulose in CNP is imparted by the hydrogen bonding between the surface chemistry of hydroxyl groups. While the long, entangled CNF offers the mechanical strength, the shorter aspect ratio of nanocellulose provides transparency with the smaller pore size [2]. The small pore in the network of hydrophilic nanofiber holds electrolytes, which gives an additional intercalation road, beneficial for the application of electrochromic CNP [39]. Yuen et al. [62] performed the ability of nanocellulose sheet to be printed with semiconducting polymer of (poly(3,4-ethylenedioxyiophene)-poly(styrenesulfonate) (PEDOT-PSS) using dimatix material printer. Nanocellulose sheets can also absorb liquids due to the high hydrophilicity allowing plating chemicals to go through the sheets and enabling the reaction of metallic plating within their matrix.

Although the presence of hydroxyl groups as tunable surface functionalities in the nanocellulose could positively enhance the mechanical properties and absorption capacity of formulated inks, which are required to develop flexible electronics, there is a fundamental weakness as they could lose their mechanical strength in humid conditions. Their hygroscopic characters can have detrimental effects due to external factors such as high relative humidity, moisture, and temperature [54,57]. Moisture behaves like a plasticizer; lowering the bonds among the fibrils; therefore, decrease the mechanical properties of CNP [57]. Likewise, Tobjörk and Österbacka [2] reported that the adsorption of water could reduce the strength of paper as the individual fiber loses the interconnection bonding strength leading to a swelling paper structure. The effect of moisture on the mechanical properties of CNP have been reported by Benitez et al. [54], Ansari et al. [63], and Wu et al. [64].

In the environment with relatively high humidity, their hydrophilic characters significantly affect its surface behavior and thus limit their application [65]. This negative effect of high humidity would be more pronounced if the base material of electrical component is made by pure nanocellulose due to the complexity of water vapor sorption or hygroscopic behavior. The saturation state of moisture content depends upon the paper structure and number of hydroxyl groups, as well as the presence of other polar chemistry groups. The more prolonged exposure to the environment with very high relative humidity can be detrimental to the performance of printed electronics devices leading to device failure. Bollstrom et al., 2014 [66] reported that very fine printed electronic lines (line widths ranging in an average of 75–685 μm) were very vulnerable to the humidity. The hydrophilicity also affects the shape instability of CNP, which is a major drawback for its application in electronic-related devices [67]. A substrate in high-resolution electronic printing needs to have excellent shape stability to avoid wrinkles, which leads to a short circuit. Hence, the challenges remain as despite having unique flexibility and porosity, tunable surface chemistry as well as optical properties, CNP is susceptible to a lower mechanical property under humid condition and shape instability due to the hydrophilic characters, and consequently, surface functionality is an imperative stage to address these challenges.

12.3.2 Surface functionality of nanocellulose for enhanced properties

The functionalization of nanocellulose, such as CNF or shorter CNC, has already been reported in the literature to a large extent compared to BNC, which is not only to address the negative effect of hydrophilic characters as previously discussed but also to tailor and modify their surface properties and satisfy the required performance for material-based electronic devices. Zhu et al. [40] used glutaraldehyde for cross-linker and HCl as a catalyst to limit the exposure of hydroxyl groups to the solvent; therefore, increasing the CNP's shape stability, suitable for the flexible radio frequency identification (RFID) antennas using gravure printing. Gao et al. [67] coated CNP with acrylic resin to obtain the waterproof CNP, making it suitable for flexible perovskite solar cell (PSC). The flexible PSC displayed excellent stability, maintaining up to 80% of its original efficiency even after 50 times of bending.

Surface functionalization also allows tailoring negative or positive electrostatic charges on the surface of nanocellulose to give a better dispersion in any solvent or polymers. The modification can control the surface energy characteristic, enhancing nanocellulose compatibility with nonpolar or hydrophobic polymer matrices [11]. The surface functionality can help to enhance the mechanical properties, as the surface of the modified hydroxyl group can make a strong covalent bond instead of only making the hydrogen bond. As a result, a more rigid hybrid network can be achieved, which limits the movement of modified nanocellulose fiber in external load and thus leading to an increase in the elastic modulus [68]. Several commonly used methods for surface functionalization are esterification, acetylation, etherification, oxidation, silylation, polymer grafting, nucleophilic substitution, carbamation, and fluorescent labeling [11,69,70]. Selected functionalization techniques to modify the characteristics of CNP depending upon the types of nanocellulose are illustrated in Table 12.2.

12.3.3 Transparent conductive cellulose nanopaper

All the substrate material either from petrochemical-based polymer such as PI, PET, polytetrafluoroethylene, or organic-based polymer as the transparent CNP, has been considered as the potential substrate to substitute glass base material in electronics. As discussed earlier, their transparency, mechanical properties, and surface smoothness have been recognized to suit the development of the next-generation flexible electronics. Nevertheless, they are nonconductive, and thus the introduction of conductive material is required to afford the ability for further implementation of CNP into the electronics industries.

Two main approaches typically are used to obtain conductive CNP, either by incorporating conductive material within the substrate structure as a composite or attaching/coating the conductive material onto the surface of CNP as a conductive layer. While much study has been devoted to this area of conductive CNP composite, most of the electronic applications require patterned or printed conducting structures onto the substrate, which is included in the latter above-mentioned approach.

Table 12.2 Selected surface modification of CNP.

Nanocelluloses	Functionalization method	Results	Reference
CNF–mechanical treatment	Mild esterification using anhydride	Increased water contact CNP angle of 118 °C	[71]
	Esterification using lactic acid	Water-resistant CNP with triple higher improvement in storage modulus	[68]
	Blending with the self-made hydrophobic substance of fluoro acrylic resin	Hydrophobic CNP with light transmittance up to 82%	[72]
CNF–TEMPO mediated oxidation	Incorporation of sepiolite using high-shear homogenization and sonomechanical treatments	Hybrid CNP with Young's Modulus of 3.4 GPa	[73]
	Silanization of methyltrichlorosilane	Self-cleaning superhydrophobic surface with contact angle 159.6° and high transparency at 90.2%	[74]
CNC–H_2SO_4 hydrolysis	Esterification with hexanoyl chloride	Free-standing CNP with a low degree of substitution	[75]
	Incorporation of ZnO/BNC via mild sonication	A decrease in water vapor permeability by 30% with improved thermal properties	[76]
	Grafting with long alkyl chain amine	Reduction in water vapor transmission rate by 40% with improved mechanical properties	[77]
BNC–mechanical treatment	Incorporation of ZnO/BNC via ultrasonication	109% improvement in tensile strength with significant moisture resistance	[78]
	Adsorption of hydrophobic polystyrene by immersion	Water-repelled CNP with contact angle ≈ 90°	[79]

There are three types of conductive particles to formulate conductive inks, including metallic nanoparticles or flakes (mainly silver, gold, and copper), carbon particles (nanotubes, graphene, and carbon black), and conductive polymers PEDOT-PSS, polypyrrole, Polyaniline (PANI), polymerizable deep eutectic solvent). Among the existing conductive inks in the IJ printing process, metallic nanoparticles inks, such as silver nanoparticle is the most frequently used and is often in the combination of other conductive particles to obtain an excellent functional layer.

As discussed earlier, transparent CNP has a unique porous structure and high ink absorption capacity required during surface modification via coating or deposition process of conductive ink [80]. The introduction of conductive material in the substrate that can improve their electrical conductivity while maintaining transparency is particularly important to open any possible plethora of applications in electronics. The fabrication of conductive CNP by the introduction of conductive layer onto the substrate surface using varied conductive material while maintaining its transparency is summarized in Table 12.3.

12.4 Printed electronics

The interest for printed electronics has increased rapidly since the first printed transistor was discovered in 2003 that marked the revolution of the electronic industry from conventional printing equipment [87,88]. The term printed electronics is defined as the printing of sensors and the circuits for the manufacturing of electronic parts and devices on paper, plastic, or textile [3,87,89]. Printed electronics combine electronic manufacturing with text printing [90]. In this context, printing means the process of attaching electronically functional materials to media surfaces or substrates such as paper and plastic [87]. Compared to conventional microfabrication techniques, printed electronics offer cost–effective, flexible, lightweight, sustainable, and more environmentally

Table 12.3 Selected studies on the fabrication of transparent conductive CNP using varied conductive materials.

Conductive components	Conductive layering method	Sheet resistance (Ω/sq)	Transparency (%)	Reference
Silver nanowire (AgNW)	Wet-lamination process by ultrafiltration	12	88	[33]
	Coating polyvinylidene fluoride mask followed by deposition	1.2	82.5	[31]
	Self-polymerization of dopamine followed by binding with AgNWs layer	14.2	90.93	[81]
	Vacuum–assisted microcontact printing (μCP)	54	82	[82]
	Pressured-extrusion papermaking process	4.43	89.56	[83]
CNT	Deposition via airbrush spray	2000	93.5	[84]
	Meyer rod coating method	–	83.5	[24]
Polymerizable deep eutectic solvent (PDES)	Screen printing PDES onto CNP under ultraviolet (UV) light source	700	94.5	[85]
AgNW and PEDOT:PSS	Aqueous pressured-extrusion paper-making process	7.32	92.56	[80]
PANI/ PEDOT:PSS	Layer-by-layer coating deep immersion	–	47.1	[86]

friendly products [91]. The field of printed electronics covers IJ, gravure, flexography, or screen printing, offset, and laser [92]. Coating processes, including bar coating, spin-coating, or sputtering that produce conductive tracks and layers, are also the common techniques in printed electronics [3]. Printed electronics provide an interesting alternative by allowing the invention of high surface area, flexible with a cost-effective and high-efficient device [93]. Several applications of printed electronics are RFID, OLED, and photovoltaic cell [87].

In the printing process, as the ink solution is directly contacted to the substrate, both properties are important to be evaluated concerning the process condition during the printing. The conditions include low-temperature deposition, high throughput, interplay with other layers including compatibility of printed layers in terms of wetting, adhesion, and dissolving as well as drying procedures after deposition of liquid layers [87]. For, substrate, the smoothness of the surface of the thermally stable material is the most critical requirement. Porosity and surface roughness also significantly affect print quality. These properties have an impact on ink penetration and ink–film continuity, which leads to the connection and interaction between conductive particles [92]. These unique properties and their challenges are discussed in the previous section. For the conductive inks, the major challenges rely on the stability of conductive material in the formulation while maintaining the required physiochemical properties of conductive printing or coating onto a respected substrate. Another crucial property of the conductive ink is the surface energy and adsorption capacity as they influence ink penetration, ink film thickness, and image resolution, which later affect the performance of functional prints [92].

12.4.1 Conductive inks in printed electronics

Printed electronics have gained more interest as a next-generation technology to fabricate electrical devices on a roll-to-roll of flexible substrates, leading to an increasing urgency for conductive inks [9]. Flexographic, gravure, offset, screen, and IJ printing are several techniques used for printed electronics [94]. Among them, IJ printing techniques have gained tremendous attention to the fabrication of highly conductive patterns [26].

IJ printing is a resourceful technique for transferring liquid ink to create a printed pattern [87]. This process allows the deposition of thin layers (0.2–2 mm) with a high resolution (up to line width of 40 mm) [5]. In this process, conductive inks are discharged through a small-diameter nozzle around 50 mm, and consequently, the lower viscosity of conductive inks is required to prevent nozzle clogging. At a low concentration of conductive ink, the morphology of the printed line is susceptible to a rough form with lateral spreading, splashing, and wavy boundaries, which leads to lower conductivity. Therefore, the surface morphology of the substrates is a significant factor in the fabrication of highly conductive and fine lines by IJ printing [26]. Also importantly, the excellent dispersion of the conductive particle in the ink formulation is crucial to ensure the homogenous, smooth, and fine printed line in the substrates. The complexity of obtaining high colloidal stability at a low concentration, either single or combined conductive particles, becomes a significant challenge. Therefore, additives and compatible solvents are usually required to prevent the agglomeration of conductive particles and thus achieve the desired performance of printed electronics.

Recently, nanocellulose has been reported to offer excellent properties as the additives or components in the formulation of conductive inks. Hoeng et al. [3] reported that nanocellulose could act as the dispersing agent as well as a binding agent to help achieving greater dispersion stability in the conductive inks due to the presence of charged density induced by tunable surface chemistry of nanocellulose. Adsorption capacity is also an important character to be fulfilled by the conductive ink to enhance the ink penetration and achieve the homogenous coating of conductive inks in the printing process. This can be managed by the presence of nanocellulose-based material in the conductive ink formulation [95]. Here, we provide the potential use of nanocellulose in conductive ink applications that could offer the desired performance of printed conductive electronics.

12.4.2 Promising characteristics and challenges of nanocellulose as conductive inks

12.4.2.1 Charged densities for dispersion stability of conductive inks

As mentioned earlier, one of the major challenges to improve the performance and functionalities of conductive inks is excellent stability and dispersion at low concentration of conductive material in the ink formulation. It is also often required to use a more complex system containing the combination of several conductive fillers to achieve the desired electrical conductivity while maintaining the transparency of the obtained conductive layer as illustrated in Table 12.3. For this purpose, the stabilizing agents are of importance in the ink formulations for improving dispersion stability and thus maintaining the homogenous ink penetration and physicochemical properties of the suspension. As reported, nanocellulose is a versatile bio-based material that can act as a stabilizing agent in ink formulation due to unique characteristics such as surface chemistry and its charged density.

The surface charge density is one of the key factors to induce the stabilized colloidal nanocellulose due to electrostatic repulsion imparted by the surface chemistry group of the nanocellulose. The presence of charged density on the surface chemistry group of the nanocellulose as the result of grafting of charged groups can be categorized into three formations: (1) native surface chemistry because of the isolation or deconstruction of microscale cellulose-based material to produce nanoscale material; (2) additional introduction or surface modification reaction of the obtained nanocellulose, and (3) adsorption process to the nanocellulose surface [96].

Among those routes, the preparation and deconstruction of microcellulose into nanocellulose offers the simultaneous surface charge benefit for a stabilizing agent. In particular, studies using TEMPO-mediated oxidation for the nanocellulose formation have been reported to convert the D-glucose cellulose units into highly charged carboxylate groups on the surface of the obtained nanocellulose backbone leading to the liberation of nanocellulose fiber and thus enhancing colloidal stability due to electrostatic repulsion [54,97,98]. Mechanical treatment via microfluidizer with the addition of mono-chloracetic acid has also been reported to produce CNF with a higher surface charge to offer greater colloidal stability [99].

A higher surface charge of nanocellulose promotes robust and more stable colloids due to higher electrostatic repulsion. Still, it should be under the magnitude of the percolation threshold of the fiber concentration. The effect of nanocellulose source, dimension, and concentration on colloidal stability is also investigated by Mendoza et al. [100]. It is interesting to note that the coloidal stability is independent from the variable of the pulp resources, yet it is significantly influenced by the concentration of fiber. The increasing cellulose concentration results in a significant increase in viscosity due to greater electrostatic repulsion and higher surface charges up to 50% improvement leading to greater colloidal stability. Nevertheless, above the percolation concentration, the reduction of surface charges is observed as a function of reduced viscosity due to the lower effective electrical interaction.

The effect of high surface charges resulting in the improved colloidal stability of conductive inks formulation has been investigated in several studies. Koga et al. [9] showed that the surface anionic of TEMPO-mediated CNF was an effective dispersing agent for carbon nanotube (CNT) even after drying, allowing the formation of highly conductive and printable nano-inks with low surface resistance (1.2 $\pm$ 0.1 kΩ/square). The presence of CNF as an excellent aqueous dispersion agent of single–walled nanotubes was also reported by Hamedi et al. [101]. Hoeng et al. [5] also observed that the CNC could act both as capping and stabilizer of conductive ink containing silver nanoparticles. The printing of the CNC–silver ink produced conductive tracks with a high resolution. The presence of nanometric rod-shaped nanocellulose created a more stable conductive suspension at a small amount of silver compared to the conventional ink containing solely silver.

12.4.2.2 Adsorption capacity

External coating using conductive materials onto substrates can modify the electrical properties of such organic substrate or conventional plastic or glasses. Surface adsorption capacity is a vital parameter that influences the ink penetration and the uniformity of ink film coating, and subsequently, significantly affects the final properties of functional prints. Several conventional methods have been established to enhance the adsorption; for example homogenous coating of PEDOT:PSS onto microcrystalline cellulose paper could be enhanced by reducing pH as well as adding salt ion due to the reduction of polyelectrolyte, which enables the absorption/penetration in the pore substrate of PEDOT:PSS and into cellulose surface [92].

Interestingly, the addition of a small amount of CNFs in the colloidal particle dispersion has also been reported recently to enhance the adsorption capabilities by suppressing the "coffee rings effect" on the solid substrate upon the drying process [95]. Coffee ring effect, also known as a pinning effect, is a typical problem when particles tend to agglomerate at the edge of the liquid dropping zone upon the drying process. It was reported that the nonspherical shape, as expected from CNF, was responsible for the capillary effect and anisotropy in Brownian motions leading to the possibility of suppressing the coffee ring effect [102]. Interestingly, such an effect was observed on the contrary impact on the use of CNC. It was observed that the use of CNC in the particle dispersion resulted in strong coffee rings [103]. Although those studies have shown that the presence of distinctive types of nanocellulose on the particle coating formulation exhibited a significant effect of adsorption, they are still a proof of concepts for understanding the suppression of coffee rings in the process of layering substrate with the colloidal inks. We would suggest that further fundamentals studies are required and be worthwhile for the further investigation on the complex system of conductive inks containing nanocellulose in the field of conductive printed electronics.

12.5 Applications of nanocellulose as green electronics
12.5.1 OLED for display

The OLED has been a very promising feature for display applications. Commercially, LEDs are fabricated on glass, which is inherently brittle, fragile, and inflexible. Multiple developments have facilitated a massive leap in the development of hi-tech display from glass substrate toward polymer-based substrate (e.g., polyamide, PET, and various fluoropolymers) that make it possible to fabricate flexible display [31]. However, CTE of plastic derived from petrochemical-based material, exceeds 200 ppm/K which is far above the restricted requirement of CTE for commercial display substrate at maximum of 20 ppm/K [104].

CNP is a highly potential substrate for OLED due to its superior thermal stability, tunable optical transmittance, lightweight, flexibility, and, more importantly, possesses a lower CTE as well as its compatibility with roll-to-roll manufacturing [15]. Okahisa et al. [105] reported that the use of CNF showed a promising result to offer extremely low CTE at only 4–6 ppm/K and excellent transparency, which is perfectly matched for display substrates. Wu et al. [106] prepared nanocellulose film, which demonstrated a high optical transmittance and diffused transmission, which is applicable as a translucent light diffuser to improve light extraction from OLEDs. The improvement of the illuminous efficacy can reach up to 15%. Tao et al. [107] developed a hybrid polymer of nanocellulose/polyarylate that offered high transparency, low haze, superior thermal properties, and excellent mechanical strength. The current efficacy is also an important characteristic of the fabrication of flexible OLED. Najafabadi et al. [108] reported that the average current efficacy of OLED fabricated on CNC substrate was similar to the average performance achieved on glass substrates. A key advantage of the fabricated OLED on the CNC substrates is that they can be easily dispersed at room temperature by simple immersion in water. These excellent results may open up a new way for flexible high-performance OLEDs. Indeed, in the creation of flexible OLED, a substrate has a significant role as the mechanical support and the primary layer for the conductive functional layers [107]. Table 12.4 shows the use of different types of nanocellulose and its functional modification as flexible OLED.

Table 12.4 Comparative studies of the application of nanocellulose as LED display.

NCs	Role	Functional material	Products (LED/OLED)	Reference
CNF	Substrate	PEDOT:PSS	Optical transparency 93%, coefficient of thermal expansion 2.7 ppm/K, maximum handling temperature 200 °C	[109]
		Indium Tin Oxide (ITO)	Optical transmission of 85%, current and luminous emission of 47 cd/A and 20 lm/W, and a maximum brightness of 10,000 cd/m².	[110]
		ITO	Optical transmittance of 90%, haze of 93%, external quantum efficiency of 17.3%, and power efficacy of 37.7 lm/W	[111]
	Additive	ITO	High transparency of 85% and a low haze of 1.75% at 600 nm, upper operating temperature up to over 220 °C	[107]
CNC	Substrate	Al/lithium fluoride (LiF) cathode and Au/MoO₃ anode	Efficient phosphorescent to-emitting inverted OLEDs with maximum current efficacy of 32.5 ± 14.1 cd/m² at 10 cd/m² and 42.7 ± 9.8 cd/A at 100 cd/m²	[108]
BNC	Substrate	ITO	Optical transmission of 88%, resistivity of 2.7×10^{-4} Ω cm, carrier concentration of —1.48×1021 cm⁻³, and mobility of 15.2 cm²/Vs	[45]
		acrylic resin	Nonflexible OLED with the coefficient thermal expansion (CTE) of 4 ppm/K	[20]

12.5.2 Electronic circuit and components

One of the basic components in electronic circuits is transistor. Table 12.5 shows the use of different types of nanocellulose applied as an electronic circuit and components, specifically as transistor and supercapacitor.

The transistor is a fundamental building block of many electronic devices to amplify and tune the electronic signals. Owing to its high transparency, low surface roughness, and flexibility, CNP as a substrate has been exploited for flexible thin-film transistors (TFT). The fabrication of TFT is mostly dominated by the use of CNF substrate. Gravure and IJ printing techniques are used to fabricate the conducting, dielectric, and semiconducting device layers onto the substrate. In addition, other printing process techniques, such as vacuum evaporation and wet etching, were also reported [122]. Huang et al. [24] reported that the TFT performance with CNP was not only mechanically stronger but also exhibited a lower CTE, and better electrochemical properties, compared to that of fabricated on a plastic substrate.

Supercapacitors are also one of the important components in electrical appliances that act as the bridge between electrolytes and rechargeable batteries. Supercapacitor is expected to exhibit higher capacitance, high surface charge, and longer life cycle suitable for energy-storage devices to support the fast development of portable electronic devices [120]. Supercapacitor based on freestanding and flexible electrodes has been reported to achieve the desired properties using types of nanocellulose such as CNF and BNC. In addition, after further carbonization process, nanocellulose can also act as electrode materials in the energy-storage device that can be directly used as a matrix, separator, binder, and electrolyte [116].

12.5.3 Antennas

An antenna enables the transmission and reception of radio waves and has been extensively used in wireless communication, such as phones, radios, and computers. To increase their portability in the next-generation of flexible and wearable electronics, the development of miniaturized and lighter and flexible antenna has been promoted [15,35]. Conventional paper has high porosity and surface roughness due to its micrometer dimension fiber, which leads to poor electrical conductivity limiting its use as antennas, whereas densely packed CNP has a smoother surface resulting in highly conductive and sensitive antenna lines at a high frequency (>1 GHz) [124]. Several applications of CNP and its modification to meet the required performances as antenna are illustrated in Table 12.6.

12.6 Future remarks and conclusion

Nanocellulose has been recognized as a material building block holding enormous applications, including electronics. Owing to its excellent mechanical properties, chemically tunable surface functionality, printability, optical transparency, excellent

Table 12.5 Selected applications of nanocellulose in transistors and supercapacitor.

NCs	Role	Functional material	Products	Reference
CNF	Substrate	Single-walled carbon nanotube (SWCNT)	Optical transmittance up to 83.5%, a coefficient of thermal expansion (CTE) of substrates is 20–100 ppm/K	[112]
	Substrate Substrate	Mo gate electrodes, fluoro-based polymer gate dielectric layer, Au source/drain electrodes	Flexible organic thin-film transistor array exhibited high mobility of up to 1 cm^2/Vs,	[113]
	Substrate	Commercial organometallic ink (Bottom (S/D) and top (gate) electrodes	Organic thin-film transistors with saturation mobility, cm^2/Vs, V_{TH} of -2.4, and On/Off ratio of 10–100	[114]
	Substrate	IGZO (In_2O_3-Ga_2O_3ZnO) and IZO (In_2O_3-ZnO)	Drain voltages of 1 V with low power, exhibiting ION/IOFF > 104 and a mobility ≈2 cm^2/Vs	[28]
	Substrate	Solid-state ionic conductive CNP	Ultra-smooth surface ~0.59 nm and high transparency above 80%, effective capacitance at 0.1 Hz is 220 nF/cm^2	[115]
	Substrate and additive inks	$NiCo_2O_4$	Asymmetric supercapacitor-based CNF porous carbon with a high electrochemical performance at 64.83 F/g at 0.25 A/g or 32.78 F/g at 4 A/g	[116]
CNC	Substrate	GIZO (Ga_2O_3–In_2O_3–ZnO layer and IZO (In_2O_3–ZnO)	Field-effect transistor with high channel saturation mobility (>7 $cm^2/V\ s$), drain–source current on/off modulation ratio higher than 105	[117]
	Substrate	CYTOP/Al_2O_3, 6,13-Bis (triisopropylsilylethynyl) (TIPS)-pentacene: Poly[bis(4-phenyl) (2,4,6-trimethylphenyl) amine] (PTAA)	Organic field-effect transistors at low operating voltage, and average field-effect mobility of 0.11 cm^2/Vs	[118]
BNC	Substrate	PEDOT: PSS	Electrolyte-sensing transistor with on/off ratio of 200. Device sensitivity up to 342 mM.	[119]
	Substrate	Multiwalled carbon nanotubes, and polyaniline (PANI)	Flexible supercapacitor with remarkable cycling capacitance degradation less than 0.5% after 1000 charge-discharge cycles	[120]
	Substrate	PEDOT:PSS/SnO_2/ reduced graphene Oxide	Flexible supercapacitor with a capacitance of 445 F/g at 2 A/g and outstanding cycling stability 84.1% over 2500 charge/ discharge cycles	[121]

Table 12.6 Several applications of CNP in antenna.

NCs	Role	Additive components	Products	Reference
CNF	Substrate	AgNW in ethylene glycol	V-shaped antenna patterns with return lose 0.5–4.0 GHz, and line resistance 0.19 Ω	[35]
CNF	Substrate	HCl and glutaraldehyde	Shape stable gravure-printed antenna with a sheet resistance of 0.3 ohms per sq.	[123]
CNF	Substrate	AgNW	Miniaturized flexible antenna on the high- k CNP with loss return around 2.60 GHz	[124]

adsorption capacity, biologically renewable, and sustainability in addition to its abundance, the advancement of nanocellulose is believed to revolutionize the development of current electronics. Although the studies in the field of electronics are reported to be significantly lower than that of other areas pertaining to nanocellulose, fundamental understanding and challenges have been identified to explore the feasibility of the use of nanocellulose in electronics, including: (1) nanocellulose has shown promising results and properties in the main features in the electronics, mostly reported as "substrate/matrixes" and limited studies as "binder/additive/component of the conductive inks," (2) several proofs of concept on the development and modification involving different types of nanocellulose and their surface functionalization as well as process fabrication have been successfully tailored and enhanced the inherited properties of nanocellulose to satisfy the required specification and performances in electronics, (3) recent potential application in electronics achievable by the nanocellulose-based material has been reported in the laboratory experiment.

However, there are still challenges to overcome before truly organic or greener electronics hit the large scale and commercial market. The most challenges are related to the relatively high production cost of nanocellulose due to high-energy consumption, as well as numerous required chemicals and processes. Table 12.7 shows the comparison of the cost of electronic substrate using different types of materials. Indeed, the cost-effective process with improved energy efficiency and lesser chemical use on the production of nanocellulose along with the high-speed fabrication of transparent CNP need to be addressed.

Moreover, although the studies of nanocellulose as an additive/filler in the electronic inks are far limited compared to substrate, the use of nanocellulose can be a promising alternative compared to other additives, given the price of functional additive; for example, AgNW is very expensive at $300–800 per gram. The shelf life of the electronics containing biodegradable cellulose material should also accurately be

Table 12.7 The cost of electronic substrates according to the varied material (adapted from [37] with permission from The Royal Society of Chemistry).

Raw materials	Substrates	Price (€/m²)	Possibility for scaling up
Polyethylene terephthalate	Plastic substrate	4–6	Yes
Microcellulose	Paper	6–7	Yes
Mechanical fibrillated CNF	Nanopaper	15–200	No
TEMPO-mediated CNF	Nanopaper	200–500	No

evaluated while considering the possible substitution with the conventional plastics. This result is of importance to understand the optimal life product while addressing the challenges related to the end-of-life disposal of electronics. Finally, the fact that the increasing awareness of the environment as well as the sustainable process and product are still the global priorities, the intensive efforts should continue to be established to produce a low-cost and high performance of "green" electronics toward industrialization.

References

[1] A. Işıldar, E.R. Rene, E.D. Van Hullebusch, P.N.L. Lens, Resources, conservation & recycling electronic waste as a secondary source of critical metals : management and recovery technologies, Resourc. Conserv. Recycl. 135 (2018) 296–312.

[2] D. Tobjörk, R. Österbacka, Paper electronics, Adv. Mater. 23 (2011) 1935–1961. https://doi.org/10.1002/adma.201004692.

[3] F. Hoeng, A. Denneulin, J. Bras, Use of nanocellulose in printed electronics: a review, Nanoscale 8 (27) (2016) 13131–13154. https://doi.org/10.1039/c6nr03054h.

[4] R. Sabo, A. Yermakov, C.T. Law, R. Elhajjar, U.F. Service, G. Pinchot, Nanocellulose-enabled electronics, energy harvesting devices, smart materials and sensors : a review, J. Renew. Mater 4 (5) (2016) 297–312. https://doi.org/10.7569/JRM.2016.634114.

[5] F. Hoeng, J. Bras, E. Gicquel, G. Krosnicki, A. Denneulin, Inkjet printing of nanocellulose-silver ink onto nanocellulose coated cardboard, RSC Adv. 7 (25) (2017) 15372–15381. https://doi.org/10.1039/c6ra23667g.

[6] DA. Gopakumar, Y.B. Pottathara, K.T. Sabu, H.P.S. Abdul Khalil, Y. Grohens, S. Thomas, Nanocellulose-Based Aerogels for Industrial Applications. Industrial Applications of Nanomaterials, Elsevier Inc, 2019. https://doi.org/10.1016/b978-0-12-815749-7.00014-1.

[7] N. Grishkewich, N. Mohammed, J. Tang, K.C. Tam, Recent advances in the application of cellulose nanocrystals, Curr. Opin. Colloid Interface Sci. 29 (2017) 32–45. https://doi.org/10.1016/j.cocis.2017.01.005.

[8] N.D. Luong, J.T. Korhonen, A.J. Soininen, J. Ruokolainen, L.-. Johansson, J. Seppala, Processable polyaniline suspensions through in situ polymerization onto nanocellulose, Eur. Polym. J. 49 (2) (2013) 335–344. https://doi.org/10.1016/j.eurpolymj.2012.10.026.

[9] H. Koga, T. Saito, T. Kitaoka, M. Nogi, K. Suganuma, A. Isogai, Transparent, conductive, and printable composites consisting of TEMPO-oxidized nanocellulose and carbon nanotube, Biomacromolecules 14 (4) (2013) 1160–1165. https://doi.org/10.1021/bm400075f.

[10] L. Valentini, M. Cardinali, E. Fortunati, L. Torre, J.M. Kenny, A novel method to prepare conductive nanocrystalline cellulose/graphene oxide composite films, Mater. Lett. 105 (2013) 4–7. https://doi.org/10.1016/j.matlet.2013.04.034.

[11] J. George, S.N. Sabapathi, Cellulose nanocrystals: synthesis, functional properties, and applications, Nanotechnol. Sci. Appl. 8 (2015) 45–54. https://doi.org/10.2147/NSA.S64386.

[12] M.C. Hsieh, H. Koga, K. Suganuma, M. Nogi, Hazy transparent cellulose nanopaper, Sci. Rep. 7 (2017) 1–7. https://doi.org/10.1038/srep41590.

[13] D. Klemm, ED. Cranston, D. Fischer, M. Gama, SA. Kedzior, D. Kralisch, F. Kramer, et al., Nanocellulose as a natural source for groundbreaking applications in materials science: today's state, Mater. Today 21 (7) (2018) 720–748. https://doi.org/10.1016/j.mattod.2018.02.001.

[14] A.J. Benítez, A. Walther, Cellulose nanofibril nanopapers and bioinspired nanocomposites: a review to understand the mechanical property space, J. Mater. Chem. A 5 (31) (2017) 16003–16024. https://doi.org/10.1039/c7ta02006f.

[15] S. Li, P.S. Lee, Development and applications of transparent conductive nanocellulose paper, Sci. Technol. Adv. Mater. 18 (1) (2017) 620–633. https://doi.org/10.1080/14686996.2017.1364976.

[16] Q.F. Guan, H.B. Yang, ZiM Han, LiC Zhou, YBo Zhu, Z.C. Ling, HeB Jiang, et al., Lightweight, tough, and sustainable cellulose nanofiber-derived bulk structural materials with low thermal expansion coefficient, Sci. Adv. 6 (18) (2020) 1–9. https://doi.org/10.1126/sciadv.aaz1114.

[17] A. Stanislawska, Bacterial nanocellulose as a microbiological derived nanomaterial, Adv. Mater. Sci. 16 (4) (2016) 45–57. https://doi.org/10.1515/adms-2016-0022.

[18] C. Sharma, NK. Bhardwaj, Bacterial nanocellulose: present status, biomedical applications and future perspectives, Mater. Sci. Eng. C 104 (2019) 109963. July. https://doi.org/10.1016/j.msec.2019.109963.

[19] C. Legnani, C. Vilani, V.L. Calil, H.S. Barud, W.G. Quirino, C.A. Achete, S.J.L. Ribeiro, M. Cremona, Bacterial cellulose membrane as flexible substrate for organic light emitting devices, Thin Solid Films 517 (3) (2008) 1016–1020. https://doi.org/10.1016/j.tsf.2008.06.011.

[20] M. Nogi, H. Yano, Transparent nanocomposites based on cellulose produced by bacteria offer potential innovation in the electronics device industry, Adv. Mater. 20 (10) (2008) 1849–1852. https://doi.org/10.1002/adma.200702559.

[21] N. Zhuravleva, A. Reznik, D. Kiesewetter, A. Stolpner, A. Khripunov, Possible applications of bacterial cellulose in the manufacture of electrical insulating paper, J. Phys. Conf. Ser. (3) (2018) 1124. https://doi.org/10.1088/1742-6596/1124/3/031008.

[22] Z. Fang, H. Zhu, Y. Yuan, D. Ha, S. Zhu, C. Preston, Q. Chen, et al., Novel nanostructured paper with ultra-high transparency and ultra-high haze for solar cells, Nano Lett. 14 (2) (2013) 765–773. https://doi.org/10.1021/nl404101p.

[23] D. Ha, Z. Fang, N.B. Zhitenev, Paper in electronic and optoelectronic devices, Adv. Electron. Mater. 1700593 (2018) 1–20. https://doi.org/10.1002/aelm.201700593.

[24] J. Huang, H. Zhu, Y. Chen, C. Preston, K. Rohrbach, J. Cumings, L. Hu, Highly transparent and flexible nanopaper transistors, ACS Nano 7 (3) (2013) 2106–2113. https://doi.org/10.1021/nn304407r.

[25] M. Nogi, S. Iwamoto, A.N. Nakagaito, H. Yano, Optically transparent nanofiber paper, Adv. Mater. 21 (16) (2009) 1595–1598. https://doi.org/10.1002/adma.200803174.

[26] M.C. Hsieh, C. Kim, M. Nogi, K. Suganuma, Electrically conductive lines on cellulose nanopaper for flexible electrical devices, Nanoscale 5 (19) (2013) 9289–9295. https://doi.org/10.1039/c3nr01951a.

[27] N. Isobe, T. Kasuga, M. Nogi, Clear transparent cellulose nanopaper prepared from a concentrated dispersion by high-humidity drying, RSC Adv. 8 (2018) 1833–1837. https://doi.org/10.1039/C7RA12672G.

[28] R. Martins, D. Gaspar, MJ. Mendes, L. Pereira, J. Martins, P. Bahubalindruni, P. Barquinha, E. Fortunato, Papertronics: multigate paper transistor for multifunction applications, Appl. Mater. Today 12 (2018) 402–414. https://doi.org/10.1016/j.apmt.2018.07.002.

[29] H. Sehaqui, Qi Zhou, O. Ikkala, LA. Berglund, Strong and tough cellulose nanopaper with high specific surface area and porosity, Biomacromolecules 12 (10) (2011) 3638–3644. https://doi.org/10.1021/bm2008907.

[30] Q. Wang, J.Y. Zhu, J.M. Considine, Strong and optically transparent films prepared using cellulosic solid residue recovered from cellulose nanocrystals production waste stream, Appl. Mater. Interfaces (2013) 52527–52534. https://dx.doi.org/10.1021/am302967m.

[31] X. Xu, J. Zhou, L. Jiang, G. Lubineau, T. Ng, B.S. Ooi, H-Yu Liao, C. Shen, L. Chen, J.Y Zhu, Highly transparent, low-haze, hybrid cellulose nanopaper as electrodes for flexible electronics, Nanoscale 8 (2016) 12294–12306. https://doi.org/10.1039/c6nr02245f.

[32] H. Fukuzumi, T. Saito, T. Iwata, Y. Kumamoto, A. Isogai, Transparent and high gas barrier films of cellulose nanofibers prepared by TEMPO-mediated oxidation, Biomacromolecules 10 (1) (2009) 162–165. https://doi.org/10.1021/bm801065u.

[33] H. Koga, M. Nogi, N. Komoda, T.T. Nge, T. Sugahara, K. Suganuma, Uniformly connected conductive networks on cellulose nanofiber paper for transparent paper electronics, NPG Asia Mater. 6 (3) (2014) 1–8. https://doi.org/10.1038/am.2014.9.

[34] U. Henniges, S. Veigel, E.-M. Lems, A. Bauer, J. Keckes, S. Pinkl, W. Gindl-altmutter, Microfibrillated cellulose and cellulose nanopaper from miscanthus biogas production residue, Cellulose (21) (2014) 1601–1610. https://doi.org/10.1007/s10570-014-0232-4.

[35] M. Nogi, N. Komoda, K. Otsuka, K. Suganuma, Foldable nanopaper antennas for origami electronics, Nanoscale 5 (2013) 4395–4399. https://doi.org/10.1039/c3nr00231d.

[36] H. Sehaqui, A. Liu, Qi Zhou, L.A. Berglund, Fast preparation procedure for large, flat cellulose and cellulose/inorganic nanopaper structures, Biomacromolecules 11 (2010) 2195–2198. https://doi.org/10.1021/bm100490s.

[37] S.R.D. Petroudy, E.R. Garmaroody, H. Rudi, Oriented cellulose nanopaper (OCNP) based on bagasse cellulose nanofibrils, Carbohydr. Polym. 157 (2017) 1883–1891. https://doi.org/10.1016/j.carbpol.2016.11.074.

[38] W. Yang, L. Jiao, D. Min, H. Dai, Effects of preparation approaches on optical properties of self-assembled cellulose nanopapers, RSC Adv. 7 (2017) 10463–10468. https://doi.org/10.1039/C6RA27529J.

[39] W. Kang, C. Yan, Ce Yao Foo, P.S. Lee, Foldable electrochromics enabled by nanopaper transfer method, Adv. Funct. Mater. 25 (27) (2015) 4203–4210. https://doi.org/10.1002/adfm.201500527.

[40] H. Zhu, Z. Fang, C. Preston, Y. Li, L. Hu, Transparent paper: fabrications, properties, and device applications, Energy Environ. Sci. 7 (1) (2014) 269–287. https://doi.org/10.1039/c3ee43024c.

[41] T. Kasuga, N. Isobe, H. Yagyu, H. Koga, M. Nogi, Clearly transparent nanopaper from highly concentrated cellulose nanofiber dispersion using dilution and sonication, Nanomaterials 8 (2) (2018) 104. https://doi.org/10.3390/nano8020104.

[42] A. Operamolla, Recent advances on renewable and biodegradable cellulose nanopaper substrates for transparent light-harvesting devices: interaction with humid environment, Int. J. Photoenergy (2019) 2019. https://doi.org/10.1155/2019/3057929.

[43] M. Wakabayashi, S. Fujisawa, T. Saito, A. Isogai, Nanocellulose film properties tunable by controlling degree of fibrillation of TEMPO-oxidized cellulose, Front. Chem. 8 (2020) 1–9. February. https://doi.org/10.3389/fchem.2020.00037.

[44] C.N. Wu, K.C. Cheng, Strong, thermal-stable, flexible, and transparent films by self-assembled TEMPO-oxidized bacterial cellulose nanofibers, Cellulose 24 (1) (2017) 269–283. https://doi.org/10.1007/s10570-016-1114-8.

[45] C. Legnani, H.S. Barud, J.M.A. Caiut, V.L. Calil, I.O. Maciel, W.G. Quirino, S.J.L. Ribeiro, M. Cremona, Transparent bacterial cellulose nanocomposites used as substrate for organic light-emitting diodes, J. Mater. Sci.: Mater. Electron. 30 (18) (2019) 16718–16723. https://doi.org/10.1007/s10854-019-00979-w.

[46] G. Chinga-Carrasco, N. Averianova, O. Kondalenko, M. Garaeva, V. Petrov, B. Leinsvang, T. Karlsen, The effect of residual fibres on the micro-topography of cellulose nanopaper, Micron 56 (2014) 80–84. https://doi.org/10.1016/j.micron.2013.09.002.

[47] A. Ferguson, U. Khan, M. Walsh, K.Y. Lee, A. Bismarck, MS.P. Shaffer, JN. Coleman, SD. Bergin, Understanding the dispersion and assembly of bacterial cellulose in organic solvents, Biomacromolecules 17 (5) (2016) 1845–1853. https://doi.org/10.1021/acs.biomac.6b00278.

[48] G.P Szekeres, Z. Nemeth, K. Schrantz, K. Hernadi, T. Graule, Insights into pore size control in cellulose nanopapers through modeling and experiments, J. Nanosci. Nanotechnol. 18 (4) (2017) 3000–3005. https://doi.org/10.1166/jnn.2018.14536.

[49] M. Henriksson, L.A. Berglund, P. Isaksson, T. Lindstro, Cellulose nanopaper structures of high toughness, Biomacromolecules 9 (6) (2008) 1579–1585. https://doi.org/10.1021/bm800038n.

[50] Q. Meng, T.J. Wang, Mechanics of strong and tough cellulose nanopaper, Appl. Mech. Rev. 71 (4) (2019) 1–30. https://doi.org/10.1115/1.4044018.

[51] M.S. Islam, L. Chen, J. Sisler, K.C. Tam, Cellulose nanocrystal (CNC)-inorganic hybrid systems: synthesis, properties and applications, J. Mater. Chem. B 6 (6) (2018) 864–883. https://doi.org/10.1039/c7tb03016a.

[52] S. Tanpichai, F. Quero, M. Nogi, H. Yano, R.J. Young, T. Lindström, W.W. Sampson, S.J. Eichhorn, Effective Young's modulus of bacterial and microfibrillated cellulose fibrils in fibrous networks, Biomacromolecules 13 (5) (2012) 1340–1349. https://doi.org/10.1021/bm300042t.

[53] L. Zhai, H.C. Kim, D. Kim, R.M. Muthoka, J. Kim, Young's Moduli of Cellulose Nanofibers Measured by Atomic Force Microscopy, Proc. Nano-, Bio-, Info-Tech Sensors, and 3D Systems II, 2018. https://doi.org/10.1117/12.2296836.

[54] A.J. Benítez, J. Torres-rendon, M. Poutanen, A. Walther, Humidity and multiscale structure govern mechanical properties and deformation modes in films of native cellulose nano fibrils, Biomacromolecules 14 (2013) 4497−4506. https://doi.org/10.1021/bm401451m.

[55] H. Fukuzumi, T. Saito, A. Isogai, Influence of TEMPO-oxidized cellulose nanofibril length on film properties, Carbohy. Polym. 93 (1) (2012) 172–177. https://doi.org/10.1016/j.carbpol.2012.04.069.

[56] Z. Fang, Bo Li, Yu Liu, J. Zhu, G. Li, G. Hou, J. Zhou, X. Qiu, Critical role of degree of polymerization of cellulose in super-strong nanocellulose films, Matter 2 (4) (2020) 1000–1014. https://doi.org/10.1016/j.matt.2020.01.016.

[57] F. Ansari, L.A. Berglund, Tensile Properties of Wood Cellulose Nanopaper and Nanocomposite Films, in: D. Puglia, E. Fortunati, J.M. Kenny (Eds.), Multifunctional Polymeric Nanocomposites Based on Cellulosic Reinforcements, Elsevier Inc, Amsterdam, 2016. https://doi.org/10.1016/B978-0-323-44248-0.00004-3.

[58] K. Syverud, Ø. Weiby Gregersen, G. Chinga-Carrasco, Ø. Eriksen, The Influence of Microfibrillated Cellulose, MFC, on Paper Strength and Surface Properties, in: S.J. I'Anson (Ed.), Advances in Pulp and Paper Research, Oxford, Trans. of the XIVth Fund. Res. Symp. Oxford, 2009, pp. 899–930. DOI:10.15376/frc.2009.2.899.

[59] W. Gindl-Altmutter, S. Veigel, M. Obersriebnig, C. Tippelreither, J. Keckes, High-modulus oriented cellulose nanopaper, ACS Symp. Ser. 1107 (2012) 3–16. https://doi.org/10.1021/bk-2012-1107.ch001.

[60] H. Sehaqui, N.E. Mushi, S. Morimune, M. Salajkova, T. Nishino, LA. Berglund, Cellulose nanofiber orientation in nanopaper and nanocomposites by cold drawing, ACS Appl. Mater. Interfaces 4 (2) (2012) 1043–1049. https://doi.org/10.1021/am2016766.

[61] X. Li, X. Zhang, S. Yao, H. Chang, Y. Wang, Z. Zhang, UV-blocking, transparent and hazy cellulose nanopaper with superior strength based on varied components of poplar mechanical pulp, Cellulose 27 (11) (2020) 6563–6576. https://doi.org/10.1007/s10570-020-03236-0.

[62] J.D. Yuen, L.C. Shriver-Lake, S.A. Walper, D. Zabetakis, J.C. Breger, D.A. Stenger, Microbial nanocellulose printed circuit boards for medical sensing, Sensors (Switzerland) (7) (2020) 20. https://doi.org/10.3390/s20072047.

[63] F. Ansari, S. Galland, M. Johansson, C.J.G. Plummer, LA. Berglund, Cellulose nanofiber network for moisture stable, strong and ductile biocomposites and increased epoxy curing rate, Compos. A: Appl. Sci. Manufact. 63 (2014) 35–44. https://doi.org/10.1016/j.compositesa.2014.03.017.

[64] M. Wu, P. Sukyai, D. Lv, F. Zhang, P. Wang, C. Liu, B. Li, Water and humidity-induced shape memory cellulose nanopaper with quick response, excellent wet strength and folding resistance, Chem. Eng. J. 392 (2020) 123673. https://doi.org/10.1016/j.cej.2019.123673.

[65] X. Guo, Y. Wu, X. Xie, Water vapor sorption properties of cellulose nanocrystals and nanofibers using dynamic vapor sorption apparatus, Sci. Rep. (2017) 14207. https://doi.org/10.1038/s41598-017-14664-7.

[66] R. Bollstrom, F. Pettersson, P. Dolietis, J. Preston, R. Osterbacka, M. Toivakka, Impact of humidity on functionality of on-paper printed electronics, Nanotechnology 25 (7) (2014). doi:10.1088/0957-4484/25/9/094003.

[67] L. Gao, L. Chao, M. Hou, J. Liang, Y. Chen, H.-.D. Yu, W. Huang, Flexible, Transparent Nanocellulose Paper-Based Perovskite Solar Cells, Flex. Electron. 3 (1) (2019) 4. https://doi.org/10.1038/s41528-019-0048-2.

[68] J. Sethi, M. Farooq, S. Sain, M. Sain, J.A. Sirviö, M. Illikainen, K. Oksman, Water resistant nanopapers prepared by lactic acid modified cellulose nanofibers, Cellulose 25 (1) (2018) 259–268. https://doi.org/10.1007/s10570-017-1540-2.

[69] H. Abushammala, J. Mao, A review of the surface modification of cellulose, Molecules 24 (2019) 1–18. https://doi.org/10.3390/molecules24152782.

[70] K.M. Chin, S.S. Ting, H.L. Ong, M.f. Omar, Surface functionalized nanocellulose as a veritable inclusionary material in contemporary bioinspired applications: a review, J. Appl. Polym. Sci. 135 (13) (2018) 46065. https://doi.org/10.1002/app.46065.

[71] H. Sehaqui, T. Zimmermann, P. Tingaut, Hydrophobic cellulose nanopaper through a mild esterification procedure, Cellulose 21 (2014) 367–382. https://doi.org/10.1007/s10570-013-0110-5.

[72] Y. Qi, H. Zhang, D. Xu, Z. He, X. Pan, S. Gui, X. Dai, J. Fan, X. Dong, Y. Li, Screening of nanocellulose from different biomass resources and its integration for hydrophobic transparent nanopaper, Molecules 25 (2020) 1–9. https://doi.org/10.3390/molecules25010227.

[73] G. del Campo, M. Mar, M. Darder, P. Aranda, M. Akkari, Y. Huttel, A. Mayoral, J. Bettini, E. Ruiz-Hitzky, Functional hybrid nanopaper by assembling nanofibers of cellulose and sepiolite, Adv. Funct. Mater. 28 (27) (2018) 1–13. https://doi.org/10.1002/adfm.201703048.

[74] S. Chen, Y. Song, F. Xu, Highly Transparent and hazy cellulose nanopaper simultaneously with a self-cleaning superhydrophobic surface, ACS Sustain. Chem. Eng. 6 (2018) 5173–5181. https://doi.org/10.1021/acssuschemeng.7b04814.

[75] A. Operamolla, S. Casalini, D. Console, L. Capodieci, F. Di Benedetto, G.V. Bianco, F. Babudri, Tailoring water stability of cellulose nanopaper by surface functionalization, Soft Matter 14 (36) (2018) 7390–7400. https://doi.org/10.1039/c8sm00433a.

[76] E. Lizundia, A. Urruchi, J.L. Vilas, L.M. León, Increased functional properties and thermal stability of flexible cellulose nanocrystal/ZnO films, Carbohydr. Polym. 136 (20) (2016) 250–258. https://doi.org/10.1016/j.carbpol.2015.09.041.

[77] C. Tan, J. Peng, W. Lin, Y. Xing, K. Xu, J. Wu, M. Chen, Role of surface modification and mechanical orientation on property enhancement of cellulose nanocrystals/polymer nanocomposites, Eur. Polym. J. 62 (2015) 186–197. https://doi.org/10.1016/j.eurpolymj.2014.11.033.

[78] H. Abral, N. Fajri, M. Mahardika, D. Handayani, E. Sugiarti, H.-. Kim, A simple strategy in enhancing moisture and thermal resistance and tensile properties of disintegrated bacterial cellulose nanopaper, J. Mater. Res. Technol. 9 (4) (2020) 8754–8765.

[79] K.S. Kontturi, K. Biegaj, A. Mautner, R.T. Woodward, B.P. Wilson, L.-. Johansson, K.-. Lee, J.Y.Y. Heng, A. Bismarck, E. Kontturi, Noncovalent surface modification of cellulose nanopapers by adsorption of polymers from aprotic solvents, Langmuir 33 (2017) 5707–5712. https://doi.org/10.1021/acs.langmuir.7b01236.

[80] Y. Su, S. Yuan, S. Cao, M. Miao, L. Shi, X. Feng, Assembling polymeric silver nanowires for transparent conductive cellulose nanopaper, J. Mater. Chem. C 7 (45) (2019) 14123–14129. https://doi.org/10.1039/c9tc03913a.

[81] Y. Su, Y. Zhao, H. Zhang, X. Feng, L. Shi, J. Fang, Polydopamine functionalized transparent conductive cellulose nanopaper with long-term durability, J. Mater. Chem. C 5 (3) (2017) 573–581. https://doi.org/10.1039/c6tc04928a.

[82] D. Kim, Y. Ko, G. Kwon, U.J. Kim, J. You, Micropatterning silver nanowire networks on cellulose nanopaper for transparent paper electronics, ACS Appl. Mater. Interfaces 10 (44) (2018) 38517–38525. https://doi.org/10.1021/acsami.8b15230.

[83] H. Zhang, L. Shi, X. Feng, Use of chitosan to reinforce transparent conductive cellulose nanopaper, J. Mater. Chem. C 6 (2) (2018) 242–248. https://doi.org/10.1039/c7tc03980h.

[84] J. Zhong, H. Zhu, Q. Zhong, J. Dai, W. Li, S.-. Jang, Y. Yao, et al., Self-powered human-interactive transparent nanopaper systems, ACS Nano 9 (7) (2015) 7399–7406.

[85] K. Zhang, G. Chen, R'Ai Li, K. Zhao, J. Shen, J. Tian, M. He, Facile preparation of highly transparent conducting nanopaper with electrical robustness, ACS Sustain. Chem. Eng. 8 (13) (2020) 5132–5139. https://doi.org/10.1021/acssuschemeng.9b07266.

[86] Xi Wang, K. Gao, Z. Shao, X. Peng, X. Wu, F. Wang, Layer-by-layer assembled hybrid multilayer thin film electrodes based on transparent cellulose nano fibers paper for flexible supercapacitors applications, J. Power Sources 249 (2014) 148–155. https://doi.org/10.1016/j.jpowsour.2013.09.130.

[87] V. Kantola, J. Kulovesi, L. Lahti, R. Lin, M. Zavodchikova, E. Coatanéa, Printed electronics, now and future, in: S. Ylönen, Y. Neuvo (Eds.), Bit Bang - Rays to the Future, Helsinki University Print, Helsinki, Finland, 2009, pp. 63–102.

[88] A. Penttilä, J. Sievänen, K. Torvinen, O. Kimmo, J.A. Ketoja, Filler-nanocellulose substrate for printed electronics : experiments and model approach to structure and conductivity, Cellulose 20 (2013) 1413–1424. https://doi.org/10.1007/s10570-013-9883-9.

[89] A. Nag, A.I. Zia, A. Babu, S.C. Mukhopadhyay, Printed electronics : present and future opportunities, Proc. Ninth International Conference on Sensing Technology, 2015. 380–389. https://doi.org/10.1109/ICSensT.2015.7438427.

[90] K. Suganuma, Introduction to Printed Electronics, Springer, New York Heidelberg Dordrecht London, 2014. https://doi.org/10.1007/978-1-4614-9625-0.

[91] A. Pammo, H. Christophliemk, J. Keskinen, T. Björkqvist, S. Siljander, M. Mäntysalo, Nanocellulose films as substrates for printed electronics*, Proc. 2019 International Conference on Manipulation, Automation and Robotics at Small Scales (MARSS), IEEE, 2019, pp. 1–6. https://doi.org/10.1109/MARSS.2019.8860931.

[92] S. Agate, M. Joyce, L. Lucia, L. Pal, Cellulose and nanocellulose-based flexible-hybrid printed electronics and conductive composites – a review, Carbohyd. Polym. 198 (June 2018) 249–260. https://doi.org/10.1016/j.carbpol.2018.06.045.

[93] A. Tang, Y. Liu, Q. Wang, R. Chen, W. Liu, Z. Fang, L. Wang, A new photoelectric ink based on nanocellulose/CdS quantum dots for screen-printing, Carbohydr. Polym. 148 (2016) 29–35. https://doi.org/10.1016/j.carbpol.2016.04.034.

[94] G. Chinga-Carrasco, D. Tobjörk, R. Österbacka, Inkjet-printed silver nanoparticles on nano-engineered cellulose films for electrically conducting structures and organic transistors: concept and challenges, J. Nanopart. Res. 14 (11) (2012) 1213. https://doi.org/10.1007/s11051-012-1213-x.

[95] Y. Ooi, I. Hanasaki, D. Mizumura, Yu Matsuda, Suppressing the coffee-ring effect of colloidal droplets by dispersed cellulose nanofibers, Sci. Technol. Adv. Mater. 18 (1) (2017) 316–324. https://doi.org/10.1080/14686996.2017.1314776.

[96] R.J. Moon, A. Martini, J. Nairn, J. Simonsen, J. Youngblood, Cellulose nanomaterials review: structure, properties and nanocomposites, Chem. Soc. Rev 40 (2011) 3941–3994. https://doi.org/10.1039/c0cs00108b.

[97] F. Martoïa, P.J.J. Dumont, L. Orgeas, M.N. Belgacem, J.-.L. Putaux, Soft matter suspensions of cellulose nanofibrils under steady state shear flow, Soft Matter 12 (2016) 1721–1735. https://doi.org/10.1039/C5SM02310F.

[98] L. Mendoza, W. Batchelor, R.F. Tabor, G. Garnier, Gelation mechanism of cellulose nanofibre gels: a colloids and interfacial perspective, J. Colloid Interface Sci. 509 (2018) 39–46. https://doi.org/10.1016/j.jcis.2017.08.101.

[99] A.B. Fall, S.B. Lindstr, O. Sundman, L. Odberg, W. Lars, Colloidal stability of aqueous nanofibrillated cellulose dispersions, Langmuir 27 (2011) 11332–11338. https://doi.org/10.1021/la201947x.

[100] L. Mendoza, T. Gunawardhana, W. Batchelor, G. Garnier, Effects of fibre dimension and charge density on nanocellulose gels, J. Colloid Interface Sci. 525 (2018) 119–125. https://doi.org/10.1016/j.jcis.2018.04.077.

[101] MM. Hamedi, A. Hajian, AB. Fall, K. Hkansson, M. Salajkova, F. Lundell, L. Wgberg, LA. Berglund, Highly conducting, strong nanocomposites based on nanocellulose-assisted aqueous dispersions of single-wall carbon nanotubes, ACS Nano 8 (3) (2014) 2467–2476. https://doi.org/10.1021/nn4060368.

[102] I. Hanasaki, Y. Isono, Detection of diffusion anisotropy due to particle asymmetry from single-particle tracking of Brownian MOTION BY THE LARGE-DEVIATION PRINCIPLE, Phys. Rev. E 85 (5) (2012) 051134.

[103] X. Mu, DG. Gray, Droplets of cellulose nanocrystal suspensions on drying give iridescent 3-D 'coffee-stain' rings, Cellulose 22 (2) (2015) 1103–1107. https://doi.org/10.1007/s10570-015-0569-3.

[104] S. Ummartyotin, J. Juntaro, M. Sain, H. Manuspiya, Development of transparent bacterial cellulose nanocomposite film as substrate for flexible organic light emitting diode (OLED) display, Ind. Crops Prod 35 (1) (2012) 92–97. https://doi.org/10.1016/j.indcrop.2011.06.025.

[105] Y. Okahisa, A. Yoshida, S. Miyaguchi, H. Yano, Optically transparent wood-cellulose nanocomposite as a base substrate for flexible organic light-emitting diode displays, Compos. Sci. Technol. 69 (11–12) (2009) 1958–1961. https://doi.org/10.1016/j.compscitech.2009.04.017.

[106] W. Wu, NG. Tassi, H. Zhu, Z. Fang, L. Hu, Nanocellulose-based translucent diffuser for optoelectronic device applications with dramatic improvement of light coupling, ACS Appl. Mater. Interfaces 7 (48) (2015) 26860–26864. https://doi.org/10.1021/acsami.5b09249.

[107] J. Tao, R. Wang, H. Yu, L. Chen, D. Fang, Y. Tian, J. Xie, et al., Highly transparent, highly thermally stable nanocellulose/polymer hybrid substrates for flexible OLED devices, ACS Appl. Mater. Interfaces 12 (8) (2020) 9701–9709. https://doi.org/10.1021/acsami.0c01048.

[108] E. Najafabadi, Y. Zhou, K.A Knauer, C. Fuentes-hernandez, B. Kippelen, Efficient organic light-emitting diodes fabricated on cellulose nanocrystal substrates, Appl. Phys. Lett. 105 (2014) 063305. https://doi.org/10.1063/1.4891046.

[109] H. Zhu, Z. Xiao, D. Liu, Y. Li, NJ. Weadock, Z. Fang, J. Huang, L. Hu, Biodegradable transparent substrates for flexible organic-light-emitting diodes, Energy Environ. Sci. 6 (7) (2013) 2105–2111. https://doi.org/10.1039/c3ee40492g.

[110] S. Purandare, EF. Gomez, AJ. Steckl, High brightness phosphorescent organic light emitting diodes on transparent and flexible cellulose films, Nanotechnology (9) (2014) 25. https://doi.org/10.1088/0957-4484/25/9/094012.

[111] J.H. Lee, S. Kang, N.-. Park, J.-. Shin, C. Woong, J. Lee, S.-. Ahn, S.-. Kang, J. Moon, Porous cellulose paper as a light out coupling medium for organic light-emitting diodes, J. Inf. Disp. 19 (4) (2018) 1–7. https://doi.org/10.1080/15980316.2018.1527260.

[112] H. Zhu, L. Hu, J. Cumings, J. Huang, Y. Chen, Highly transparent and flexible nanopaper transistor, ACS Nano 7 (3) (2013) 2106–2113. https://doi.org/10.1021/nn304407r.

[113] Y. Fujisaki, H. Koga, Y. Nakajima, M. Nakata, H. Tsuji, Transparent nanopaper-based flexible organic thin-film transistor array, Mater. Views (2014) 1657–1663. https://doi.org/10.1002/adfm.201303024.

[114] T. Hassinen, A. Alastalo, K. Eiroma, T. Tenhunen, V. Kunnari, T. Kaljunen, U. Forsström, T. Tammelin, All-printed transistors on nano cellulose substrate, MRS Adv. 1 (2015) 645–650. https://doi.org/10.1557/adv.2015.31.

[115] S. Dai, Y. Chu, D. Liu, F. Cao, X. Wu, J. Zhou, B. Zhou, Y. Chen, J. Huang, Intrinsically ionic conductive cellulose nanopapers applied as all solid dielectrics for low voltage organic transistors, Nat. Commun. 9 (1) (2018) 1–10. https://doi.org/10.1038/s41467-018-05155-y.

[116] Qi Zhang, C. Chen, W. Chen, G. Pastel, X. Guo, S. Liu, Q. Wang, et al., Nanocellulose-enabled, all-nanofiber, high-performance supercapacitor, ACS Appl. Mater. Interfaces 11 (2019) 5919–5927. https://doi.org/10.1021/acsami.8b17414.

[117] D. Gaspar, S.N. Fernandes, A.G. de Oliveira, J.G. Fernandes, P. Grey, R.V. Pontes, L. Pereora, R. Martins, M.H. Godinho, E. Fortunato, Nanocrystalline cellulose applied simultaneously as the gate dielectric and the substrate in flexible field effect transistors, Nanotechnology 25, (2014) 094008. https://doi.org/10.1088/0957-4484/25/9/094008.

[118] C.-y. Wang, C. Fuentes-hernandez, J.-c. Liu, A. Dindar, S. Choi, J.P. Youngblood, R.J. Moon, B. Kippelen, Stable low-voltage operation top-gate organic field-effect transistors on cellulose nanocrystal substrates., Appl. Mater. Interfaces 7 (8) (2015) 4804–4808. https://doi.org/10.1021/am508723a.

[119] J.D. Yuen, S.A. Walper, B.J. Melde, M.A. Daniele, D.A. Stenger, Electrolyte-sensing transistor decals enabled by ultrathin microbial nanocellulose, Sci. Rep. 7 (2017) 1–9. https://doi.org/10.1038/srep40867.

[120] S. Li, D. Huang, B. Zhang, X. Xu, M. Wang, G. Yang, Y. Shen, Flexible supercapacitors based on bacterial cellulose paper electrodes, Adv. Energy Mater. 4 (10) (2014) 1301655. https://doi.org/10.1002/aenm.201301655.

[121] K.-. Liu, Q. Jiang, C. Kacica, H.G. Derami, P. Biswas, S. Singamaneni, Flexible solid-state supercapacitor based on tin oxide/reduced graphene oxide/bacterial nanocellulose, RSC Adv. 8 (2018) 31296–31302. https://doi.org/10.1039/C8RA05270K.

[122] T. Hassinen, A. Alastalo, K. Eiroma, T.M. Tenhunen, V. Kunnari, T. Kaljunen, U. Forsström, T. Tammelin, All-printed transistors on nano cellulose substrate, MRS Adv. 1 (10) (2016) 645–650. https://doi.org/10.1557/adv.2015.31.

[123] H. Zhu, B. Baby Narakathu, Z. Fang, A.T. Aijazi, M. Joyce, M. Atashbar, L. Hu, A gravure printed antenna on shape-stable transparent nanopaper, Nanoscale 6 (15) (2014) 9110–9115. https://doi.org/10.1039/c4nr02036g.

[124] T. Inui, H. Koga, M. Nogi, N. Komoda, K. Suganuma, A miniaturized flexible antenna printed on a high dielectric constant nanopaper composite, Adv. Mater. 27 (6) (2014) 1112–1116. https://doi.org/10.1002/adma.201404555.

CHAPTER 13

Nanocellulose in paper and wood industry

Mansi Chugh[a], Tulsi Chandak[a], Shruti Jha[b], Deepak Rawtani[c]
[a]School of Engineering and Technology, National Forensic Sciences University (Ministry of Home affairs, GOI), Gandhinagar, Gujarat, India
[b]Sardar Vallabhbhai National Institute of Technology, Surat, Gujarat, India
[c]School of Pharmacy, National Forensic Sciences University (Ministry of Home affairs, GOI), Gandhinagar, Gujarat, India

13.1 Introduction

Industrialization in current world is contributing as one of the important parts of the economic growth. There is a great impact on the overall growth and development of wood-based trades, which reflects emerging trend in the world. Its high biodegradable property being a major reason for its worldwide commercialization as the world is aiming toward no plastic era. Wood industry basically comprises traditional forest products that exist in society since ancient times like timber, furniture, paper, and other functional material (wood coatings, wood preservatives, wood composites). Broadly wood industry is classified into three types: sawn wood, wood-based panels, and paper boards of which worldwide production had been recorded by forestry statistics 2018 from year 1990 to 2016 as mentioned in Figs. 13.1, 13.2, and 13.3 [1–3].

However, the manufacturing of main wood-based products, that is paper, have faced numerous hurdles such as (1) good quality consumer expectation to meet overall quality properties (physical, mechanical, printing) of paper, (2) limitations toward production cost, (3) increasing recycling rate leads to damage of recycled fibers [4]. To overcome these obstacles, nanocellulose (NC) has been introduced because of its characteristic properties such as high surface area, unique optical properties, lightweight, stiffness, high strength, etc. This results in getting the spotlight in paper industry and gave a new wave in research and development in growing to new heights. Thus, providing importance in forming a new efficient quality product [4,5].

13.2 Application in wood and paper industry

Nowadays, there is an increasing interest in the production of cellulose at a nanoscale (known as NC). In recent years, its involvement has been observed in the paper and wood industry. Wide applications of NC are based on extracted NC from wood and nonwood NCs [6].

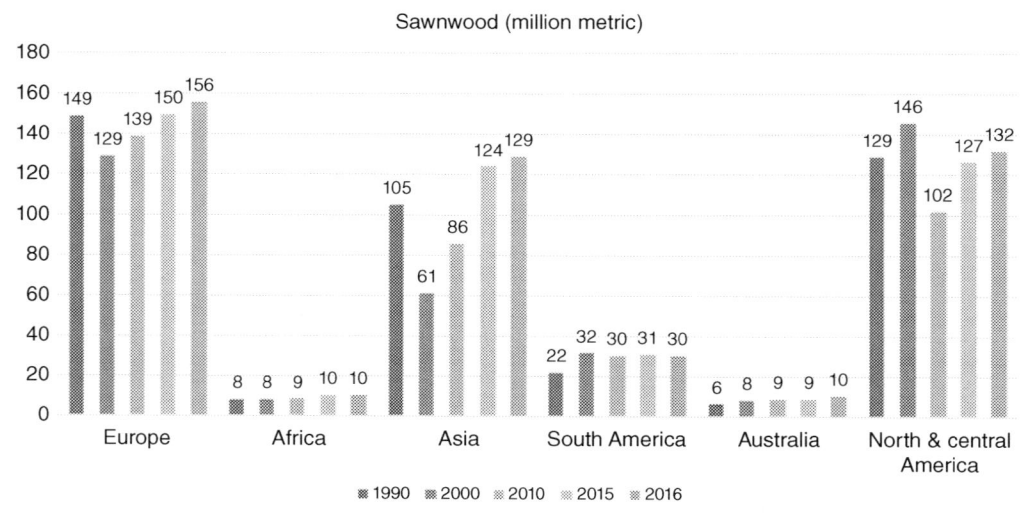

Fig. 13.1 Worldwide production of sawn wood (source - Production of wood products by region, 1990–2016 (Forestry Statistics 2018 Chapter 9: International Forestry).

- *Extracted wood NC*

NC is usually extracted by wood cellulose through various methods such as chemical method (strong acid hydrolysis), thermochemical (pyrolysis, liquefication, and gasification), enzymatic hydrolysis. Among this enzymatic hydrolysis is the most preferred methodology because it eases the production of nanostructure with high stability [2,5,7,8]. By applying further processes on extracted NC, variety of other NCs of different dimensions can be obtained.

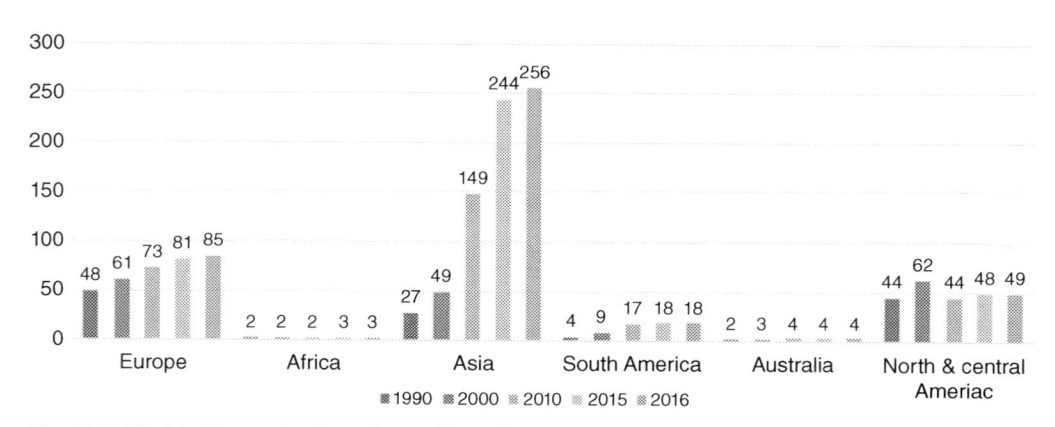

Fig. 13.2 Worldwide production of wood-based panels (source - Production of wood products by region, 1990–2016 (Forestry Statistics 2018 Chapter 9: International Forestry).

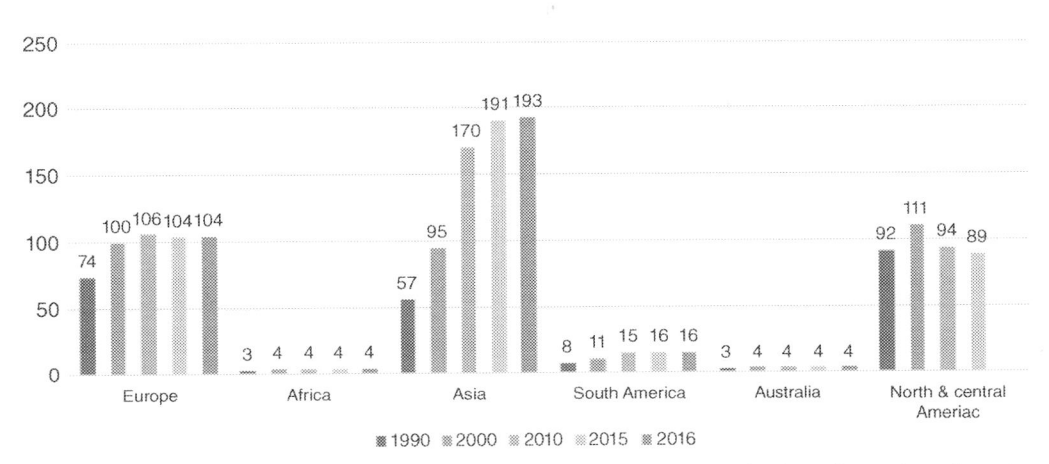

Fig. 13.3 Worldwide production of paper and paperboard (source - Production of wood products by region, 1990–2016 (Forestry Statistics 2018 Chapter 9: International Forestry).

13.3 Papermaking

Paper, a thin material, obtained by basic raw materials that is, two fiber sources—primarily as cellulose fibers, generally wood, agricultural residues, recycled paper; and secondary from nonwood-based fibers such as jute, flax, bagasse (sugarcane fibers), bamboo, cereal straw, reeds, esparto grass, and sisal; water; coal; and Chemicals (chlorine, limestone sulfur). Papermaking is a two-step process in which cellulose pulp is obtained first from basic raw material and then flexible sheets are produced by pressing moist fibers followed by drying. The papermaking process starts with chipping of debarked (elimination of bark) logs followed by pulping obtained by various methods (mechanical, chemical, and sulfate), and bleaching is carried out for increasing the brightness of the pulp. The bleached pulp is then diluted to form a thin fiber mixture, pressure is applied to increase material density which leads to fiber formation and paper is obtained by introducing fiber in paper feed machine, then the paper is known as jumbo rolls when it is reeled up into large reels before to trimming, finishing, and sale (Fig. 13.4) [9–13].

13.4 Application of NC in papermaking

NC in their various forms is utilized in different steps in papermaking process due to its characteristic of high mechanical strength (increase of interfiber H-bonds), biocompatibility, biodegradability, adjustable surface chemistry, good biocompatibility, adjustable surface chemistry, high specific area, biodegradability, modifiable physical and chemical

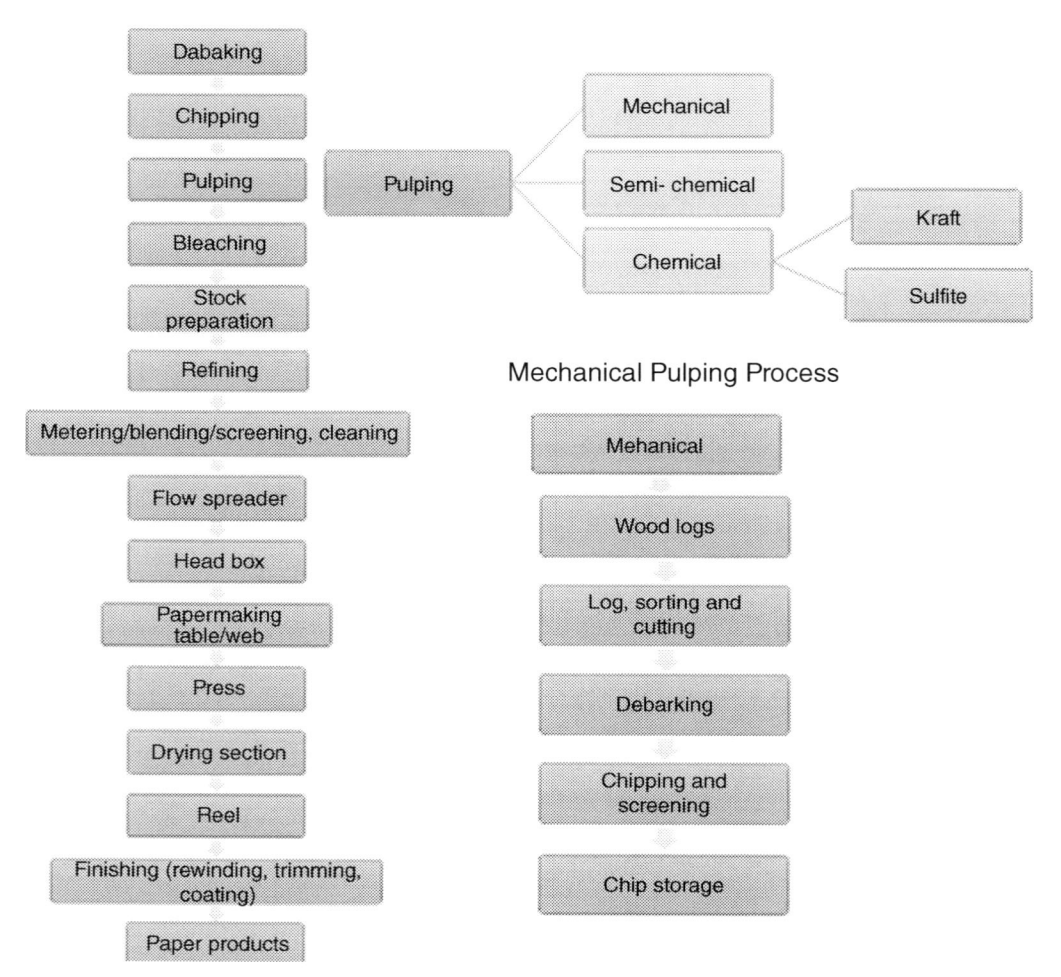

Fig. 13.4 Flow sheet describing outline of steps in manufacturing of paper in paper industry (Source - [14]).

properties. This implies every step in papermaking affects, directly or indirectly, the paper's surface properties. Factors including sizing treatment, fine and filler content, fiber orientation, coating, and calendaring nip pressure can affect brightness, opacity, smoothness, and printability of paper [15–17].

13.5 Pulping

Pulp is obtained by various methods (mechanical, chemical, and sulfate) in process of paper manufacturing. Different types of methods used for pulping give different paper quality including high optical transmittance, superior smoothness, etc. NC in the form of cellulose nanofibrils (CNFs) and cellulose nanocrystals (CNCs) can be possibly added at two steps either during the start of pulp slurries in papermaking process or at the end as a paper coating on the paper. It is established that they are widely used when mixed with fibers, cationic polymers, fillers, and other additives for its wide range of advantages in properties of paper. It has been introduced in papermaking slurries for making strong bonds with the pulp fiber through hydrogen bonding. Various properties of paper have been observed after adding NC in different types of pulping process [9,18–21].

1. *Mechanical pulping process*

Two methods can be used under process of mechanical pulping first by mechanical grinders and refiner during mechanical pulping. Mechanical grinder is used for reduction of woodchips to fibers and pulp is obtained by abrasive action on a rotating stone. Mechanical refiner is recently opted for pulping, in this method wood chips pass through two rotating discs and pulp is formed. The former method leads to very less amount of lignin, which directly impacts quality of paper with respect to chemical pulp. Mechanical pulps are generally designed for tissues, news print, cardboard, paper towels [18,20].

Advantage of CNF added to pulp suspension by mechanical method results in highest tensile strength and rigidity in comparison to any of the methods. However, drastic reduction in porosity takes place [18]. Tensile strength is affected when CNF is added as it depends on quantity of CNF, fibrillation intensity, degree of refining pulp, methods and ways of addition of retention agents. As stated, the quantity of CNF is proportional to high tensile strength that is also proportional to degree of fibrillation, this means more the amount of CNF, higher the tensile strength and higher the degree of fibrillation [22–24].

2. *Thermo-mechanical pulping (TMP) process*

Mechanical disintegration takes place under controlled high pressure of stem when heated with woodchips, which leads to the formation of thermomechanical pulp [18,20].

Advantages of CNF added to pulp suspension by the TMP method are as follows:

- Good resistance to air by increasing compactness and fibrillar bonding of fibers of TMP.
- Stronger adhesion between the clay fillers and fibers by evolving of good amount of contact particles and sites of bounding.
- Overall structure of pores is maintained in a suitable way of scattering.
- Enhance the paper quality promoting property including density, surface area, and surface charge [18].

As stated, pulp density increases by two different methods. One of the methods is lowering radius of water meniscus, which refers to increment in difference of pressure in phase of water and surroundings during sheet dewatering process. This results in compacting of fibers by pressing of sheets. Other method includes attachment of CNF layer on fiber surface, which directly impacts improving area of contact and number of hydrogen bonds. Therefore, by bond establishment contact between fibers becomes stronger [23].

In addition, paper obtained from the TMP provides high mechanical strength paper as compared with mechanical pulping process. Along with mechanical properties it reduces fine content, shows good fiber length as well as high energy consumption [18,24].

3. *Chemical pulping process*

The process is developed when raw material is cooked with kraft (sulfate) and sulfite process, it releases components of wood (lignin). Basically, this method is used for high-grade paper fabrication [18].

Advantages of CNF added to pulp suspension by chemical methods are as follows:

- Improvement in mechanical and physical properties followed by increasing in specific surface area.
- Enhancement in the overall paper strength including wet and dry strengths with higher fibrillation yield as well as higher density and lower porosity.
- Suppressing the mechanical beating by preventing the overall fibrillar structure offered by added nanofibers.
- Supports the reinforcement properties of paper including degree of fibrillation, charges on surface, and morphology.
- Addition with mineral fillers (calcium carbonate, kaolin, or talc) to overcome the filler retention and reinforcement property [18,23].

From the engineering point-of-view, pulp refining is also done to favor the improved printing properties and smoothen the sheets of paper by using CNF as additive. Refining results in modification in fibers toward softness, length shortening, elasticity, etc. Increase in surface area of fibers leads to strong fiber bonding in manufacturing of paper. However, a drastic decrease in tensile strength is observed in refined pulp compared to unrefined pulp due to its interaction with fibrillated surface [18].

It is found that with respect to mechanical pulping process, chemical pulp offers supreme mechanical and optical properties that is, brightness [18,22,23].

4. *Other pulping process*

- Unbleached pulping process: For advancement in unbleached eucalyptus pulp, CNF is added as a bulk additive with retention agents such as cationic starch and colloidal silica. By doing so it helps in increasing tensile strength. Moreover, adsorption of CNF is low with addition in large amounts of fibers on surface. Therefore, tensile strength fails as it depends on the bonded area and tensile strength of fiber [18,22,25].

5. *Deinked pulping process*

These pulps are used for recycling the old magazines and papers. CNF is majorly added to deinked pulp as it forms strong fiber network that is bonded by hydrogen bonding; therefore, it improves the property of final product by excluding mechanical beating process that includes high mechanical and tensile strength and denser, less porous, and stiffer paper. CNF utilization provides advantage in preventing fiber structure and increasing its shelf life. Besides, CNC also has impact on the process with or without combining it with cationic starch and cationic polyacrylamide in a deinked process, which mainly focus on increasing the retention of fillers and fines and properties of strengthening [18,19,26].

Effect of NC in agricultural produced pulp: It has been established that optional to wood as a source for paperboard and paper fabrication, agriculture and forest waste such as wheat, rice straw, sugarcane bagasse, and rapeseed is used. One of the main reasons for using CNF in pulping for papermaking process from agriculture residues is due to broad availability and low economic price of the residues.

The addition of CNF in agriculture waste (bagasse pulp) leads to improved quality of paper including overall strength. It also enhances wet and dry pulp strength by improving the nanofibers and fibers interaction. Hence, as a reinforcing agent, it is adopted as the most used method in papermaking industry of Egypt [18].

13.6 Wet-end chemistry

During the wet-end process, wet sheet is known to have adequate amount of water before passing it to the press section of paper machine. Performance and production efficiency plays a leading role in the wood and paper industry, which is completely achieved by wet-end chemicals.

These water-soluble additives are added to not only quicken water removal from foaming sheet but also increases wet web functional and structural sheet properties and hence lessen machine down time, drainage aids, and production loss. In addition, NC has been widely used as a chemical additive for wet end of papermaking process [16].

Hence, its few applications in papermaking industry have been recently adapted which include the void filling between cellulose strands, improving primary properties (mechanical strength, elasticity, wet strength and printability), and overcoming the problems related to high cost, drainability, and drying efficiency [16].

CNF is known to have high interfiber bonding, therefore its addition in wet end of paper machine provides good tensile strength to paper. Hence, it is replaced by the other fillers used in papermaking industry such as (precipitated calcium carbonate (PCC), grounded calcium carbonate (GCC), powder, kaolin, and TiO_2) and improves printability at a greater extent [16].

Impact on paper strength is affected by few characteristics that mainly include length of fiber, bond strength specificity, bonded area between fibers, formation of sheets, and stress concentration. Henceforth implementation of NC as wet strength additive for enhancing z-strength and stiffness in paper and papers products (paper towels, tissue paper, paperboard, and other paper grades) has been done. However, more than micro fibrillated cellulose (MFC), CNF is known to have strong strength property [27,28].

Moreover, in filler, CNF works as retention aid support strengthening by connecting the filler-induced void in the bonding domain between fibers. This also acts as high filler binder and strengthens the fiber-filler and filler-filler contacts [29].

This shows strong intermolecular hydrogen bonding between cellulosic fibrils and hence increases the fibers' bonded area [29].

In paper manufacturing, drainage is considered as a crucial parameter because of its effects directly and indirectly on efficiency and paper quality, respectively. This property is highly recommended for improving the paper visual properties, which is mainly affected by factors such as pH, ionic strength, fiber dimensions, type of polyelectrolyte dimensions [28–30].

CNF added as wet-end chemical reduces drainability, which depends on the method used for a papermaking pulp suspension process. This leads to no change in behavior of drainage [28].

Though use of CNF increases the drainage time as increase in time for formation of fiber mat, the bad drainage characteristics yield and followed by showing great impact on high specific energy of industrial usage [16,28]. The solution to this problem is using attaching of CNF with good retention agents such as cationic polymers (cationic polyacrylamide) on the surface of fibers [29].

While NC upgrades the elasticity of paper, the drainability diminishes in light of the fact that the pores of the wet paper web are filled by CNFs, bringing about a decrease in diameter of pores [16]. As industry majorly focused on economic status in terms of productivity, wet-end additives help in saving the cost, increasing efficiency which reflects overall productivity [6,30].

13.7 Paper coating

Paper coating develops uniformity on surface of sheet. Cellulose fiber of the outer boundaries is covered and then some materials such as binders, pigments, agents, etc. are added between the spaces in the sheets [16]. CNFs are majorly used as paper coating agent among all the other types of NC.

Since a few years ago, the use of CNF in paper industry has been established. CNF is used in coating formulation, applying CNF with different strategies (size press coating, foam coating, spray coating, and bar coating) has a great impact on quality of paper sheet [31–33].

CNF when applied with spray coating uniformly distributed, a very thin coating layer is obtained [31]. Foam coating has recently established for use of CNF for obtaining more uniformity on surface layer with less coating weight. Besides, the other methods, such as bar coating and size press coating, give high coating weight [31,34]. Various applications of NC as paper coating agent have been discussed in Fig. 13.5.

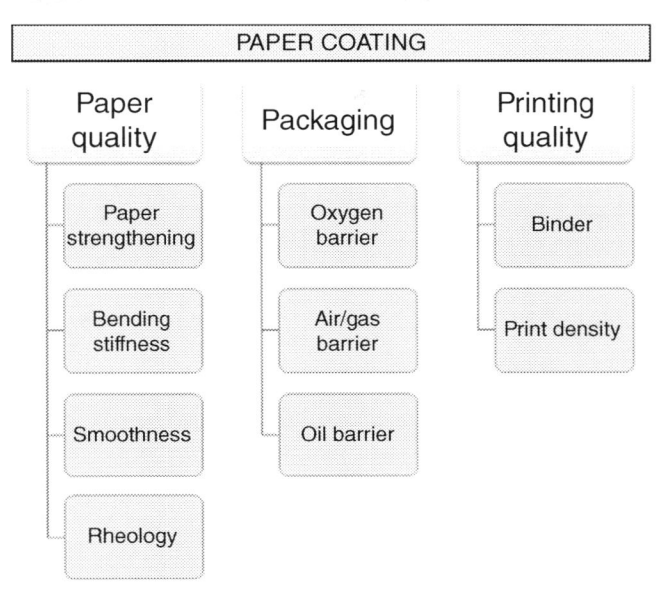

Fig. 13.5 Application of NC as a paper coating agent.

13.7.1 Paper quality

• Paper strengthening

Due to the weakest links in the structure of base paper, it gets crack and breaks by applying forces. This problem is overcome by covering the surface of paper with strong CNF layer to enhance the tensile index and strength property of paper [11,25,31].

CNF in coating composition contributes to strengthening of paper, which can be described broadly in two mechanisms (Fig. 13.6) [18].

In the above mechanism, hydrogen bonding formation is the main reason for strengthening the paper interfibrils bond [16,33].

Following are the advantages of CNF as paper coating composition compared to base paper [18].

• Favors fracture toughness of paper
• Reduce roughness and increase smoothness of coated paper

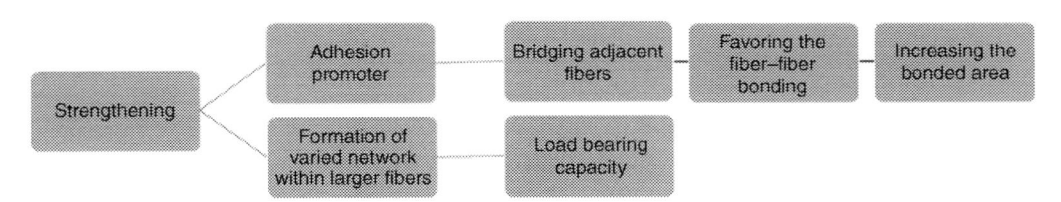

Fig. 13.6 Mechanisms of CNF in strengthening of paper [18].

- Quality of paper improvement for printing and wrapping
- Avoids fold cracking
- Resist the mechanical defects of coated paper including tearing, spallation, blistering, and substrate peeling
- Bending stiffness

CNF as a precoating agent is known to increase bending stiffness of paperboard and cardboard than the reference material that is, successive rewetting—drying cycles of cardboard. This is so because adding CNF helps in decreasing water uptake and changes structure of base paper [31,35].

- Smoothness

After the process of precoating, CNF is recoated to decrease the paper surface roughness specially when applied TMP and clay-based paper sheets. It smoothens the paper by filling the fiber to fiber crossing that is produced by material pores. For improved smoothness, there is a requirement of smaller and narrower size distributer. And thus, providing flexibility in packaging and printing applications [23,31,35].

- Rheology

At a low solid content, CNF in coating formulation can increase its rheology as it helps in transferring the coating to the paper surface in a controlled way. Increase in high viscosity layer of CNF is due to the end of shear forces on paper surface [31,32,35].

13.7.2 Packaging

Coating NC on the paper has established its role in improving barrier properties and there by its wide application has been observed as transparent and biodegradable packaging [18]. The two main varieties of NC, that is, CNFs and CNCs have major effect individually as well as in combination in providing the barrier against oil, air, and grease. Both CNF and CNC are widely used in combination as multilayer coating due to its individual negative and positive impacts as a barrier [16,36–38].

- Oxygen barrier

CNF films can be replaced by many synthetic packaging films (ethylene vinyl alcohol) at 0% relative humidity (RH), as it turns out to be a barrier to oxygen permeability because of decrease of oxygen transmission rate (OTR) [31]. However, there is swelling of fibrillar network at high relative humidity, which is responsible for increasing OTR [31,39].

- Air/gas barrier

Air barrier property is attained by using CNF as coating or film over a base paper. After applying over base paper, it closes the open pores and lowers the air permeability up to 10 nm/pas; moreover, its effects decrease with a multilayer coat of CNF [23,31,39].

- Oil barrier

Oil resistance is a major issue in field of packaging which can be recovered by coating base paper by CNF. Oil barrier property is more effective than lower air permeability with increase in layer of CNF [31,39].

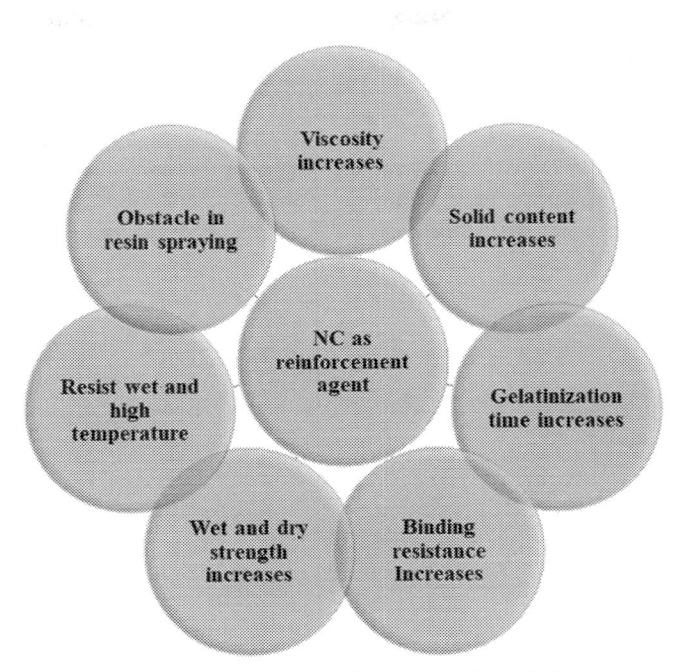

Fig. 13.7 Addition of NC (CNC and CNF) in wood adhesives results in following changes [40].

In addition, CNF as a coating agent promotes the quality of packaging by inhibiting growth of microorganisms such as Gram-positive bacteria and provides a standard coating product. Therefore, as a current need for having an environmentally friendly packaging material, NC has been recent replacement to give uniformity with antimicrobial and barrier property [16].

13.7.3 Printing quality

Due to very strong binding property of CNF as a paper coating agent, it has established its use in application of printing field. The characteristic of high-quality printing paper is due to its strong strength on surface, and as a result CNF is suitable for binder for many printing techniques including inkjet printing, flexograph. Petro-based binder has been ruled out for its use as CNF because it has good water absorption which weakens coated paper [16,31].

CNF is known to increase print density to a great extent as high quantity of ink pigments are very strongly captured by the small pores of porous CNF-coated layer [16,31].

CNF as a binder is also added with coating color in low content (less than 10% of CNF of total solids) to increase printing quality, in addition it avoids runnability difficulties at a pilot scale [16,31].

CNF is used in the following combinations with other materials for improved properties:

- CNF with clay in coating formulation provides good printing density and surface smoothness with low proportions of CNF in comparison to clay because high proportion of CNF with clay does not affect the properties to a greater extend [31].
- Alkyl ketene dimer (AKD) is added with NC to enhance the printability by lowering print-through and upgrading print density during ink-jet printing [16].
- The paper coating formulation includes nanocrystalline cellulose with cationic styrene-acrylic and starch emulsion boosting printing property of paper [16].
- The combining of CNF with mineral pigment such as kaolin in paper coating formulation provides strong print density property. However, rate of ink absorption decreases [16].

Using NC for increasing printability can be considered by paper industry. Moreover, choosing the ink and printing system affects the CNF printing surface [31].

13.8 Other application of NC in papermaking

13.8.1 Wood—adhesives (reinforcement agent)

NC has been introduced as a reinforced adhesion between wood components, which helps to modify the wood adhesive chemistry to improve mechanical and performance properties and adhesive strength by adding filler [40,41]. Utilization of NC provides advantage as a reinforcing agent and binder for wood adhesives results in final product quality in the terms of physical and mechanical characteristics as represented in Fig 13.7. Moreover, it decreases the liberation of formaldehyde when used with synthetic adhesives [41,42].

NC has major impact on adhesion property of wood-reconstituted products for example, plywood panels, flakeboard, particle board that affects the end product quality [41].

Usually, synthetic adhesives (phenol-formaldehyde and urea-formaldehyde (UF)) are used for bonding the overlapped woods under controlled pressure and temperature, this implies crossing of their fibers perpendicular (90°) to each other to form plywood panels [41]. One of the broadly used adhesives in wood industry is UF. It is known to have low prize, which reflects it economic feasibility in industry. But UF alone lowers the overall mechanical strength by developing microcracks, which leads to CNC enhancement to overcome the weaknesses of UF adhesive that is the resistance to moisture and its resistance glue line to the dry. Increase in higher frequency of effective hydrogen bonds between UF-CNC-wood and UF-CNC in plywood panels results in increasing binder resistance of the panels. Other adhesives including water-based polyvinyl acetate (PVAc) and polyurethane in mixture with CNF promote the mixture's rheological stability [40].

Addition of CNF as a reinforcement agent in flaked panels results in enhancing the properties such as flexural strength and internal bonding as well as lowering swelling after a day in water [40].

Commonly synthetic adhesives including melamine-formaldehyde, phenol-formal–dehyde, and UF are used in production of particleboard wood panels, which includes the haphazardly arrangement of small particles bond with these adhesives, followed by gluing under controlled heat and pressure [41]. One of the major advantages of CNF as an adhesive is known to have nonemission of volatile organic compounds as well as fixation of carbon by the trees producing cellulose [15,40,43].

13.8.2 Recycled paper

Recycling of paper is one of the processes of paper manufacturing industry as it benefits both business and environment. For paper industry, it acts as a major source of raw material. Paper manufacturing sector is known for developing economically feasible processes [4,44]. In Europe, 52.4% raw material for papermaking industry is obtained from reused paper and this contributes to a rate of 72.3% of recycled paper [4].

In addition to the advantages, one of the major disadvantages of reused paper that it is produced from recycled fibers having poor properties such as resistance with respect to original fibers of paper. Reused fibers have variation in morphology compared to original fibers, which includes average length shortening, lower elasticity, low capacity of hydration, and formation of interbonds of fibers [26,44,45]. In addition, the use of natural and artificial strength additives in recycling paper processing is also one of the major limitations of getting low tensile strength [4].

To overcome these limitations, NC is added as a secondary fiber for improving interfiber bonding and hence providing a new recycled product with improved strength quality [4,44,46].

The addition of NC has the following few advantages:
- Structural: good configuration of fibers, more closely packed, uniform, and resistant
- Properties: mechanical (strength increases), physical (increase density, thickness reduction)
- Others: sustainability, biodegradability, increase in surface area, and more availability [4,44].

One of the major advantages is the cost benefits and easy transportation. Considering these two major advantages paper industries have started adapting to the use of NC in paper manufacturing process, also with consideration of current situation this helps in good environmental health as it is biodegradable [4,44].

13.9 Challenges and future aspects in wood and paper industry

Since over past few years, NC has become a subject of high importance to mainstream researchers in paper and wood industry [47]. As in recent times, it is highly recommended to move toward sustainable sources as an optional raw material to reduce the utilization of natural resources NC has been introduced [16]. With lots of advantages of using NC which mainly include biodegradable, renewable, and low-cost material that provides benefits to not only business economy but also bioeconomy as a whole, there come few drawbacks and challenges for its usage [44].

NC is mainly used in its two forms CNF and CNC in paper industry, high cost of production of CNC and CNF stands as a major drawback [16,48]. In addition, difficulties in transportation and higher requirement of energy as well as difficulties associated in getting uniformity in NC particles and problems associated with pumping and dewatering process bring restriction in its usage as well as high production of NC [4].

As stated above, these drawbacks bring lime light on the future development to this sector in regard to its eco-friendly nature. In upcoming time, there are possibilities of *in-situ* CNF production along with number of studies and findings of alternatives to the disadvantages being considered, and researchers have been working on different aspects to overcome this and generalize its usage. However, there has been lack of research in this field [4,16,34]. The advantages to this (*in situ*) approach are increment in the production yield, prevention from drying process followed by significantly reduced cost of transportation and generation waste [4]. Moreover, this approach will also benefits paper manufacturers in terms of determining the correlation between the minimum quality of CNF and need of certain requirement in paper recycling, also permitting digital control of the CNF characteristics and fulfilling the needs of production demand [4,26]. However, the scientists have known to be working since a couple of decades ago but still no findings have been concluded; large-scale production of CNF is still a major issue for paper manufacturers and hence its usage has been limited [4,28].

Various other additives mainly cationic starches combined with CNF and CNC for its wide approach toward effective utilization has emerged a new branch in advance development in paper industry, which is known to affect the overall efficiency of paper [16,49]. This major principle stands behind using nanomaterials for enhancing the functional, mechanical, and optical characteristics [49].

Recent advancement has also started toward the using of recycling of paper and obtaining the desirable qualities in the end product, this gives a major importance as it benefits the manufacturers and reduces the environmental load [4].

Considering both the sides of the coin, it is concluded that because of the outstanding characteristics of NC in wood and paper industry around the globe more research and development regarding its detailed studies and application has been carried out and voids between the raw material, processing, production, packaging, and commercialization are gradually filled [42,44,50].

References

[1] IFOS-Statistics, Forestry Statistics 2018 A Compendium of Statistics About Woodland, Forestry and Primary Wood Processing in the United Kingdom, first ed., Forestry Commission, Edinburgh, 2018.

[2] I. Kajanto, M. Kosonen, The potential use of micro-and nanofibrillated cellulose as a reinforcing element in paper, J. Sci. Technol. For. Prod. Processes 2 (6) (2012) 42–48.

[3] L. Jasmani, R. Rusli, T. Khadiran, R. Jalil, S. Adnan, Application of nanotechnology in wood-based products industry: a review, Nanoscale Res. Lett. 15 (1) (2020) 1–31.

[4] A. Balea, J.L. Sanchez-Salvador, M.C. Monte, N. Merayo, C. Negro, A. Blanco, In situ production and application of cellulose nanofibers to improve recycled paper production, Molecules 24 (9) (2019) 1800.

[5] S.J. Eichhorn, A. Dufresne, M. Aranguren, N.E. Marcovich, J.R. Capadona, S.J. Rowan, C. Weder, W. Thielemans, M. Roman, S. Renneckar, W. Gindl, Current international research into cellulose nanofibres and nanocomposites, J. Mater. Sci. 45 (1) (2010) 1–33.

[6] A. Isogai, Wood nanocelluloses: fundamentals and applications as new bio-based nanomaterials, J. Wood Sci. 59 (6) (2013) 449–459.

[7] D. Klemm, F. Kramer, S. Moritz, T. Lindström, M. Ankerfors, D. Gray, A. Dorris, Nanocelluloses: a new family of nature-based materials, Ang. Chemie Int. Ed. 50 (24) (2011) 5438–5466.

[8] M. Michelin, D.G. Gomes, A. Romaní, M.D.L. Polizeli, J.A. Teixeira, Nanocellulose production: exploring the enzymatic route and residues of pulp and paper industry, Molecules 25 (15) (2020) 3411.

[9] A. Latha, M. Arivukarasi, C. Keerthana, R. Subashri, V. Vishnu Priya, Paper and pulp industry manufacturing and treatment processes—a review, Int. J. Eng. Res. Technol 6 (2) (2018) 6.

[10] T. Lindstrom, A. Naderi, A. Wiberg, Large scale applications of nanocellulosic materials-a comprehensive review, J. Korea TAPPI 47 (6) (2015) 5–21.

[11] A.F. Lourenço, J.A. Gamelas, P. Sarmento, P.J. Ferreira, Enzymatic nanocellulose in papermaking–the key role as filler flocculant and strengthening agent, Carbohydr. Polym. 224 (2019) 115200.

[12] G. Thompson, J. Swain, M. Kay, C.F. Forster, The treatment of pulp and paper mill effluent: a review, Bioresour. Technol. 77 (3) (2001) 275–286.

[13] O. Tünay, E. Erdeml, I. Kabdaşli, T. Ölmez, Advanced treatment by chemical oxidation of pulp and paper effluent from a plant manufacturing hardboard from waste paper, Environ. Technol. 29 (10) (2008) 1045–1051.

[14] P. Bajpai, Environmentally Friendly Production of Pulp and Paper, John Wiley & Sons, New Jersy, USA, 2011.

[15] Y. Habibi, L.A. Lucia, O.J. Rojas, Cellulose nanocrystals: chemistry, self-assembly, and applications, Chem. Rev. 110 (6) (2010) 3479–3500.

[16] M. Sharma, R. Aguado, D. Murtinho, A.J. Valente, A.P.M. De Sousa, P.J. Ferreira, A review on cationic starch and nanocellulose as paper coating components, Int. J. Biol. Macromol 162 (2020) 578–598.

[17] S. Wang, Q. Cheng, A novel process to isolate fibrils from cellulose fibers by high-intensity ultrasonication, part 1: process optimization, J. Appl. Polym. Sci. 113 (2) (2009) 1270–1275.

[18] S. Boufi, I. González, M. Delgado-Aguilar, Q. Tarrès, M.À. Pèlach, P. Mutjé, Nanofibrillated cellulose as an additive in papermaking process: a review, Carbohydr. Polym. 154 (2016) 151–166.

[19] S. Xie, X. Zhang, M.P. Walcott, H. Lin, Applications of cellulose nanocrystals: a review, Eng. Sci. 2 (14) (2018) 4–16.

[20] R. Guimond, B. Chabot, K.N. Law, C. Daneault, The use of cellulose nanofibres in papermaking, J. Pulp Paper Sci. 36 (1-2) (2010) 55–61.

[21] H. Kargarzadeh, M. Ioelovich, I. Ahmad, S. Thomas, A. Dufresne, Methods for extraction of nanocellulose from various sources, in: H. Kargarzadeh, I. Ahmad, S. Thomas, A. Dufresne (Eds.), Wiley-VCH Verlag GmbH & Co. KGaA, Wenheim, Germany, 2017, pp. 1–49.

[22] E. Duker, E. Brannvall, T. Lindstrom, The effects of CMC attachment onto industrial and laboratory-cooked pulps, Nord. Pulp Paper Res. J. 22 (3) (2007) 356.

[23] A.K. Das, M.N. Islam, M. Ashaduzzaman, M.M. Nazhad, Nanocellulose: its applications, consequences and challenges in papermaking, J. Packag. Technol. Res. 4 (2020) 253–260.

[24] H.A. Khalil, Y. Davoudpour, M.N. Islam, A. Mustapha, K. Sudesh, R. Dungani, M. Jawaid, Production and modification of nanofibrillated cellulose using various mechanical processes: a review, Carbohydr. Polym. 99 (2014) 649–665.

[25] Ø. Eriksen, K. Syverud, Ø. Gregersen, The use of microfibrillated cellulose produced from kraft pulp as strength enhancer in TMP paper, Nord. Pulp Paper Res. J. 23 (3) (2008) 299–304.

[26] M. Delgado-Aguilar, I. González, M.A. Pèlach, E. De La Fuente, C. Negro, P. Mutjé, Improvement of deinked old newspaper/old magazine pulp suspensions by means of nanofibrillated cellulose addition, Cellulose 22 (1) (2015) 789–802.

[27] M. Diab, D. Curtil, N. El-shinnawy, M.L. Hassan, I.F. Zeid, E. Mauret, Biobased polymers and cationic microfibrillated cellulose as retention and drainage aids in papermaking: comparison between softwood and bagasse pulps, Ind. Crops Prod. 72 (2015) 34–45.

[28] S.H. Osong, S. Norgren, P. Engstrand, Processing of wood-based microfibrillated cellulose and nanofibrillated cellulose, and applications relating to papermaking: a review, Cellulose 23 (1) (2016) 93–123.

[29] G.H. Yang, G.R. Ma, M. He, X. Ji, H.J. Youn, H.L. Lee, J. Chen, Application of cellulose nanofibril as a wet-end additive in papermaking: a brief review, Paper Biomater. 5 (2) (2020) 76.

[30] I. Siró, D. Plackett, Microfibrillated cellulose and new nanocomposite materials: a review, Cellulose 17 (3) (2010) 459–494.

[31] F.W. Brodin, Ø.W. Gregersen, K. Syverud, Cellulose nanofibrils: challenges and possibilities as a paper additive or coating material–a review, Nord. Pulp Paper Res. J. 29 (1) (2014) 156–166.

[32] R.S. Ribeiro, B.C. Pohlmann, V. Calado, N. Bojorge, N. Pereira Jr, Production of nanocellulose by enzymatic hydrolysis: trends and challenges, Eng. Life Sci. 19 (4) (2019) 279–291.

[33] F. Rol, B. Karakashov, O. Nechyporchuk, M. Terrien, V. Meyer, A. Dufresne, M.N. Belgacem, J. Bras, Pilot-scale twin screw extrusion and chemical pretreatment as an energy-efficient method for the production of nanofibrillated cellulose at high solid content, ACS Sustain. Chem. Eng. 5 (8) (2017) 6524–6531.

[34] M. Delgado Aguilar, I. González Tovar, J.A. Tarrés Farrés, M. Alcalà Vilavella, M.À. Pèlach Serra, P. Mutjé Pujol, Approaching a low-cost production of cellulose nanofibers for papermaking applications., Bioresources 10 (3) (2015) 5435–5455 2015núm.

[35] I. González, F. Vilaseca, M. Alcalá, M.A. Pèlach, S. Boufi, P. Mutjé, Effect of the combination of biobeating and NFC on the physico-mechanical properties of paper, Cellulose 20 (3) (2013) 1425–1435.

[36] N. Lavoine, I. Desloges, A. Dufresne, J. Bras, Microfibrillated cellulose–its barrier properties and applications in cellulosic materials: a review, Carbohydr. Polym. 90 (2) (2012) 735–764.

[37] N. Lavoine, J. Bras, I. Desloges, Mechanical and barrier properties of cardboard and 3D packaging coated with microfibrillated cellulose, J. Appl. Polym. Sci. 131 (8) (2014a).

[38] N. Lavoine, I. Desloges, B. Khelifi, J. Bras, Impact of different coating processes of microfibrillated cellulose on the mechanical and barrier properties of paper, J. Mater. Sci. 49 (7) (2014b) 2879–2893.

[39] H. Fukuzumi, T. Saito, T. Iwata, Y. Kumamoto, A. Isogai, Transparent and high gas barrier films of cellulose nanofibers prepared by TEMPO-mediated oxidation, Biomacromolecules 10 (1) (2009) 162–165.

[40] E.C. Lengowski, E.A.B. Júnior, M.M.N. Kumode, M.E. Carneiro, K.G Satyanarayana, Nanocellulose-reinforced adhesives for wood-based panels, in: M. Inamuddin, S. Thomas, R.K. Mishra, A.M. Asiri (Eds.), Sustainable Polymer Composites and Nanocomposites, Springer, Cham, 2019, pp. 1001–1025.

[41] S.K. Vineeth, R.V. Gadhave, P.T. Gadekar, Nanocellulose applications in wood adhesives, Open J. Polym. Chem. 9 (4) (2019) 63–75.

[42] M.L. Hassan, A.P. Mathew, E.A. Hassan, N.A. El-Wakil, K. Oksman, Nanofibers from bagasse and rice straw: process optimization and properties, Wood Sci. Technol. 46 (1-3) (2012) 193–205.

[43] S. Iwamoto, A.N. Nakagaito, H. Yano, M. Nogi, Optically transparent composites reinforced with plant fiber-based nanofibers, Appl. Phys. A 81 (6) (2005) 1109–1112.

[44] L.C. Viana, D.C. Potulski, G.I.B.D. Muniz, A.S.D. Andrade, E.L.D. Silva, Nanofibrillated cellulose as an additive for recycled paper, Cerne 24 (2) (2018) 140–148.

[45] H. Wang, D. Li, R. Zhang, Preparation of ultralong cellulose nanofibers and optically transparent nanopapers derived from waste corrugated paper pulp, BioResources 8 (1) (2013) 1374–1384.

[46] Q. Tarrés, E. Saguer, M.A. Pèlach, M. Alcalà, M. Delgado-Aguilar, P. Mutjé, The feasibility of incor-
      porating cellulose micro/nanofibers in papermaking processes: the relevance of enzymatic hydrolysis,
      Cellulose 23 (2) (2016) 1433–1445.
[47] A. Winter, B. Arminger, S. Veigel, C. Gusenbauer, W. Fischer, M. Mayr, W. Bauer, W. Gindl-Altmutter,
      Nanocellulose from fractionated sulfite wood pulp, Cellulose 27 (16) (2020) 9325–9336.
[48] A.M. Springer, Industrial Environmental Control: Pulp and Paper Industry. Georgia, USA 1985.
[49] J. Kasmani, A. Samariha, Effect of nano-cellulose on the improvement of the properties of paper
      newspaper produced from chemi-mechanical pulping, BioResources 14 (4) (2019) 8935–8949.
[50] D. Klemm, D. Schumann, F. Kramer, N. Heßler, M. Hornung, H.P. Schmauder, S. Marsch, Nanocelluloses as
      innovative polymers in research and application, in: D. Klemm (Ed.), Nanocelluloses as innovative polymers in
      research and application, Polysaccharides II. Advances in Polymer Science, vol 205 (2006) 49–96.

CHAPTER 14

Environmental, legal, health, and safety issue of nanocellulose

Gurudatta Singh[a], Syed Saquib[a], Ankita Gupta[a], Swati[b]
[a]Institute of Environment & Sustainable Development Banaras Hindu University Varanasi, Uttar Pradesh, India
[b]Department of Botany, Institute of Science, Banaras Hindu University Varanasi, Uttar Pradesh, India

14.1 Introduction

Cellulose is a homopolysaccharide polymer and is ubiquitous and renewable in nature to prepare materials with a much smaller diameter (<10–100 nm). Cellulases are used to produce cellulose, produced by bacteria, pillows, and some protozoans [1]. Nanocellulose is readily available, biocompatible, and biodegradable, making it suitable for a wide range of applications from packaging to pharmacological use [2]. Nanocellulose is widely used in adsorption studies to remove various pollutants. There are four primary sources of nanocellulose, namely bacteria, plants, algae, and lower animals. They have an excellent property of incorporating different metals and minerals that ultimately demonstrate the catalytic, optical, and electronic capabilities of such nanomaterials [3,4]. The use of green, renewable, and recycled materials has become increasingly necessary to manufacture various high-value goods with low environmental impact [5]. Cellulose is one of the most important natural polymers, almost inexhaustible raw material, and a primary industrial source of sustainable materials [6]. Economic crops such as wood and agricultural byproducts, seaweed, bacteria, fungi, are traditional cellulose sources and can also be harvested from tunicates [7].

As cellulose chains are clustered together, the advantage of cellulose can be further expanded, creating highly organized regions that can be isolated as nanoparticles, known as cellulose nanomaterials or nanocelluloses, considered as a good class of futuristic materials due to their physicochemical characteristics [5]. To extract noncellulosic materials such as lignin, hemicellulose, lipid, wax, pectin, alginate, and then nanocellulose 1–100 nm in diameter, the raw material is subjected to a series of pretreatments. There are several techniques for developing nanocellulose from purified cellulose, mainly in three categories: mechanical, chemical, and biological [8]. Although nanocellulose is an environmentally friendly composite material, it is highly debatable and reported by various studies in the literature. This chapter will highlight the general procedure for obtaining different cellulose nanocomposites along with their properties, applications, and possible health effects. Nanocellulose is classifiable into two major groups [4,8]:

1. Nanostructured materials (cellulose microcrystals and cellulose microfibrils) and
2. Nanofibers (cellulose nanofibrils, cellulose nanocrystals, and bacterial cellulose (BC))

While all forms are similar in chemical composition, they are different in morphology, particle size, crystallinity, and some properties due to the variation in sources and extraction methods.

Nanocellulose is an excellent candidate material for composite functional materials and has the advantages of large specific surface area, high transparency, high strength, low density, comprehensive chemical modification, and biocompatibility [6]. The use of various types of nanocellulose provides many advantages, including recyclability, reproducibility, biocompatibility, and surface tunable chemistry, as stated in the literature [8]. These advantages have convinced several researchers to explore the use of nanocellulose in different fields, such as water and wastewater treatment, tissue engineering, 3D bioprinting, packaging, biosensing, and drug delivery [7,9].

Nanocellulose is a nanomaterial, despite these useful properties. The behavior and physicochemical properties of materials change at the nanoscale, giving them new characteristics even while having the capacity to adversely affect human health and the environment [10]. Due to a broad range of consumer and industrial applications, the following various issues have been raised for this purpose [11]:

1. Small particles may be introduced into the air and inhaled in the workplace when released.
2. The long-term effects of exposure remain mostly unknown.

Therefore, while nanocellulose is widely considered nontoxic, there are still gaps in knowledge about its impacts on the environment and human health, and information remains scarce [12,13].

14.1.1 Cellulose to nanocellulose

Cellulose in a nanometer range or nanocellulose has attracted much attention from researchers because of its unique properties. Nanocellulose can be obtained by acid hydrolysis of cellulose. Conversion processes involved in nanocellulose production are discussed in Fig. 14.1.

14.1.2 Cellulose: the raw material

Nanocellulose is generated mainly from native cellulose, which is found in plant cells. These cells' walls consist of lignocellulose, which consists mostly of cellulose, lignin, and hemicellulose. The main contribution to the production of nanocellulose is the cellulosic content of this wall, and other impurities are removed by pretreatment methods [8]. The most abundant polymeric raw material on the earth is cellulose, an attractive and renewable feedstock. Cellulose is essentially constituted in the 4C1-chain structure

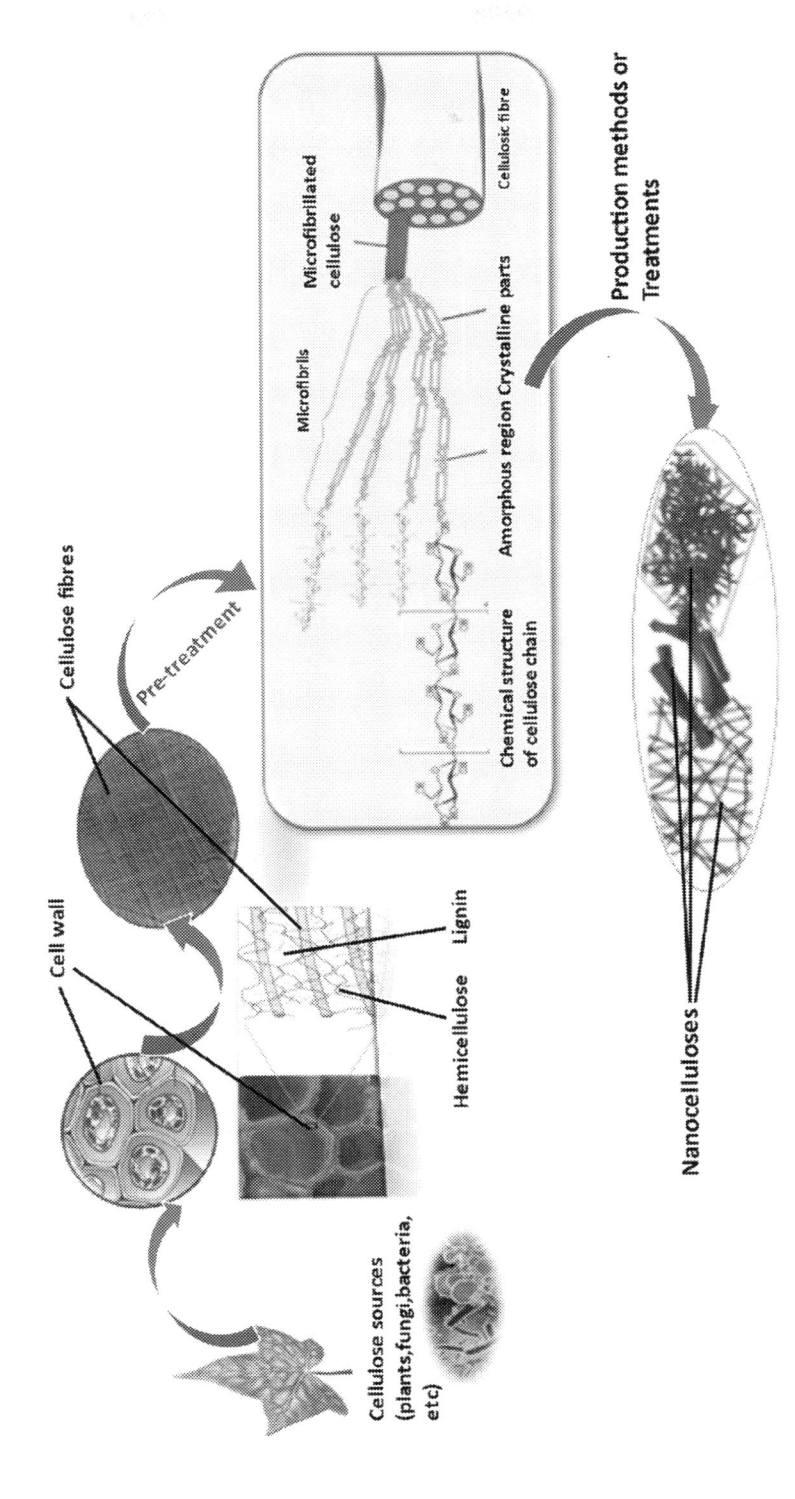

Fig. 14.1 Diagrammatic view of cellulose to nanocellulose conversion (modified from [7]).

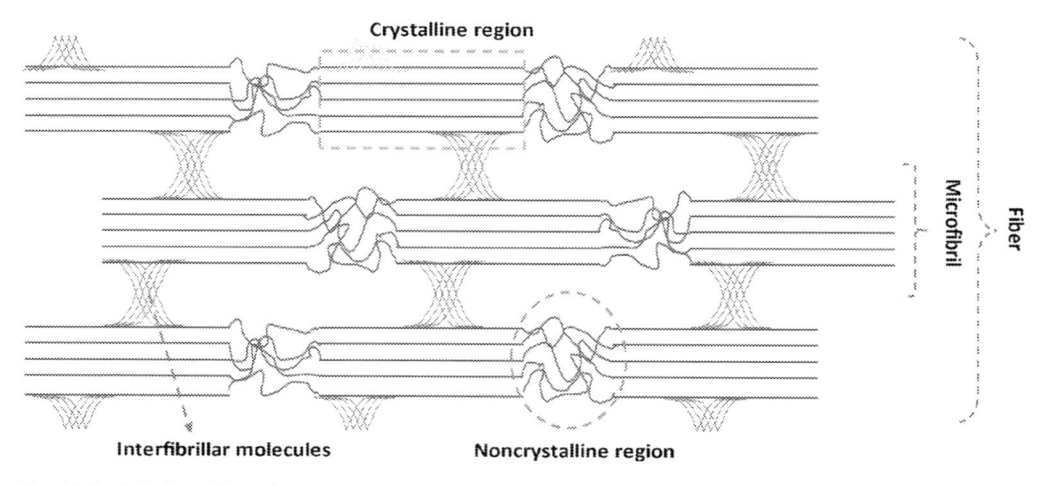

Fig. 14.2 Cellulose fiber showing crystalline and noncrystalline or amorphous regions.

by repeating $\beta(1,4)$-bound units of D-glucopyranosyl in which each monomer unit is corkscrewed at 180° relative to its neighbors [5,12]. The produced cellobiose units are bound together to produce a crystalline cellulose structure known as elementary fibrils. These latter are merged to produce microfibrils that in turn form macrofibrils or cellulosic fibers. Cellulose fiber showing crystalline and noncrystalline or amorphous regions are presented in Fig.14.2.

14.2 Characteristics of nanocellulose

Nanocellulose is an attractive, renewable, and sustainable product that attracts attention in academia and industry to unlock its potential [6]. Cellulosic materials with one dimension on the nanometer scale are commonly referred to as nanocellulose [5,14]. Nanocellulose blends essential cellulose properties with nanoscale materials' basic characteristics, such as high specific strength, modulus, and hydrophilicity. This is mainly due to the extensive surface area of these materials. It also has a modifiable surface, excellent mechanical strength, and a high aspect ratio [9]. Nanocellulose also has high rigidity (one order of magnitude less than that of single-walled carbon nanotubes), low thermal expansion, and low density, characteristics comparable to those of carbon nanotubes, an excellent material for a wide range of biomedical engineering and materials science applications. It has a high induction potential [15].

14.3 Properties of nanocellulose materials

Nanocellulose has various properties such as mechanical, optical, and thermal properties.

14.3.1 Mechanical properties

Physical and mechanical properties of cellulose are evident and are used by the society from ages in different purposes. These properties however strongly depend on many other factors including chemical composition and source location. The mechanical properties of cellulose microfibrils should be higher, homogenized than those of parent lignocellulose, and free from impurities [16]. The distribution of particles is polydisperse with average tensile strength ranging from 2 to 6 GPa and 2 to 4 GPa in Cellulose nanocrystal (CNCs) and MicroFibrillated Cellulose (MFCs), respectively [17]. The utility of nanocellulose composites (NC) becomes great as a reinforced polymer due to such impressive physical properties. The Young modulus of CNCs and cellulose nanofibrils/nanofibers (CNFs) were reported around 15–150 GPa or 1.5–1.6 g/cm^3 which is much higher than glass fiber (70 GPa). Microfibrils and nanocrystals are much stronger than steel [18]. Khan et al. [19] studied the production and properties of methylcellulose-based biodegradable reinforced nanocellulose film. They reported that 1% of nanocellulose content increases the puncture strength by 126%. Another study also reported similar positive trends in mechanical properties except elongation [20].

14.3.2 Optical properties

Using UV-Vis spectroscopy, regular light transmission is considered for investigating the optical properties of nanocellulose materials. Measurements are carried out in the 200–1000 nm range, with an ideal transmittance reported at 600 nm. When cellulose fibrils are closely packaged, microfibrillated cellulose (MFC) films are transparent. Freeze-dried films are opaque, however. The deformation-reformation under loading and unloading is the reason behind transparency [13]. MFC transparency was found to be around 71%, and 2,2,6,6-tetramethylpiperidine-N-oxyl (TEMPO) oxidized MFC films were found to be around 90%. [21].

On the other hand, CNCs produce transparent and iridescent films. CNCs perform the phase transition from liquid to the liquid crystal at various concentrations under suspension. To produce a semi-translucent film, the suspension may slowly evaporate [22].

14.3.3 Thermal properties

In polymer and material science, thermal properties are significant, particularly in industrial packaging and biomedical use. Thermal stability and thermal conductivity are mainly involved in these properties. The degradation behavior of nanocomposites depends on thermal stability. As nanocellulose decomposition occurs between 200 and 300 °C, higher stabilized nanocomposites are produced by a controlled production process lower than this limit [23]. CNF and CNC thermal decomposition has been reported at 200 and 220 °C, respectively. Various factors, including the source, manufacturing and isolation methods of nanocellulose, and the crystallinity of nanocellulose, influence the thermal stability of nanocellulose.

Another important property for the electronics and insulation industries is thermal conductivity, particularly at higher operating temperatures. Previous studies have identified various factors that affect thermal conductivity, including crystalline structure, cross-sectional area, pore size, and surface layer resistance [24]. CNCs have demonstrated significant thermal conductance of 2.5 W/m K, three to five times that of other common plastic films used in electronic appliances.

14.4 Preparation and types of nanocellulose

The source of cellulose raw materials and the difference in production conditions are the two main factors that affect the size, morphology, and properties of nanocellulose. Higher crystallinity raw materials, such as corncob and pineapple leaf, led to nanocrystals with a high aspect ratio. Shorter nanorods, such as rice straw, result from higher amorphous cellulose raw materials [5,15]. The key difference between CNC and CNF is the proportion of the material's amorphous stage and size. There are distinct forms of nanocellulose based on characteristics and preparation. Nanocellulose can be divided mainly into three categories as shown in Fig. 14.3.

1. CNFs, commonly attributed as nanofibrillated cellulose or MFCs.
2. Nanocellulosic crystals (NCC/CNCs), commonly called whiskers.
3. BCs.

14.4.1 Cellulose nanofibrils

CNFs consist primarily of plant-separated microfibrils through multiple mechanical and chemical actions. CNFs are usually produced by high-pressure homogenization,

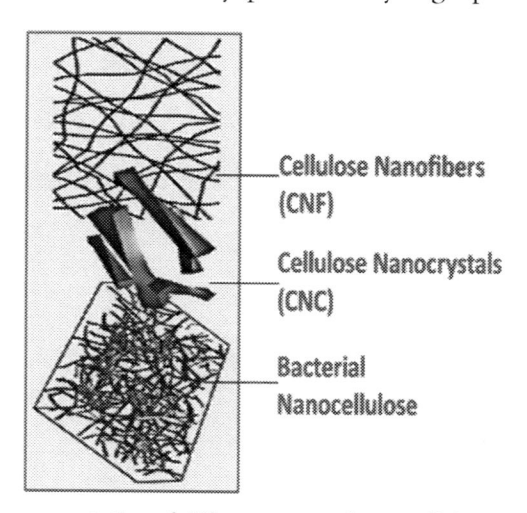

Fig. 14.3 Diagrammatic representation of different types of nanocellulose.

grinding, and milling of cellulosic pulp. Chemical treatments consist of oxidation mediated by TEMPO [25]. In an individual fiber with a length of 4 mm and a diameter of 10–50 nm, CNF consists of both amorphous and crystalline structures. Wood-based MFCs have been widely documented in the literature. They range in diameter from 10 to 30 nm.

14.4.2 Nanocellulose crystals

By the action of acid hydrolysis on cellulosic materials, cellulose nanocrystals or CNCs are made. MFCs undergo transverse cleavage along the amorphous regions, resulting in rod-like CNCs combining centrifugation and sonication. In comparison with microfibrils, cellulose whiskers have a relatively low aspect ratio. For cellulosic degradation, hydrochloric and sulfuric acids are used prominently. However, other acids, that is, nitric, phosphoric, and hydrobromic acids, have also been cited [26,27]. In general, CNCs range in length from 200 to 500 nm and in diameter from 3 to 35 nm [28]. CNCs are highly durable, strong materials, and are light [27].

14.4.3 Bacterial cellulose

As CNFs and CNCs are derived from plants (either wood, pulp, or straw), BCs are directly produced from different bacterial strains belonging to Gram-negative genera including *Acetobacter, Aerobacter, Agrobacterium, Alkaligenes, Azotobacter, Pseudomonas, Rhizobium, Rhodobacter, Salmonella,* and *Gluconacetobacter* [29]. Multiple species of *Gluconacetobacter* were widely used by different researchers that significantly include *Gluconacetobacter xylinus, Gluconacetobacter europaeus,* and *Gluconacetobacter medellinensis* [30]. BC is chemically identical with plant cellulose but is free of contaminants such as lignin, pectin, and hemicelluloses. Glucose molecules when combined with cell wall of microbes, ribbon-shaped BCs are formed. These nanofibers generally range from 20 to 100 nm in length [31]. Papers on nanocellulose published in different journals from 2010 to 2020 are presented in Fig. 14.4.

14.4.4 Other sources of nanocellulose

Nanocellulose is also derived from other natural sources, which mainly include algal and animal-based nanocellulose materials, in addition to plant and bacterial synthesis. Two significant genera of algae are *Cladophora* and *Cystoseira.* For their biomedical and pollutant removal capabilities, both of them were extensively researched. In biomedical and therapeutic applications for endotoxins and heavy metal toxicity, and cell culture and the removal of toxic dyes such as Congo red by adsorption, *Cladophora* was useful [32,33]. In combination with Fe_3O_4, nanocellulose produced from *Cystoseira myrica* also plays an important role in combating environmental pollutants, particularly removing mercury ions from the aqueous medium [2,34].

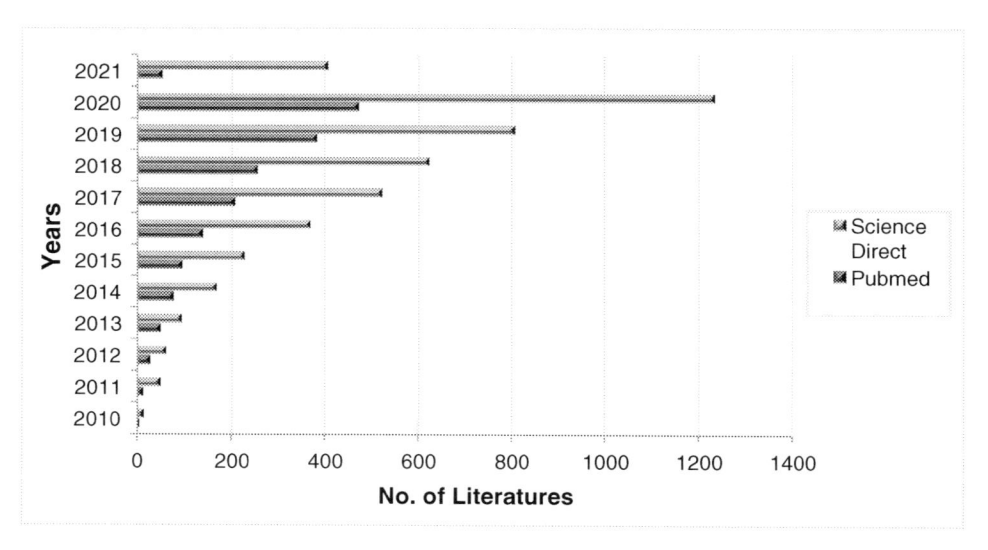

Fig. 14.4 Number of publications on nanocellulose between years 2010 and 2021 based on two databases PubMed and Science direct.

A possible source of nanocellulose is also a certain primitive higher animal. Tunicates belonging to the phylum *Chordata* are usually included. Examples of tunicates are *Styela clava* or *Halocynthia roretzi Drasche*. *Clava* is used as wound dressing and as fiber for stitching [35,36]). *Halocynthia roretzi Drasche* is primarily used to remove oil spills when combined with TiO_2 nanoparticles [36].

14.5 Production and extraction nanocellulose

Techniques used for the production of nanocellulose are presented in Fig. 14.5.

As mentioned before, CNCs are produced by acid hydrolysis technique, which is a chemical technique, or CNFs produced by mechanical technique. Similarly, other techniques are used to extract nanocellulose from different sources [8]. The extraction of nanocelluloses from the raw material includes the following various steps:

14.5.1 Selection of sources

The raw materials with high yielding cellulose are preferable. Nanocelluloses can be extracted from plants (e.g., wood, wheat straw, rice straw, corn husk, potato pulp, etc.) or by bacterial processes (genus such as *Achromobacter, Alcaligenes,* and *Gluconacetobacter* (previously known as *Acetobacter*) [4]. Cellulose content in some plant materials is as follows: Cotton: 85–90%, bast plants: 65–75%, wood: 40–50%, bamboo: 23–46%. It can be produced using various waste materials such as sugarcane bagasse, waste newspaper,

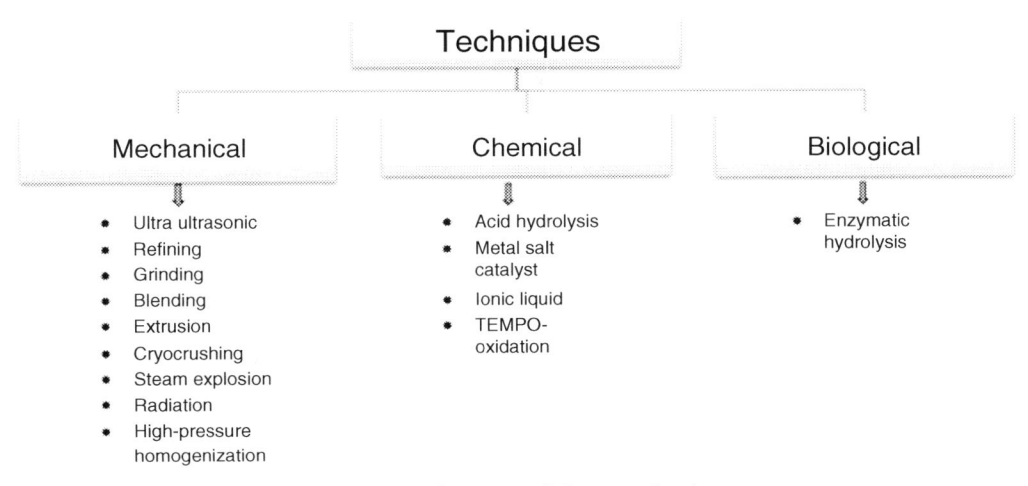

Fig. 14.5 Represents different techniques for nanocellulose production.

municipal solid waste, etc. It is also produced from pulpwood, bleached plants, other sources such as alpha cellulose, cotton linter, kenaf pulp, etc. [8].

14.5.2 Pretreatment

It is required after choosing a suitable source to remove impurities (such as hemicellulose, lignin, wax); impurities such as lignin act as a physical barrier. Typical pretreatments are as follows: physical, physicochemical, chemical, and biological. Majorly used treatments are *alkaline pretreatment* and *bleaching pretreatment* [8,37].

14.5.3 Extraction

This step influences the characteristics of nanocellulose, such as the aspect ratio, crystallinity, thermal stability, and stability of the dispersion, in addition to sources and pretreatment techniques. Extraction technique can be divided into biological, chemical, and mechanical [8,38]. The extraction process of nanocellulose is presented in Fig. 14.6.

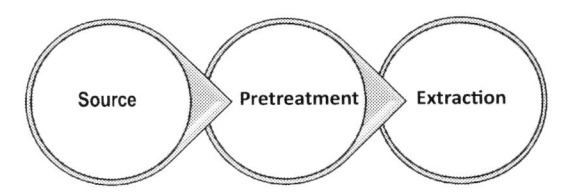

Fig. 14.6 Representation of the extraction process of nanocellulose.

14.6 Application of nanocellulose

14.6.1 Nanocellulose in paper and film industry

One of the prominent sinks for nanocellulose use is the paper industry. The extraction of more than 100 MT/annum of cellulose is commercially used for paper and cardboard production. During cellulose fiber refining, nanocellulose particles increase the tensile strength of paper [39]. Such sheets' mechanical characteristics are at least two to five times greater than those of popular papers. The fibers' diameter and the porosity are the sole differences between nanocellulose paper and common cellulosic paper. The resulting paper is more transparent due to the much lower width. By filtration of 0.2% MFC suspension through a 0.65 μm pore size filter membrane, Sehaqui et al. [40] produced nanocellulose documents. In a Rapid-Kothen semiautomatic sheet former, the wet gel cake was clamped dry at 363 K and 70 mbar. BC, aside from MFCs, is an alternative material for the production of nanocellulose paper. BC naturally forms membranes, so the step of filtration can be omitted, and by simply pressing the BC hydrogel, BC papers can be obtained.

14.6.2 Nanocellulose in the biomedical industry

Due to its biodegradability from drug delivery to tissue repair, nanocellulose is now widely used in biomedical applications. SH–SY5Y neuroblastoma cells were reported to be cultured on 3D bacterial nanocellulose (BNC) scaffolds in neural tissue engineering; adhered to, proliferated, and also differentiated into mature neurons [41]. BC nanocellulose is also used in cartilage reconstruction due to its high water-retention capacity and mechanical strength. Studies have also reported bioinclusion based on nanocellulose in 3D bioprinting with living cells. Alginate and chondrocyte plant-based MFCs were used to construct the human ear and meniscus [42]. Nanocellulose can be administered for drug delivery as it is nano-sized in structure. A carrier drug can penetrate and treat skin diseases through skin pores. For cellular uptake, CNCs combined with folic acid have been synthesized and favorably facilitated cancer-targeting chemotherapy [43]. The binding capabilities of namely tetracycline and doxorubicin with CNCs for rapid release were reported in another study [44].

14.6.3 Environmental application of nanocellulose

14.6.3.1 Water contaminant removal

Nanocellulose is an excellent, sustainable material that can be used to remove water from particular contaminants. Textile dyes, heavy metals, and pesticides are important aqueous contaminants. Adsorption is one of the efficient phenomena commonly used to combat these contaminants. Due to its high aspect ratio, cost efficiency, high natural abundance, cellulose nanomaterials are an effective adsorbent [45]. The ions La(II),

Pb(II), and Cu(II) can be removed efficiently by CNF. Another significant study shows up to 98% efficacy of adsorption and removal of toxic chromate ions on CNCs [46]. A large number of dyes are used by most textile, paper, and plastics industries to color their products. They are water soluble and pose a significant threat to bodies of water. This contamination of the dye can lead to serious problems with the environment and health. By TEMPO oxidation, cellulose nanocrystals incorporate the carboxylic group and are effective for cationic dye adsorption [47]. By using positively charged cellulosic materials, anionic dyes can be removed. Through successive sodium periodate oxidation and reaction with ethylenediamine, the cationic CNCs were prepared.

14.6.3.2 Pollutant sensors

Various researchers studied the possible application of nanomaterials for biosensor capabilities. Au-doped nanocellulose nanocomposites are extensively studied for pollutant sensing [48,49]. The biocompatibility of BC and conductive nature of AuNPs are incorporated to detect chemicals electrochemically. Synthesized AuNP/BC nanocomposite are formed and used as sensors for H_2O_2 and glucose detection [50]. Surface-enhanced Raman spectroscopy (SERS) based nanocellulose biosensors are extensively used for the monitoring of waterborne pathogens and organic contaminants. One such study used gold-biosensor for the detection of waterborne pathogens namely *Cryptosporidium parvum* and *Giardia lamblia*.

14.6.4 Nanocellulose in the energy application

In energy production and storage devices such as solar cells, fuel cells, lithium–ion battery production, nanocellulose materials are also used. Fuel chemical energy generally from H_2 catalytically oxidized at the anode and oxygen at the cathode is catalytically reduced to H_2O in a fuel cell to generate electrons flowing through an external circuit from the anode to cathode. This results in the creation of electricity. The substrate for several nano-sized anode catalysts is BNC [75]. Hu et al. [51], based on transparent nanocellulose paper, designed a solar cell. The transparent nanopaper increased the visible light path length, thereby resulting in greater absorption of solar light. This solar cell showed 0.4% power conversion efficiency. Nanocomposites are also used in Li–ion battery manufacturing, in addition to the above applications. Nanocellulose paper provides a chance for thin, flexible, and high-performance batteries to be designed. The sources and the application of nanocellulose are presented in Fig. 14.7.

14.7 Health issues of nanocellulose

There is not much research conducted on the health-related issues of nanocellulose. Due to the lack of sufficient data on the risk of nanocellulose, the negative effects of nanocellulose on human health and the environment are little known. Regarding the

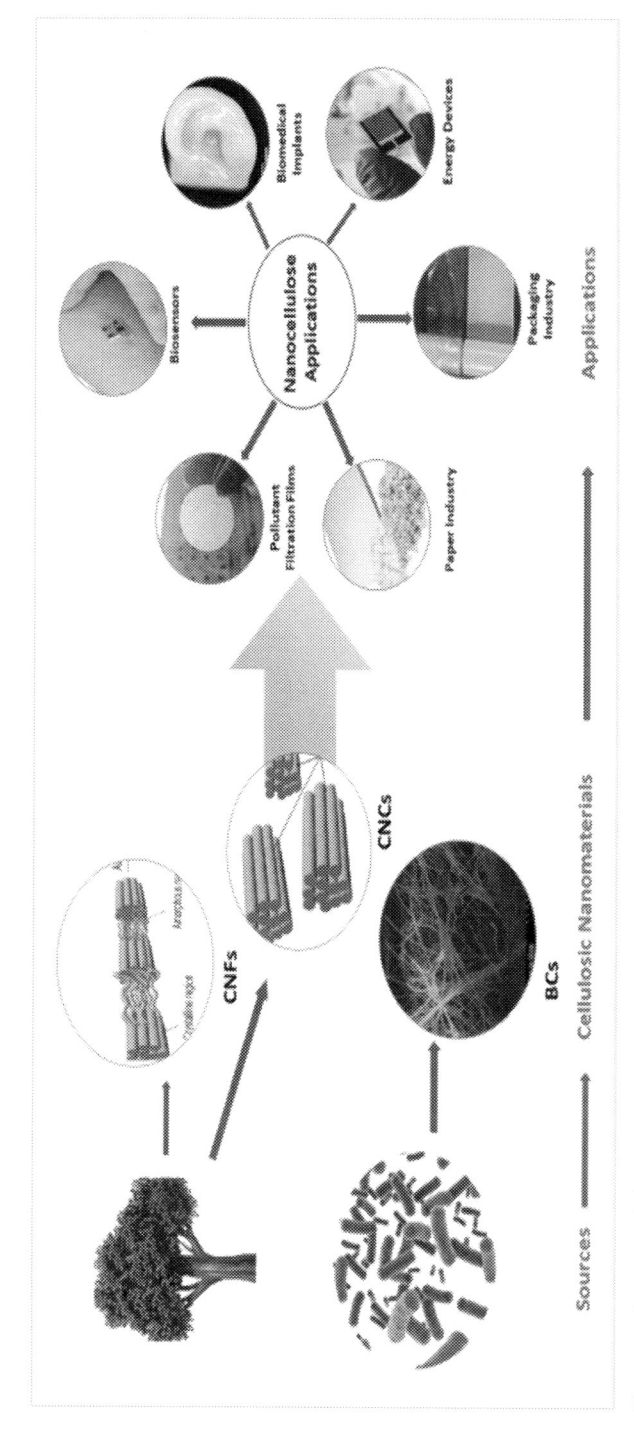

Fig. 14.7 Conceptual diagram depicting major sources and resulting from nanocellulose along with their possible applications.

effect of nanocellulose on the human body, long-term research needs to be carried out to prevent any potential threat beforehand. Cellulose itself is not harmful, but it can penetrate deep into the cells and persist for longer periods of time when used on a nanoscale level. Use of nanocellulose in various sectors, especially in healthcare and as food additives and packaging, has the potential of negative biological impacts [52]. Occupational exposure is the main entry route of such commercially manufactures NCs including gastrointestinal tract (GIT), dermal, and respiratory exposure through lungs. Inhalation of nanocellulose fibers, particularly in industries where nanocellulose is present at a very high level, is also the main way through which nanocellulose can enter the human body [53]. The deeper parts of the lungs can be reached by very small nanocellulose particles, while the comparatively larger particles can be cleared by entrapment in the upper airways. Administered cellulose powder of about 15 mg intratracheally and found severe lung disorders in test organisms, that is, fibrosis and alveobronchiolitis. Another significant study shows accumulation and higher durability of cellulosic fibers in rats' lungs administered during health risk assessment [54]. Cytotoxic studies, conducted in previous years, reported both negative as well as no significant concern on health. Kovacs et al. [55] carried out a cytotoxic study of effluent terminated from a CNC production plant with dose ranges from 0.03 to 10 g/L of fibers on various aquatic life forms, and reported no such concern on health. Negative aspects of nanocellulose were investigated for uptake in human embryonic kidney cells (HEK 293) and Sf9 insect cells and it was reported that exposure to 0.1 mg/mL of negatively charged CNCs and labeled with fluorescein isothiocyanate, led to membrane rupture [56]. The exposure of nanocellulose materials also reported inflammatory and oxidative stress. Not much research is conducted on the long-term human health effects of nanocellulose, but the available information on the effects of nanocellulose on the human health can be divided in to *in-vivo* and *in-vitro* research.

14.7.1 *In-vivo* studies

Very few *in-vivo* tests have been performed regarding the negative effects of nanocellulose. When mice were exposed to CNC in two forms, gel, and powder (50–200 μg/mouse), they showed variations in inflammatory response and tissue damage. Both the types of materials showed negative effects on mice worse than asbestos in the form of severe acute inflammatory response, the gel showed more severe effect than the powder form of CNC [57]. However when the mice were exposed to CNF and CNC (40 and 80 μg/mouse), no significant inflammatory response was seen in the mouse, even though the materials remained in the lungs for 14 days after the exposure [58].

Long-term exposure however negatively affected the mice health as seen in the following studies. When mice were exposed to CNC for 3 weeks continuously, chronic pulmonary inflammation was observed in the mice. Also, the histology of the respiratory airways was found to be abnormal after the long-term exposure of CNC [59]. Also, in a study conducted by Farcas et al. [60], the reproductive capacity was affected, the testes became abnormal pathologically, inflammatory response was seen after 3 months of exposure of nanocellulose for 3 weeks (40 μg/mouse) to the mice.

14.7.2 *In-vitro* studies

No significant genotoxicity in the tested cell lines was observed in the case of CNC. No variations in genotoxicity in human bronchial epithelial cells were observed at a dosage exposure of 2.5–100 μg/mL [61]. However, significant genotoxicity was observed in different studies after CNF exposure. It has been observed that the genotoxicity of CNF is different based on the source from which it is derived. In a study conducted by de Lima et al. [62], the brown cotton and curaua CNF were found to be genotoxic at all exposed concentrations (0.01%, 0.1%, and 1%), while the CNF caused no genotoxic effects derived from white cotton, ruby cotton, and green cotton.

Two sources (cotton and tunicates) were used for deriving CNC materials in a study conducted by Endes et al. [52]. With regard to stiffness, surface charge, zeta potential, etc., the physiochemical aspect was also different for both CNC materials. When tested on A549 cells, these two CNC materials found no significant oxidative stress in the cell lines. Many studies have been performed with regard to the inflammatory response caused by CNF materials. A comparable conclusion reached by many authors is that the surface chemistry of the nanocellulose material may trigger the inflammatory response. Two types of CNF materials were used in a comparative study conducted by Ilves et al. [63]. There were two materials with unmodified surface chemistry, carboxymethylation, and carboxylation modifying the other three materials. In comparison to the modified CNF materials, the nonmodified materials showed a higher inflammatory response.

14.7.3 Nanocellulose effect on GIT

Because of the advanced use of nanocellulose in the food industry, the effect it can cause after entering the human body must be understood as it can affect the GIT, thereby affecting digestion, absorption, and ingestion of different food products. It can be exposed to various factors in the body when the nanocellulose fibers enter the human gut, such as intestinal microorganisms; various internal surfaces such as the stomach, intestine, etc.; fatty acids, proteins, lipids, and various other GIT components [64,65]. Like every other nanoparticle, the effect of nanocellulose in the GIT is determined by the intrinsic and extrinsic variables. The fate of nanocellulose is presented in Table 14.1.

14.7.3.1 *Intrinsic factors*

The different intrinsic factors that can cause various effects of the nanocellulose particles in the human gut are surface area, particle concentration, shape, surface chemistry, particle size, aggregation, or agglomeration state [66]. Smaller nanocellulose particles size has an increased effect on the GIT components. The smaller the size, the larger the

Table 14.1 The fate of nanocellulose during the whole digestion process in the different parts of GIT (modify from Liu and Kong et al. [65]).

Type of Nanocellulose	Mouth	Stomach	Small intestine	Large intestine
	No significant change was observed	No significant change was observed	No significant change was observed	Biodegradation was observed but only partially
Modified by carboxylation (TEMPO-CNF)	No significant change was observed	The particle size was increased, and the viscosity was increased due to the aggregation and form phase separation	Aggregation was persistent	Biodegradation was observed but only partially
CNC (modified by sulfation)	No significant change was observed	Development of the hydrogel network, but no significant change was observed in the particle size	The hydrogel network was maintained	Biodegradation was observed but only partially

surface area [64]. Also, as nanocellulose is derived from the materials made up of cellulose, there is a very high possibility that it may behave in the same manner like dietary fiber [67–69].

14.7.3.2 Extrinsic factors

The extrinsic factors responsible for the effects of nanocellulose on human health consist of the GIT effect and the food matrix effect. The effects of nanocellulose are highly dependent on the pH of the GIT because the zeta potential of the nanocellulose is highly dependent on the pH of the gut [66]. According to a study done by Liu and Kong [67], the stability of various types of nanocellulose was analyzed and it was observed that the CNF remained stable during the whole digestion process whereas the TEMPO-oxidized CNF (TEMPO-CNF) aggregated and the CNC formed a network of hydrogel during the digestion process. Also, the particle size varied only in the case of TEMPO-CNF, its size increased, whereas the particle size was not affected in the case of the cellulose nanofibers and cellulose nanocrystals. The presence of a food matrix may also become the cause of the various effects of the nanocellulose particles on the human health.

In a study conducted by Nsor-Atindana et al. [70], it was seen that when CNC was added to the protein isolate-starch setup before heating at 70 °C in an experiment, the viscosity of the digested material was observed to be increased and the diffusion of glucose and its release was found to be decreased. It was also observed that with the aspect ratio higher of the CNC, this result was more enhanced. For the case of lipid digestion, in a study conducted by DeLoid et al. [71] 0.75 wt% nanocellulose was added to the experimental setup, and it was observed that the addition of nanocellulose can affect the hydrolysis of triglycerides. The fatty acids used in the experimental setup were heavy cream, corn oil, mayonnaise, and coconut oil. In the case of protein digestion effects on the addition of nanocellulose, *in-vitro* studies have been performed. In a previous study, when 3 wt% cellulose nanofibrils were added to the experimental setup, the digestion of proteins was affected when the protein was analyzed using the SDS-PAGE technique [72].

The toxicity of the nanocellulose is majorly dependent on the shape and size of the nanocellulose particles, because smaller the size of the particle, greater will be its surface area, thereby more interaction with the internal components of the body. Also, the surface chemistry plays a very important role in the effect of nanocellulose on the various parts in the GIT. Many *in-vitro* studies have also shown that at different concentration ranges, CNF and CNC did not exhibit any kind of cytotoxicity, but CNC induce cytotoxicity at different doses [73,74].

14.7.4 Cytotoxic effects of nanocellulose

To test the cytotoxicity of nanocellulose materials, several other studies have been conducted. In a study conducted by Colic et al. [75], CNF at a 1000 µg/mL concentration was used (obtained from wood using mechanical fibrillation). The murine fibroblast L929 cell lines, rat thymocytes and human peripheral blood mononuclear cells did not show any type of cytotoxicity to the cellulose nanofibrils. The problem with these kinds of cellular models is that, depending on the different physiochemical conditions in the GIT, they do not consider the internal environmental conditions in the GIT that can affect the condition of the nanocellulose materials. And there have been very few animal models used to test the effect of nanocellulosic materials on the animals' health. Therefore, further research needs to be done to understand the different types of effects that could be caused by nanocellulose materials inside the body of any living organism.

The extent of nanocellulose biodegradability is slightly greater than that of macroscale cellulosic materials. In their experimental setup, Kümmerer et al. [76] used cellulose nanowhiskers and macroscopic cellulose to see the extent of degradation for 28 days. The cellulose nanowhiskers were 54% degraded after 28 days, and the macroscopic cellulose was only 46% degraded. The larger surface area it has due to its small size could be the reason why the nanocellulose degrades more than its macroscopic counterpart.

Not much research is done on the interaction between nanocellulose and the gut microbiota. One study conducted by Nsor-Atindana et al. [77] showed that by improving the production of fatty acids and the count of Bifidobacterium bacteria in both *in-vivo* and *in-vitro* experimental setups, cellulose nanocrystals might be beneficial for prebiotic effects in the gut microbiome. Similar studies have been carried out on the nanocellulose-intestinal microbiota interaction, but further studies need to be done to better understand the human gut's actual interaction between the NC and the intestinal microbiota and all the factors that can affect this interaction.

We need more and more studies on the various physicochemical factors of nanocellulosic materials such as particle size, surface area, surface chemistry, rheological properties, etc. Their interaction with the various GIT components to understand the effects that nanocellulose could cause on human health can avoid any adverse effects, nanocellulose cooperatives, etc.

14.8 Environmental issues related to nanocellulose

As discussed earlier, NC becomes a reliable material of interest in research to industries due to their environment compatible properties. Such interest leads to higher consumption and eventually results in the release of nanocellulose into the environment. Being nanoscale in size, the properties of nanocellulose materials rapidly change, which might have the potential to impart negative effect on human and environment. However, nanocellulose is thought to be nontoxic, still few but important studies were conducted to quantify their environmental risk assessment and exposure and hazard assessment. However, due to uncertainties these studies were not able to give a direct conclusion regarding health and safety issues with such materials. Shatkin and Kim, [53], carried out an life cycle assessment (LCA) assessment for the possible environmental risk and exposure of nanocellulose. They also pointed out the required research and data gap related to environmental issues to nanocellulose. Another study conducted by Endes et al. [12] identifies a major lack of dose–response and exposure of data related to various human and environmental health.

Due to lack of techniques for the direct measurement of concentration for such engineered nanocellulose materials, modeling approaches were efficient in assessing the environmental risk associated with them [78]. One of the significant modeling studies was done to assess environmental exposure and risk in Europe [10]. They quantify the exposure, hazard and risk assessment throughout the life cycle of a material from manufacturing to end of life. The exposure was modeled using dynamic probabilistic material flow analysis (DPMFA). The model includes various environmental regimes, from and into which nanocellulose flows. They were air, surface water, subsurface water, and soil. Hazard assessment was done using probabilistic species sensitivity distribution (PSSD). Risk of nanocellulose materials was quantified using predicted environmental concentration (PEC) from hazard and exposure. The result shows that NC in end-of-life stage from the paper industry: 82% were reprocessed and 18% going to mixed municipal solid waste, ending up in landfills. The model predicts that almost 3% of NCs from textiles are released into environment, of which 50% goes in air and wastewater. Remaining 97% ended up in landfills as MSWs. Hazard assessment shows the extensive aquatic toxicity among 8 species and 76 data points collected from various studies for modeling.

14.9 Legal issues of nanocellulose

The use of nanocellulose is widely reported in different forms and in fields at commercial/industrial level, researcher globally from past few years. There is no such significant literature available till now, which points out the legal constraints related to their production and usage. Although in European Union nanocellulose are considered under nanomaterial which came under the scanner of certain regulatory assumptions guided by risk assessments and regulatory discretion. In the European Union, nanomaterials' production and use are coming under the control of the REACH regulation, which sets conditions for the market access to chemical substances. Kautto and Valve, [79] in their paper, discussed the fitting of REACH program for nanocellulose. If the regulatory and legal assumptions are adopted on the terms, a regulatory fit could be achieved. The study conducted by Kautto and Valve [79] also point out the discussions and interviews of researchers, innovation of scholars, focusing on the positive and negative aspects of nanocellulose composited on the environment and society during the formation of the regulatory framework.

14.10 Safety issues of nanocellulose

The safety issues of nanocellulose majorly contain safety of the highly exposed people to the nanocellulose particles during the production of various products using nanocellulose in the industries and the potential environmental safety risks associated with the high usage of cellulose at the nanoscale level. There is no official occupational exposure limit (OEL) value for this material, but as a precautionary measure, the recommended value of OEL is 0.01 fibers/cm^3 due to the risk of inhalation of nanocellulose fibers in the industries [80]. More research needs to be done regarding the potential safety issues of nanocellulose to prevent any possible ill-effects of nanocellulose in the near future. Exposure and safety issue of nanocellulose are presented in Table 14.2 [81].

14.11 Conclusion

Nanocellulose obtained from cellulose—the most abundant biopolymer on Earth—is an emerging renewable polymeric product. The nanocellulose is a very useful material that is used in many industries such as petroleum industry, pharmaceutical industry, textile industry, in making industrial molecules such as herbicides, chlorinated phenolic compounds, toluene, in making biomolecules such as immunoglobulin etc., in the removal of heavy metals. Various applications in biomedical, environmental industrial domain and properties make them ideal alternative among other nanomaterials.

The major way by which it can enter the human body is via inhalation in the occupational setups and via food products. There are no proper rules and regulations regarding the health hazard of the nanocellulosic materials for the humans because not

Table 14.2 Possible conditions regarding the exposure and safety issues of nanocellulose (Adapted from [69]).

Conditions	Life cycle	Potential risk	Exposure
A facility employee can inhale the dried material	Production stage	Inhalation	Occupational
Using the nanocellulose powder with other materials	Manufacturing stage	Inhalation	Occupational
The dumping of nanocellulose particles in the open landfills	After the life cycle ends of nanocellulose production	Possible risk of contamination of the environment by nanocellulose particles	Environmental
Ingestion of nanocellulose via the food products	Application	Possible health effects caused by nanocellulose	At consumer level
The flow of nanocellulose particles in the sewer system in the residential areas	After the life cycle ends of nanocellulose production	Degradation of nanocellulose	Environmental

much research has been conducted in this prospect. But more research needs to be done for the effects that nanocellulose could cause in the long-term exposure period. Also, as the nanocellulose effects are majorly dependent on their various physiochemical properties, these properties also need to be taken into consideration before coming to any conclusion. Such detrimental factors also lead to various environmental, legal, and health issues. However lack of proper studies in these areas challenges the proper implementation of regulations regarding production, usage, and waste management of nanocellulose and their byproducts.

References

[1] G. Singh, A.K. Patel, A. Gupta, D. Gupta, V.K. Mishra, Current advancements in recombinant technology for industrial production of cellulases: Part-II. In: M. Srivastava, N. Srivastava, P.W. Ramteke, P.K. Mishra (Eds.), Approaches to Enhance Industrial Production of Fungal Cellulases, Springer, Cham, 2019, pp. 177–201.

[2] L. Bacakova, J. Pajorova, M. Bacakova, A. Skogberg, P. Kallio, K. Kolarova, V. Svorcik, Versatile application of nanocellulose: from industry to skin tissue engineering and wound healing, Nanomaterials 9 (2) (2019) 164.

[3] N. Lavoine, L. Bergström, Nanocellulose-based foams and aerogels: processing, properties, and applications, J. Mater. Chem A 5 (31) (2017) 16105–16117.

[4] H. Wei, K. Rodriguez, S. Renneckar, P.J. Vikesland, Environmental science and engineering applications of nanocellulose-based nanocomposites, Environ. Sci. 1 (4) (2014) 302–316.

[5] D. Trache, A.F. Tarchoun, M. Derradji, T.S. Hamidon, N. Masruchin, N. Brosse, M.H. Hussin, Nanocellulose: from fundamentals to advanced applications, Front. Chem. 8 (392) (2020) 1–33.

[6] D. Klemm, F. Kramer, S. Moritz, T. Lindström, M. Ankerfors, D. Gray, A. Dorris, Nanocelluloses: a new family of nature-based materials, Angew. Chem. Int. Ed 50 (24) (2011) 5438–5466.

[7] J. Wang, X. Liu, T. Jin, H. He, L. Liu, Preparation of nanocellulose and its potential in reinforced composites: a review, J. Biomater. Sci. 30 (11) (2019) 919–946.

[8] S. Salimi, R. Sotudeh-Gharebagh, R. Zarghami, S.Y. Chan, K.H. Yuen, Production of nanocellulose and its applications in drug delivery: a critical review, ACS Sustain. Chem. Eng. 7 (19) (2019) 15800–15827.

[9] T. Abitbol, A. Rivkin, Y. Cao, Y. Nevo, E. Abraham, T. Ben-Shalom, S. Lapidot, O. Shoseyov, Nanocellulose, a tiny fibre with huge applications, Curr. Opin. Biotechnol. 39 (2016) 76–88.

[10] N. Stoudmann, B. Nowack, C. Som, Prospective environmental risk assessment of nanocellulose for, Eur. Environ. Sci. 6 (8) (2019) 2520–2531.

[11] J.A. Shatkin, B. Kim, Environmental Health and safety of cellulose nanomaterials and composites, Handbook of nanocellulose and cellulose nanocomposites 2, 2017, pp. 683–729.

[12] C. Endes, S. Camarero-Espinosa, S. Mueller, E.J. Foster, A. Petri-Fink, B. Rothen-Rutishauser, C. Weder, M.J.D. Clift, A critical review of the current knowledge regarding the biological impact of nanocellulose, J. Nanobiotechnol. 14 (1) (2016) 1–14.

[13] N. Stoudmann, M. Schmutz, C. Hirsch, B. Nowack, C. Som, Human hazard potential of nanocellulose: quantitative insights from the literature, Nanotoxicology 14 (9) (2020) 1241–1257.

[14] T. Yi, H. Zhao, Q. Mo, D. Pan, Y. Liu, L. Huang, H. Xu, B. Hu, H. Song, From cellulose to cellulose nanofibrils—a comprehensive review of the preparation and modification of cellulose nanofibrils, Materials 13 (22) (2020) 5062.

[15] A. Isogai, Emerging nanocellulose technologies: Recent developments, Adv. Mater. (2020) 2000630.

[16] A. Dufresne, Nanocellulose: a new ageless bionanomaterial, Mater. Today 16 (6) (2013) 220–227.

[17] S. Thomas, D. Pasquini, S.Y. Leu, D.A. Gopakumar (Eds.), Nanoscale Materials in Water Purification, Elsevier, Amsterdam, 2018.

[18] J. Lemaitre, J.L. Chaboche, A. Benallal, R. Desmorat, Mécanique des matériaux solides, 3e éd, Dunod, 2020.

[19] R.A. Khan, S. Salmieri, D. Dussault, J. Uribe-Calderon, M.R. Kamal, A. Safrany, M. Lacroix, Production and properties of nanocellulose-reinforced methylcellulose-based biodegradable films, J. Agric. Food Chem. 58 (13) (2010) 7878–7885.

[20] H.M. Azeredo, L.H.C. Mattoso, D. Wood, T.G. Williams, R.J. Avena-Bustillos, T.H. McHugh, Nanocomposite edible films from mango puree reinforced with cellulose nanofibers, J. Food Sci. 74 (5) (2009) N31–N35.

[21] H. Fukuzumi, T. Saito, T. Iwata, Y. Kumamoto, A. Isogai, Transparent and high gas barrier films of cellulose nanofibers prepared by TEMPO-mediated oxidation, Biomacromolecules 10 (1) (2009) 162–165.

[22] R.H. Marchessault, F.F. Morehead, N.M. Walter, Liquid crystal systems from fibrillar polysaccharides, Nature 184 (4686) (1959) 632–633.

[23] N.C. Nepomuceno, A.S. Santos, J.E. Oliveira, G.M. Glenn, E.S. Medeiros, Extraction and characterization of cellulose nanowhiskers from Mandacaru (*Cereus jamacaru* DC.) spines, Cellulose 24 (1) (2017) 119–129.

[24] K. Uetani, T. Okada, H.T. Oyama, Crystallite size effect on thermal conductive properties of non-woven nanocellulose sheets, Biomacromolecules 16 (7) (2015) 2220–2227.

[25] A. Isogai, T. Saito, H. Fukuzumi, TEMPO-oxidized cellulose nanofibers, Nanoscale 3 (1) (2011) 71–85.

[26] S.Y. Lee, D.J. Mohan, I.A. Kang, G.H. Doh, S. Lee, S.O. Han, Nanocellulose reinforced PVA composite films: effects of acid treatment and filler loading, Fiber Polym. 10 (1) (2009) 77–82.

[27] D. Liu, T. Zhong, P.R. Chang, K. Li, Q. Wu, Starch composites reinforced by bamboo cellulosic crystals, Bioresour. Technol. 101 (7) (2010) 2529–2536.

[28] M. Nasir, R. Hashim, O. Sulaiman, M. Asim, Nanocellulose: Preparation methods and applications. In: M. Jawaid, S. Boufi, A. Khalil H.P.S (Ed.), Cellulose-Reinforced Nanofibre Composites, Woodhead Publishing, 2017, pp. 261–276.

[29] Y. Qiu, L. Qiu, J. Cui, Q. Wei, Bacterial cellulose and bacterial cellulose-vaccarin membranes for wound healing, Mater. Sci. Eng. C 59 (2016) 303–309.

[30] M. Osorio, P. Fernández-Morales, P. Gañán, R. Zuluaga, H. Kerguelen, I. Ortiz, C. Castro, Development of novel three-dimensional scaffolds based on bacterial nanocellulose for tissue

engineering and regenerative medicine: effect of processing methods, pore size, and surface area, J. Biomed. Mater. Res.A 107 (2) (2019) 348–359.

[31] V. Raghavendran, E. Asare, I. Roy, Bacterial cellulose: biosynthesis, production, and applications, Adv. Microb. Physiol. 77 (77) (2020) 89.

[32] J. Liu, S. Willför, A. Mihranyan, On importance of impurities, potential leachables and extractables in algal nanocellulose for biomedical use, Carbohydr. Polym. 172 (2017) 11–19.

[33] C.Q. Ruan, M. Strømme, J. Lindh, Preparation of porous 2, 3-dialdehyde cellulose beads crosslinked with chitosan and their application in Congo red dye's adsorption, Carbohydr. Polym. 181 (2018) 200–207.

[34] S. Zarei, M. Niad, H. Raanaei, The removal of mercury ion pollution by using Fe_3O_4-nanocellulose: synthesis, characterizations and DFT studies, J. Hazard. Mater. 344 (2018) 258–273.

[35] S.H. Song, K.Y. Seong, J.E. Kim, J. Go, E.K. Koh, J.E. Sung, H.J. Son, Y.J. Jung, H.S. Kim, J.T. Hong, D.Y. Hwang, Effects of different cellulose membranes regenerated from *Styela clava* tunics on wound healing, Int. J. Mol. Med. 39 (5) (2017) 1173–1187.

[36] H. Zhan, N. Peng, X. Lei, Y. Huang, D. Li, R. Tao, C. Chang, UV-induced self-cleanable TiO_2/nano-cellulose membrane for selective separation of oil/water emulsion, Carbohydr. Polym. 201 (2018) 464–470.

[37] A.F. Jozala, L.C. de Lencastre-Novaes, A.M. Lopes, V. de Carvalho Santos-Ebinuma, P.G. Mazzola, A. Pessoa-Jr, D. Grotto, M. Gerenutti, M.V. Chaud, Bacterial nanocellulose production and application: a 10-year overview, Appl. Microbiol. Biotechnol. 100 (5) (2016) 2063–2072.

[38] J. Pennells, I.D. Godwin, N. Amiralian, D.J. Martin, Trends in the production of cellulose nanofibers from non-wood sources, Cellulose 27 (2) (2020) 575–593.

[39] M. Ioelovich, Preparation and application of nanoscale cellulose biocarriers, CIBTech J. Biotechnol. 4 (2015) 19–24.

[40] H. Sehaqui, A. Liu, Q. Zhou, L.A. Berglund, Fast preparation procedure for large, flat cellulose and cellulose/inorganic nanopaper structures, Biomacromolecules 11 (9) (2010) 2195–2198.

[41] M. Innala, I. Riebe, V. Kuzmenko, J. Sundberg, P. Gatenholm, E. Hanse, S. Johannesson, 3D Culturing and differentiation of SH-SY5Y neuroblastoma cells on bacterial nanocellulose scaffolds, Artific. Cells Nanomed. Biotechnol. 42 (5) (2014) 302–308.

[42] K. Markstedt, A. Mantas, I. Tournier, H. Martínez Ávila, D. Hagg, P. Gatenholm, 3D bioprinting human chondrocytes with nanocellulose–alginate bioink for cartilage tissue engineering applications, Biomacromolecules 16 (5) (2015) 1489–1496.

[43] S. Dong, H.J. Cho, Y.W. Lee, M. Roman, Synthesis and cellular uptake of folic acid-conjugated cellulose nanocrystals for cancer targeting, Biomacromolecules 15 (5) (2014) 1560–1567.

[44] J.K. Jackson, K. Letchford, B.Z. Wasserman, L. Ye, W.Y. Hamad, H.M. Burt, The use of nanocrystalline cellulose for the binding and controlled release of drugs, Int. J. Nanomed. 6 (2011) 321.

[45] Y. Zhou, S. Fu, L. Zhang, H. Zhan, M.V. Levit, Use of carboxylated cellulose nanofibrils-filled magnetic chitosan hydrogel beads as adsorbents for Pb (II), Carbohydr. Polym. 101 (2014) 75–82.

[46] K. Singh, J.K. Arora, T.J.M. Sinha, S. Srivastava, Functionalization of nanocrystalline cellulose for decontamination of Cr (III) and Cr (VI) from aqueous system: computational modeling approach, Clean Technol. Environ. Policy 16 (6) (2014) 1179–1191.

[47] R. Batmaz, N. Mohammed, M. Zaman, G. Minhas, R.M. Berry, K.C. Tam, Cellulose nanocrystals as promising adsorbents for the removal of cationic dyes, Cellulose 21 (3) (2014) 1655–1665.

[48] W. Wang, T.J. Zhang, D.W. Zhang, H.Y. Li, Y.R. Ma, L.M. Qi, Y.L. Zhou, X.X. Zhang, Amperometric hydrogen peroxide biosensor based on the immobilization of heme proteins on gold nanoparticles–bacteria cellulose nanofibers nanocomposite, Talanta 84 (1) (2011) 71–77.

[49] M. Park, H. Chang, D.H. Jeong, J. Hyun, Spatial deformation of nanocellulose hydrogel enhances SERS, BioChip J. 7 (3) (2013) 234–241.

[50] T. Zhang, W. Wang, D. Zhang, X. Zhang, Y. Ma, Y. Zhou, L. Qi, Biotemplated synthesis of gold nanoparticle–bacteria cellulose nanofiber nanocomposites and their application in biosensing, Adv. Funct. Mater. 20 (7) (2010) 1152–1160.

[51] L. Hu, G. Zheng, J. Yao, N. Liu, B. Weil, M. Eskilsson, E. Karabulut, Z. Ruan, S. Fan, J.T. Bloking, M.D. McGehee, Transparent and conductive paper from nanocellulose, Energy Environ. Sci. 6 (2) (2013) 513–518.

[52] C. Endes, O. Schmid, C. Kinnear, S. Mueller, S. Camarero-Espinosa, D.Vanhecke, E.J. Foster, A. Petri-Fink, B. Rothen-Rutishauser, C. Weder, M.J. Clift, An in vitro testing strategy towards mimicking the inhalation of high aspect ratio nanoparticles, Part. Fibre Toxicol. 11 (1) (2014) 1–13.

[53] J.A. Shatkin, B. Kim, Cellulose nanomaterials: life cycle risk assessment, and environmental health and safety roadmap, Environ. Sci. 2 (5) (2015) 477–499.

[54] H. Muhle, H. Ernst, B. Bellmann, Investigation of the durability of cellulose fibres in rat lungs, Ann. Occup. Hyg. 41 (1997) 184–188.

[55] T. Kovacs, V. Naish, B. O'Connor, C. Blaise, F. Gagné, L. Hall, V. Trudeau, P. Martel, An ecotoxicological characterization of nanocrystalline cellulose (NCC), Nanotoxicology 4 (3) (2010) 255–270.

[56] K.A. Mahmoud, J.A. Mena, K.B. Male, S. Hrapovic, A. Kamen, J.H. Luong, Effect of surface charge on the cellular uptake and cytotoxicity of fluorescent labeled cellulose nanocrystals, ACS Appl. Mater. Interfaces 2 (10) (2010) 2924–2932.

[57] N. Yanamala, M.T. Farcas, M.K. Hatfield, E.R. Kisin, V.E. Kagan, C.L. Geraci, A.A. Shvedova, In vivo evaluation of the pulmonary toxicity of cellulose nanocrystals: a renewable and sustainable nanomaterial of the future, ACS Sustain. Chem. Eng. 2 (7) (2014) 1691–1698.

[58] E.J. Park, T.O. Khaliullin, M.R. Shurin, E.R. Kisin, N. Yanamala, B. Fadeel, J. Chang, A.A. Shvedova, Fibrous nanocellulose, crystalline nanocellulose, carbon nanotubes, and crocidolite asbestos elicit disparate immune responses upon pharyngeal aspiration in mice, J. Immunotoxicol. 15 (1) (2018) 12–23.

[59] A.A. Shvedova, E.R. Kisin, N. Yanamala, M.T. Farcas, A.L. Menas, A. Williams, P.M. Fournier, J.S. Reynolds, D.W. Gutkin, A. Star, R.S. Reiner, Gender differences in murine pulmonary responses elicited by cellulose nanocrystals, Part. Fibre Toxicol. 13 (1) (2015) 1–20.

[60] M.T. Farcas, E.R. Kisin, A.L. Menas, D.W. Gutkin, A. Star, R.S. Reiner, N. Yanamala, K. Savolainen, A.A. Shvedova, Pulmonary exposure to cellulose nanocrystals caused harmful effects to reproductive system in male mice, J. Toxicol. Environ. Health A 79 (21) (2016) 984–997.

[61] J. Catalán, M. Ilves, H. Järventaus, K.S. Hannukainen, E. Kontturi, E. Vanhala, H. Alenius, K.M. Savolainen, H. Norppa, Genotoxic and immunotoxic effects of cellulose nanocrystals in vitro, Environ. Mol. Mutagen. 56 (2) (2015) 171–182.

[62] R. de Lima, L.O. Feitosa, C.R. Maruyama, M.A. Barga, P.C. Yamawaki, I.J. Vieira, E.M. Teixeira, A.C. Corrêa, L.H.C. Mattoso, L.F. Fraceto, Evaluation of the genotoxicity of cellulose nanofibers, Int. J. Nanomed. 7 (2012) 3555.

[63] M. Ilves, S. Vilske, K. Aimonen, H.K. Lindberg, S. Pesonen, I. Wedin, M. Nuopponen, E. Vanhala, C. Højgaard, J.R. Winther, M. Willemoës, Nanofibrillated cellulose causes acute pulmonary inflammation that subsides within a month, Nanotoxicology 12 (7) (2018) 729–746.

[64] J. Magalhães, A. Vieira, S. Santos, M. Pinheiro, S. Reis, Oral administration of nanoparticles-based TB drugs. In: A.M. Grumezescu (Ed.), Multifunctional Systems for Combined Delivery, Biosensing and Diagnostics, Elsevier, Amsterdam, 2017, pp. 307–326.

[65] L. Liu, F. Kong, The behavior of nanocellulose in gastrointestinal tract and its influence on food digestion, J. Food Eng. 292 (2021) 110346.

[66] I.L. Bergin, F.A. Witzmann, Nanoparticle toxicity by the gastrointestinal route: evidence and knowledge gaps, Int. J. Biomed. Nanosci. Nanotechnol. 3 (1-2) (2013) 163–210.

[67] L. Liu, F. Kong, Influence of nanocellulose on in vitro digestion of whey protein isolate, Carbohydr. Polym. 210 (2019) 399–411.

[68] L. Liu, W.L. Kerr, F. Kong, Characterization of lipid emulsions during in vitro digestion in the presence of three types of nanocellulose, J. Coll. Interface Sci. 545 (2019) 317–329.

[69] L. Liu, W.L. Kerr, F. Kong, D.R. Dee, M. Lin, Influence of nano-fibrillated cellulose (NFC) on starch digestion and glucose absorption, Carbohydr. Polym. 196 (2018) 146–153.

[70] J. Nsor-Atindana, H.D. Goff, W. Liu, M. Chen, F. Zhong, The resilience of nanocrystalline cellulose viscosity to simulated digestive processes and its influence on glucose diffusion, Carbohydr. Polym. 200 (2018) 436–445.

[71] G.M. DeLoid, I.S. Sohal, L.R. Lorente, R.M. Molina, G. Pyrgiotakis, A. Stevanovic, R. Zhang, D.J. McClements, N.K. Geitner, D.W. Bousfield, K.W. Ng, Reducing intestinal digestion and absorption of fat using a nature-derived biopolymer: interference of triglyceride hydrolysis by nanocellulose, ACS Nano 12 (7) (2018) 6469–6479.

[72] A. Sarkar, S. Zhang, B. Murray, J.A. Russell, S. Boxal, Modulating in vitro gastric digestion of emulsions using composite whey protein-cellulose nanocrystal interfaces, Coll. Surf. B Biointerfaces 158 (2017) 137–146.

[73] Y. Chen, Y.J. Lin, T. Nagy, F. Kong, T.L. Guo, Subchronic exposure to cellulose nanofibrils induces nutritional risk by non-specifically reducing the intestinal absorption, Carbohydr. Polym. 229 (2020) 115536.

[74] G.M. DeLoid, X. Cao, R.M. Molina, D.I. Silva, K. Bhattacharya, K.W. Ng, S.C.J. Loo, J.D. Brain, P. Demokritou, Toxicological effects of ingested nanocellulose in in vitro intestinal epithelium and in vivo rat models, Environ. Sci. 6 (7) (2019) 2105–2115.

[75] M. Čolić, D. Mihajlović, A. Mathew, N. Naseri, V. Kokol, Cytocompatibility and immunomodulatory properties of wood-based nanofibrillated cellulose, Cellulose 22 (1) (2015) 763–778.

[76] K. Kümmerer, J. Menz, T. Schubert, W. Thielemans, Biodegradability of organic nanoparticles in the aqueous environment, Chemosphere 82 (10) (2011) 1387–1392.

[77] J. Nsor-Atindana, Y.X. Zhou, M.N. Saqib, M. Chen, H.D. Goff, J. Ma, F. Zhong, Enhancing the prebiotic effect of cellulose biopolymer in the gut by physical structuring via particle size manipulation, Food Res. Int. 131 (2020) 108935.

[78] B. Nowack, M. Baalousha, N. Bornhöft, Q. Chaudhry, G. Cornelis, J. Cotterill, A. Gondikas, M. Hassellöv, J. Lead, D.M. Mitrano, F. von der Kammer, Progress towards the validation of modeled environmental concentrations of engineered nanomaterials by analytical measurements, Environ. Sci. 2 (5) (2015) 421–428.

[79] H. Stockmann-Juvala, P. Taxell, T. Santonen, Formulating occupational exposure limits values (OELs)(Inhalation & Dermal). Scaffold Public Documents-Ref, ScaffoldSPD7, Finnish Institute of Occupational Health (FIOH), 2014.

[80] P. Kautto, H. Valve, Cosmopolitics of a regulatory fit: the case of nanocellulose, Sci. Cult. 28 (1) (2019) 25–45.

[81] J.A. Shatkin, B. Kim, Environmental Health and safety of cellulose nanomaterials and composites, In: H. Kargarzadeh, I. Ahmad, S. Thomas, A. Dufresne (Eds.), Handbook of Nanocellulose and Cellulose Nanocomposites 1, Wiley online library, Wiley-VCH Verlag GmbH & Co. KGaA, 2017, pp. 683–729, Print ISBN:9783527338665, Online ISBN:9783527689972, doi:10.1002/9783527689972©.

CHAPTER 15

Integration of geospatial technology for mapping of algae: an economical perspective for assessing nanocellulose

Anamika Shalini Tirkey[a], Shashikant Shivaji Vhatkar[b], Ramesh Oraon[b]
[a]Department of Geoinformatics, School of Natural Resource Management, Central University of Jharkhand, Brambe, Ranchi, Jharkhand, India
[b]Department of Nanoscience and Technology, Central University of Jharkhand, Brambe, Ranchi, Jharkhand, India

15.1 Introduction

The emanation of advanced, new nanomaterials is redefining many industries and blurring traditional research boundaries. The increasing use of renewable, biodegradable, environmentally friendly materials as a substitute for nonrenewable resources has generated great interest worldwide. Besides, emerging environmental concerns warrant the use of biorenewable polymers in the manufacture of nanomaterials. The high reactivity and large surface area of nanomaterials are some of the notable features that provide an advantage in environmental remediation over other conventional alternatives [1]. Promising materials such as graphitic carbon nitride [2,3], carbon nanodots [4], and 2D carbon-based nanocomposites [5–7] are some of the most popular nanomaterials that have extensive applications in both environmental remediations as well as in energy production.

Cellulose, one of the most abundant natural polymers, has the potential to overcome challenges related to material biodegradability, renewability, cost, and energy. Besides, cellulose found in most lignocellulosic biomass, derived of agricultural waste, energy crops, and forestry residues provides a better option over food crops as a renewable resource for the production of new materials due to its availability in large quantities and low cost [8]. In recent decades, nanocellulose has shown good performance in numerous applications, including paper making [9], water [10] and wastewater treatment [11], biomedical engineering [12], and energy production [13]. New cellulose products such as cellulose nanocrystal (CNC) and microfibrillated cellulose or cellulose microfibril were mentioned for the first time by Ranby and Turbak et al. [14] and since then it is being produced by several different organizations. As an example, the potential applications of nanocellulose are gaining a significant level of attention and have been reviewed extensively.

Nanocellulose Materials: Fabrication and Industrial Applications
DOI: https://doi.org/10.1016/B978-0-12-823963-6.00015-6

15.1.1 Sources of nanocellulose and their impact on economy

Cellulose has to be extracted from a source before nanocellulose can be produced. Microorganisms including algae, bacteria, and tunicates can produce nanocellulose directly [15]. Of these, algae have a great impact on the environment. The green alga, known by its scientific name *Cladophora*, is known for its use in the production of nanocellulose. Recent research on artificial photosynthesis such as CO_2 fixation and water splitting has increased the appeal of algae and hence much attention has been paid to its use for environmental remediation. Filters made from *Cladophora* algae cellulose (also known as Shiogusa seaweed in Japan) have been tested and proven to trap swine influenza virus particles with a retention efficiency that matches that of industrial filters [16]. Besides, adsorbent beads made with *Cladophora* cellulose for rapid adsorption of palladium (II) ions (of up to 80% maximum capacity in 2 h) [17] could develop an alternative for e-waste management by recovering and recycling palladium through low-cost and efficient technology [18]. So, algae as a supply for nanocellulose production could increase cellulose production significantly economically and reduce greenhouse gases by absorbing CO_2 compared to bacteria. The filamentous algae *Cladophora glomerata* L. (Kuetzing) however is a worldwide nuisance algae species that due to its cultural eutrophication negatively affects both dissolved oxygen and pH in lakes and rivers. Widespread algal blooms also affect property values, tourism, and recreation. Therefore, mapping spatial algal cover, including *Cladophora*, is an important component of water-quality science and management that can be used to characterize the ecological well-being and aesthetic perception of a water body. The use of these green filamentous algae for the production of nanocellulose is attractive as it can help resolve problems related to water pollution in coastal areas.

The CNCs are considered an emerging nanomaterial based on their commercial production [19] and their potential to solve new and existing materials problems [20]. The escalating climate change will tremendously affect freshwater and marine environments and may undergo nutrient pollution, causing harmful algal blooms (HABs) to occur more often. This may endanger human health, the environment, and economies globally. Therefore, in the drive for sustainable development, CNCs can sustainably be used for water purification systems such as using hydrogel beads for the removal of aqueous dyes [21], water filtration membranes [10,22,23], nanocomposite heavy metal sensors and absorbents [24,25], aerogels [26], flocculants, and nanocomposite filters for groundwater remediation [27], which can significantly reduce the costs incurred in cleaning and water treatment.

15.1.2 Remote sensing

Various remote sensing techniques have been developed to map the spatial coverage of *Cladophora*. Satellite or aerial measurement of spectral reflectance (ocean color) is an effective method to monitor phytoplankton for its proxy, concentration of chlorophyll–a, the green pigment present in all algae [28]. Remote sensing offers a practical and inexpensive means of discriminating and estimating the biochemical and biophysical parameters of wetland species and can make field sampling more focused and efficient, which is otherwise labor-intensive, costly, time-consuming, and sometimes inapplicable due to poor accessibility, and therefore impractical on relatively small areas [29]. Repetitive coverage offers temporal datasets for analyzing change detection over time, which can be integrated into geographic information system (GIS) for more analysis [30]. Several researchers have used both multispectral data such as Landsat TM and SPOT images to identify general vegetation classes or to attempt to discriminate broad vegetation communities [31], whereas the hyperspectral satellite datasets are being used to discriminate and map wetland vegetation at species level [32,33].

15.2 Remote sensing applications for mapping spatial algal cover and harmful algal blooms (HABs)

When light enters water, it is absorbed and scattered by pure water, dissolved organics, and organic/inorganic particles [34]. A typical spectral reflectance curve of clear water and algae-laden water is shown in Fig. 15.1. Clear water scatters the blue wavelength of the electromagnetic spectrum, making the ocean to appear blue, while chlorophyll-a strongly absorbs the blue wavelengths. As the chlorophyll concentration increases, more of the blue wavelength is absorbed, and the green wavelength is reflected back by the phytoplankton [35].

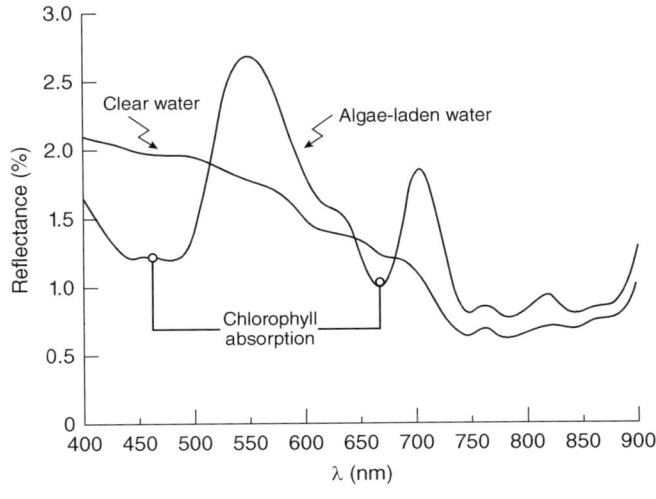

Fig. 15.1 Spectral reflectance curve for clear and algae-laden water (Source: [35]).

Thus, the open ocean regions with low primary productivity appear blue in satellite images, and the coastal upwelling regions and areas near river mouths, appear green as nutrients are added from the sea bottom and river runoffs with high plankton concentrations and high primary productivity [36,37]. The wind-induced upwelling in coastal regions brings nutrients to the surface and creates patchy zones with high biological productivity accompanied by high concentrations of chlorophyll that can be detected by satellite sensors [38,39]. Furthermore, marine phytoplanktons exhibit high spatial and temporal variability, the size of which is difficult and tedious to map using conventional methods especially at regional and global scales, which is why remote sensing datasets are advantageous for ocean-color mapping. The temporal resolution of satellite datasets enables frequent coverage on a basin-scale and monitoring of HABs and therefore has been employed by scholars for their detection and forecasting [40–43]. Various multispectral sensors using visible bands have been used to detect of HAB species (Table 15.1) [44].

Table 15.1 HABs for which remote sensing is currently being used [44,45].

HAB species	Region (example)	Sensing type	Impact
Diatoms (Bacillariophyceae)			
Pseudo-nitzschia spp.	Upwelling regions (West Coast, USA)	Sea surface temperature (SST), chlorophyll	Amnesic shellfish poisoning (ASP), harmful to copepods, invertebrates, marine mammals, fish, birds
Dinoflagellates (Dinophyceae)			
Karenia brevis (=Gymnodinium breve, Ptychodiscus breve)	Open ocean (Gulf of Mexico, USA)	Chlorophyll, optics, SST (in some cases)	Neurotoxic shellfish poisoning (NSP), respiratory irritation, toxic to marine organisms.
Karenia mikimotoi (=*Gymnodinium aureolum*)	Coastal ocean (Hong Kong, English channel, Ireland)	SST, chlorophyll	NSP, ichthyotoxic, harmful to zooplankton and marine invertebrates
Gymnodinium catenatum	Estuaries, coastal ocean, upwelling regions (Portugal, Spain, South China, Mexico)	SST, chlorophyll	Paralytic shellfish poisoning (PSP)
Alexandrium spp.	Coastal ocean (Gulf of Maine, USA)	SST	PSP, toxic to marine organisms
Dinophysis acuminata	Estuaries, coastal ocean (South Africa, Spain)	SST	Diarrheic shellfish poisoning (DSP)
Gonyaulax spinifera	Upwelling regions (British Columbia)	AVHRR band ratio	Harmful to marine organisms
Lingulodinium polyedra (=*Gonyaulax polyedra*)	Coastal ocean, upwelling regions (California)	UV absorption chlorophyll	Toxic to marine organisms

HAB species	Region (example)	Sensing type	Impact
Cochlodinium polykrikoides	Coastal ocean (British Columbia)	SST	Ichthyotoxic, molluscicidal
Noctiluca scintillans	Estuaries (Bohai Bay-China)	AVHRR visible, SST	Ichthyotoxic
Prymesiophyte			
Phaeocystis pouchetti	Estuaries (Loch Striven-Scotland)	Chlorophyll	Ichthyotoxic
Cyanobacteria			
Nodularia spumigena	Enclosed sea (Baltic Sea)	Color, AVHRR visible	Hepatotoxic, toxic to zooplankton
Microcystis spp.	Estuary, lakes (Lake Erie, Lake Pontchartrain, USA)	Color	Hepatotoxic, ichthyotoxic
Pelagophyte			
Aureococcus anophagefferens	Estuaries, upwelling regions (South Africa)	Chlorophyll from aircraft	Harmful to invertebrates, ecosystem impacts

It is evident that thermal infrared imagery can complement the visible band multispectral data by identifying upwelling areas, large river plumes, and water temperatures conducive to algal bloom development. Satellite sensors measure the spectral radiances at the upper end of the atmosphere from which, after atmospheric and radiometric corrections, the spectral radiances emerging from the ocean surface are extracted [46,47]. The surface radiances are converted to reflectances, providing scientists with the spectral signatures required for identifying chlorophyll and other water constituents. To produce valid data, such as global ocean chlorophyll concentrations, a meticulous calibration and validation approach must be used [34,48,49]. Instrumented ships, buoys, and ocean gliders are used to calibrate and validate maps of chlorophyll-a and total suspended sediment obtained with ocean-color sensors. In coastal and estuarine waters, this data must often be collected very close to the satellite passage time and must be statistically representative of the prevailing conditions. The water samples are usually taken from the top half meter of the water column. Sites for calibrating remotely sensed data, such as chlorophyll concentrations in coastal waters, must be located at well-known points representing the entire range of variables to be measured [28]. Remotely sensed concentrations of chlorophyll-a and total suspended sediments can be used as indicators of the severity of the eutrophication and turbidity, respectively. Satellite sensors can give satisfactory results even with fewer spectral bands than the hyperspectral images used to measure precise concentration levels [50–52]. Table 15.2 shows the criteria for estuarine water quality according to the National Coastal Assessment Program of the Environmental Protection Agency.

In their study, Wang and Shi [54] demonstrated the use of Moderate Resolution Imaging Spectroradiometer (MODIS) on the National Aeronautics and Space Administration

Table 15.2 United States Environmental Protection Agency National Coastal Assessment Program criteria for assessing chlorophyll-a conditions [53].

Parameter	Value
Water quality vs chlorophyll–a concentration (μg/L)	
Oligotrophic (good)	<5
Mesotrophic (fair)	5–20
Eutrophic (poor)	>20
Water quality vs total suspended sediments (mg/L)	
Clear	<10
Moderately turbid	10–50
Highly turbid	>50

(NASA) Aqua satellite to monitor a massive blue-green algae bloom (Microcystis) in Lake Taihu, China, which caused an environmental crisis and caused officials to limit the tap water supply to several million residents. The MODIS data (with a resolution of 1 km, 0.25 km, and 0.5 km) provided synoptic observations of the entire bloom event. Fig. 15.2 shows the MODIS image of chlorophyll-a concentration and normalized water-leaving radiance at 443 nm. The spatial variation of the lake's optical and

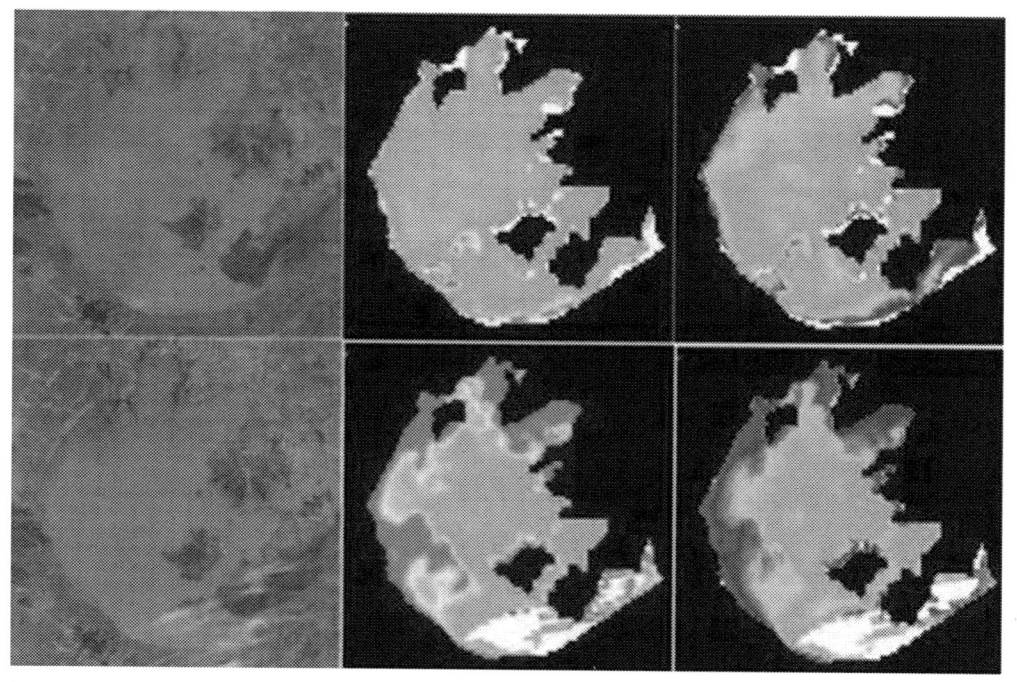

Fig. 15.2 MODIS true color images (first column), derived chlorophyll-a concentration (second column), and normalized water-leaving radiance at 443 nm (third column) for Lake Taihu, before (top) and during (bottom) the peak of the bloom event (Source: [54]).

Table 15.3 Some satellite remote sensing systems used to measure ocean color (Modified from [56]).

Sensor	Agency	Satellite	Operating dates	Spatial resolution (m)	Number of bands	Spectral coverage (nm)
CZCS	NASA	Nimbus-7	1978–1986	825	6	433–12,500
SeaWiFS	NASA	OrbView-2	1997–2010	1100	8	402–885
MODIS-Terra	NASA	Terra	Launched 1997	250/500/1000	36	405–14,385
MODIS-Aqua	NASA	Aqua	Launched 2002	250/500/1000	36	405–14,385
MERIS	ESA	Envisat-1	Launched 2002	300/1200	15	412–1050

Note: MODIS instrument is carried aboard two platforms (Terra and Aqua). NASA denotes the National Aeronautics and Space Administration; ESA denotes the European Space Agency. SeaWiFS denotes Sea-viewing Wide Field-of-view Sensor.

biological properties is depicted for both before and during the peak of the bloom event (Fig. 15.2). Due to the significant radiance contributions from the highly turbid lake at near-infrared wavelengths, longer short-wave infrared bands had to be used for atmospheric corrections [55].

Algal blooms that have high concentration of pigment result in reduced radiance in blue band (443 nm) due to the strong absorption by algae at that wavelength. A description of the remote-sensing satellites used to measure ocean color is given in Table 15.3. The ocean-color products obtained from these satellites define the optical properties of the water for a wide range of applications, such as bathymetry, euphotic depth estimation, water quality (turbidity), and diver visibility [57].

Although cloud cover hinders the mapping of species in all optical sensors, the disadvantage of using Sea-viewing Wide Field-of-view-Sensor (SeaWiFS) is that there are very few species-specific algorithms that can be used to determine the concentrations of the most important harmful algae from the water-leaving radiance. High concentrations of algal cells can be present deeper in water columns than can be passively mapped. Even if they are on or near the surface, the spatial resolution of the sensor is limited to 1 km as the bloom approaches the coast, where tracking may be most important. SeaWiFS, MODIS, and Medium Resolution Imaging Spectrometer (MERIS) were developed for monitoring the optical properties of marine waters [58–60]. Their spectral coverage and resolution (bandwidths of 20 nm) and spatial resolution (250 m to 1200 m pixel size) are mainly useful for covering large areas of open water, with color deviations mainly depending on the presence of planktonic pigments. Coastal and estuarine blooms are often characterized by optically complex, patchy waters with highly reflective components and colors that are strongly influenced by dissolved organic matter and suspended sediments. The mapping of estuarine or coastal eutrophic areas requires high spatial resolution. Therefore, sensors with high spatial resolution such as Landsat Thematic Mapper (30 m) have proven useful for studying eutrophic conditions along the coast, despite their lower spectral resolution and sensitivity [35,61,62]. A more promising application of earth observation is to identify environmental conditions that are suitable for bloom development, such as increasing water temperature and nutrients supply, and tracking the progress of a bloom as it moves onshore. This technology could make it possible to predict when and where a coastal region would be affected by an HAB. One such example is tracking sea surface temperature (SST) features, such as fronts, where HAB species are likely to accumulate, using Advanced Very High Resolution Radiometer (AVHRR) data [63,64]. Alternatively, variations in ocean color can be used to detect abnormally high levels of chlorophyll, which may indicate an impending bloom [65]. Many river and bay plumes, oceanic debris, and coastal fronts can be detected remotely due to their strong surface signatures caused by high turbid color, and temperature gradients [66]. The dynamic behavior of coastal plumes and fronts has been tracked with Landsat and SeaWiFS imagery [67–70]. A relatively small number of multi-spectral bands can be used to study the dynamics of coastal plumes. However, it is difficult to determine the composition and concentration of their content even with hyperspectral images that have more than a hundred spectral bands.

15.2.1 Remote sensing–based algal spectral indices

Several remote sensing–based spectral indices (band ratio algorithms) assist in identification and mapping of phytoplankton blooms. The University of California, Santa Cruz, and the NASA are some institutions that have applied various indices to extract algal bloom information. Table 15.4 describes some of the remote sensing–based spectral indices used to identify and map algal blooms. Although each of these indices is specifically designed for the aquatic environment in question, the NGRDI is particularly useful in detecting algal blooms in coastal environments. The red and near infra-red (NIR) band incorporated indices perform fairly successfully in river and estuarine environments, whereas indices that incorporate the green or blue bands are utilized more frequently for lake and coastal bloom identification. Other algorithms for delineating areas affected by blooms are the Adaptive Cosine Estimator (ACE), Spectral Angle Mapper (SAM), Fast Non–Negative Least Squares Solution (FNNLS), Vertex Component Analysis (VCA), and Harsanyi–Farrand–Chang (HFC).

In addition, researchers can create spectral library repositories using hyperspectral sensors [78] for algal bloom research. Spectral libraries based on hyperspectral data can help remote-sensing experts and biologists to identify specific species of harmful blooms, thereby enhancing the ability of public health officials to address and predict disease outbreaks caused by these events.

15.2.1.1 Cellulose absorption index (CAI)

Plant cellulose fibers are one of the most intensively studied and important sources for the production of nanocellulose. Plant fibers used as sources of cellulose can be classified

Table 15.4 Remote sensing–based spectral indices for identification and mapping of algal blooms [71].

Sl. no	Algal bloom index	Wavelength/formulae employed	Reference
1.	Cyanobacteria index (CI)	681 nm, 665 nm, and 709 nm	[72]
2.	*Aphanizomenon-Microcystis* index (AMI)	640 nm, 510 nm, 642 nm, and 625 nm	[72]
3.	Scattering line height (SLH) index	714 nm and 654 nm	[72]
4.	Normalized difference vegetation index (NDVI)	(NIR − Red)/(NIR + Red)	[73]
5.	Normalized green red difference index (NGRDI	(Green − Red)/(Green + Red)	[74]
6.	Blue normalized difference vegetation index (BNDVI)	(NIR − Blue)/(NIR + Blue)	[75]
7.	Normalized green blue difference index (NGBDI)	(Green − Blue)/(Green + Blue)	[74]
8.	Green normalized difference vegetation index (GNDVI)	(NIR − Green)/(NIR + Green)	[73]
9.	Green leaf index (GLI)	(2 × Green − Red − Blue)/ (2 × Green + Red + Blue)	[74]
10.	Excess green (EXG)	2 × Green − Red − Blue	[74]
11.	Algal bloom detection index (AI)	$((R_{850nm} - R_{660nm})/(R_{850nm} + R_{660nm})) + ((R_{850nm} - R_{625nm})/(R_{850nm} + R_{625nm}))$	[76,77]

into six groups: bast fibers, core fibers, grass and reed fibers, leaf fibers, seed fibers, and other fibers [79]. The CAI is a commonly used vegetation index in the field of remote sensing that discriminates plant materials from pure surfaces. The determination of this index is based on the reflectance characteristics of the spectral bands of 2000 nm, 2100 nm, and 2200 nm [80,81]. The CAI ($0.5(R_{2.0} + R_{2.2}) - R_{2.1}$) describes the average depth of the cellulose absorption characteristic at 2.1 μm in reflectance spectra. Positive CAI values indicate the presence of cellulose.

The imaging spectrometer CAI and the multispectral lignin–cellulose absorption index are reflectance band height indices that use three spectral bands between 2000 and 2400 nm to estimate crop residue cover [80,81]. Table 15.5 lists the sensors and the spectral bands used in the evaluation of CAI.

15.3 Unmanned aerial vehicles (UAVs)

UAVs have recently emerged in algal bloom studies, providing users with on-demand high spatial and temporal resolution at a lower cost when needed. Due to the challenges of processing water images, payload costs and restrictions, and a lack of standardized

Table 15.5 Sensors used in evaluating cellulose absorption index [82–85].

S. No	Name	Spectral range	Bands	Specific formula
1.	AISA	400–2506.7	498	$100(0.5(420+449)-431)$
2.	ARIES-1	400–2587	105	$100(0.5(68+80)-73)$
3.	AVIRIS	355.15–2515	224	$100(0.5(176+194)-183)$
4.	DAIS-7915	400–12,600	79	$100(0.5(42+51)-45)$
5.	EnMap	420–2450	244	$100(0.5(202+220)-209)$
6.	HYDICE	400–2500	210	$100(0.5(163+181)-170)$
7.	HyMap	429–2493.8	126	$100(0.5(99+109)-103)$
8.	Hyperion	349.896–2582.28	242	$100(0.5(B188+B206)-B195)$
9.	Orbview-4	450–2450	200	$100(0.5(158+176)-165)$
10.	SASI600	950–2450	100	$100(0.5(72+84)-77)$

methods, UAV-based algal bloom studies have not achieved critical traction. The details of the UAV platforms, onboard sensors, and the indices used in algal bloom mapping are given in Table 15.6.

15.3.1 Benefits of UAVs in algal bloom research

UAVs can be used in places where it is difficult to reach areas affected by algal blooms. They save both money and time [75,77,86–88]. They reduce the cost of *in-situ* water quality sampling. In addition, UAV-based algal bloom research can reduce the amount of fieldwork required to perform water-quality measurements and can reduce the environmental and health impacts of traversing protected, fragile, or toxic ecosystems [89]. In addition, UAV-based spectral identification (via spectral indices) can increase the accuracy of detection of surface blooms with high biomass due to the increased temporal and spatial resolution and the ability to see what the naked eye cannot (via thermal imaging and hyperspectral classifications) [74,88,90,91]. UAV-based algal research can be more flexible as users can choose their own flight paths, spatial, spectral, temporal resolutions, and revisit times, which is critical as changes in phytoplankton biomass in space and time have high variability and may have revisit times as frequent as everyday [74,87,88,90,91]. Repeated flights over specific blooms at high spatial resolution can help researchers understand seasonal patterns of phytoplankton dynamics or identify potential nutrient inputs to a water system, helping to identify contributors to algal proliferations [92]. Another advantage of using UAVs for algal bloom research is that its image acquisition is not prohibited by cloud cover, which can be a major limitation to satellite imagery analysis of algal proliferations [93]. However, in their current states, satellite and especially airborne sensors used to quantify chlorophyll content or algal biomass generally provide more narrow-band capabilities than do UAVs, [94] with the exception of UAV-based thermal [95] or hyperspectral studies [96].

Table 15.6 UAVs used in algal bloom mapping and the parameters/indices mapped (Modified from [71]).

Target, location	UAV platform	Sensor(s)	Parameters and indices	Validation
Lake cyanobacteria, Chapultepec Park, Mexico	Multirotor: DJI Phantom 3	12-megapixel red green blue (RGB) camera	Chlorophyll-a at 544 nm, phycocyanin absorption at 619 nm	*In-situ* upwelling and downwelling irradiances with hyperspectral spectroradiometer (GER-1500), cyanobacteria sampling, meteorological data
Cyanobacterial mats, Antarctica	Fixed-wing: modified UAV (Kevlar fabric, Skycam UAV Ltd. airframe) "Polar fox"	Sony NEX 5 RGB and near-infrared (NIR) cameras	Not specified	*In-situ* spectral reflectance using multi and hyperspectral cameras. Field microscopy of cyanobacterial mats.
Shallow river algae (*Cladophora glomerata*) in Clark Fork River, Montana, USA	Multirotor: DJI (unspecified)	GoPro Hero 3 with 12MP RGB camera	ACE, SAM	*In-situ* data of total suspended sediment samples, light attenuation profiles, daily streamflow
Cyanobacteria biomass, Rhode Island, USA	Multirotor: 3DR X8+ Multirotor: 3DR Iris+	Tetracam Mini-MCA6 multispectral camera 4 MAPIR Survey 1 cameras	NDVI, GNDVI	In-situ water quality data (not specified)
Shallow lake algae, Finland	Helicopter UAV: not specified	Fabry-Perot interferometer (FPI)	500–900 nm filter	*In-situ* radiometric water measurements. Signal-to-noise ratios.
River algal blooms, Daecheong Dam, South Korea	Fixed-wing: Ltd'seBee	Canon S110 RGB and NIR cameras	AI	*In-situ* field spectral reflectance
River algal blooms, Nakdong River, South Korea	Fixed-wing: Ltd'seBee	Canon Powershot S110 RGB and NIR camera	AI	*In-situ* water quality, microscopy analysis of phytoplankton species field spectral reflectance
Pond Cyanobacteria, Kingston, Canada	Octorotor: UAV not specified with Pixhawk as autopilot	Sony ILCE-6000 mini SLR camera (RGB) with MTI-G-700 gimbal Sony HX60v used for test flights	Not specified	Only technical validation performed

(continued)

Table 15.6 (Cont'd)

Target, location	UAV platform	Sensor(s)	Parameters and indices	Validation
Lake blue-green algal distribution, water turbidity, Finland	Helicopter UAV: not specified	Fabry-Perot interferometer (FPI) CMV4000 4.2 Megapixel CMOS camera	SAM, HFC, VCA, FNNLS	*In-situ* water quality measurements (Temp, conductance, turbidity, chlorophyll a, blue-green algae distribution)
Estuarine phytoplankton bloom (*Phaeocystis globosa*), Western Taiwan Strait and Weitou Bay, China	Fixed-wing: LT-150, from TOPRS Technology Co., Ltd.	AvaSpec-dual spectroradiometer with 2 sensors that span 360–1000 nm	Not specified	*In-situ* chlorophyll a sampling and spectroradiometer measurements via ship
Algal cover in Tain-Pu Reservoir, Taiwan, China	Fixed-wing: Ltd'seBee	Canon Powershot S110 RGB & Canon Powershot S110 NIR cameras	Chlorophyll a regression model	*In-situ* total phosphorus, chlorophyll a, Secchi disk depth
Lake cyanobacterial (Microcystis) buoyant volume, Kansas, USA	Fixed-wing: Zephyr sUAS from RitewingRC; Controller: Ardupilot Mega 2.6 Multirotor: DJI F550; Controller: NAZA V2	Canon Powershot S100 NDVI camera Multirotor gimbal (Gaui Crane II) and real-time video (ReadymadeRC)	BNDVI	*In-situ* microscopy identification of cyanobacterial genus
Green tide algae (on *Pyropia yezoenis* rafts) biomass in aquaculture zones, Yellow Sea, China	Multirotor: DJI Inspire 1	DJI X3 RGB camera	NGRDI, NGBDI, GLI, EXG	*In-situ* samples of green algae attached to *P. yezoenis* in the aquaculture zone

ACE: Adaptive cosine estimator; SAM: Spectral angle mapper; GNDVI: Green normalized difference vegetation index; AI: Algal bloom detection index; HFC: Harsanyi-Ferrand-Chang; VCA: vertex component analysis; FNNLS: Fast non-negative least squares solution; BNDVI: Blue normalized difference vegetation index; NGRDI: Normalized green red difference index; NGBDI: Normalized green blue difference index; GLI: Green leaf index; EXG: Excess green

Table 15.7 Advantages of fixed-wing UAVs and rotorcraft UAVs for algal bloom studies [71].

Fixed-wing UAVs	Rotorcraft UAVs
Has longer flight time than rotorcraft due to aerodynamic factors that enable survey of larger blooms with extended battery power and greater potential to fly multiple sensors to measure chlorophyll-a concentrations	Convenient for observing small or microscopic colonies of algae due to their higher spatial resolution potential
Suitable for algal blooms that affect larger spatial extents	Advantageous for closer analysis to help identify phytoplankton groups
Difficult to find substantial areas of flat, dry regions to safely launch and land adjacent to water bodies	Vertical take-off and landing abilities enable for surveying closely

15.3.2 UAV platforms used in algal bloom researches

Two main platforms used in UAV-based algal bloom studies are fixed-wing and rotor-craft UAVs (Table 15.6). Most fixed-wing vehicles are eBee platforms, while the majority is multirotor Da-Jiang innovations (DJI) platforms. UAV platforms used for algal bloom studies can also benefit from custom waterproofing or surface water landing capabilities demonstrated in UAV-based water-quality sampling research. Some key advantages of fixed-wing UAVs and rotorcraft UAVs are described in Table 15.7.

15.4 Sensors and cameras used in remote sensing for algal bloom mapping

The list of sensors and cameras for algal bloom studies is described in Table 15.8.

Table 15.8 Sensors and cameras used in algal bloom mapping [71].

Sensors	Cameras
Hyperspectral sensors: i. Fabry–Perot interferometer (FPI) (range: 400–1000 nm), ii. OCI-1000 (measures in the visible and near-infrared and has a payload of 180 g iii. Headwall's Nano-Hyperspec sensor (range: 380–2500 nm, weight: 680 g) iv. microHSI (range: 400–2400 nm, weight: 450 g–2.6 kg) v. Ocean Optics USB4000 (range: 350–1000 nm, weight: 190 g) AvaSpec-dual spectroradiometer with two sensors (range: 360–1000 nm)	RGB cameras, or RGB + infrared (IR) cameras (multispectral cameras) Hyperspectral cameras offering greater precision in regard to species or genus compositions of algal blooms (especially for discerning phytoplankton function types and toxicity potential)

15.5 Quantitative analyses

High surface concentrations of algal biomass can be quantified by producing results such as absorption value maps of *in-situ*-validated species, such as *Microcystis aeruginosa* [90], or chlorophyll-a concentration maps using precise reflectance spectra from radiometric UAV data [97]. The surface area or biomass of the bloom can also be quantified using the math toolbox available in ArcGIS software in the GIS environment. Hyperspectral sensors with more than a hundred spectral bands help with quantitative research of algal blooms. Currently, most UAV-based algal bloom studies provide first-level assessments of the spatial and temporal extents of blooms.

15.6 Future opportunities

15.6.1 Hyperspectral UAVs

Airborne platforms that use hyperspectral sensors to detect and map algal blooms provide good spectral information and are advantageous for identifying specific algal species. There are thousands of species of phytoplankton; ~300 species that cause blooms or red tides; and approximately 80 species that produce toxins that can negatively affect fish and shellfish consumers [98]. Hyperspectral images provide more continuous spectral information about the reflectivity of algae, and this is important because species have their own unique spectral signatures. The use of hyperspectral sensors on UAVs increases the ability to detect different groups of phytoplankton and offers higher spatial and temporal resolution. This information can be integrated into a program for predicting HABs [99]. Successfully used algae indices from airborne and satellite sensors will augment HAB species monitoring and their management in a sustainable way.

15.7 Socioeconomic impact

Anthropogenic activities have negatively affected socioeconomic aspects globally in regard to algal ecosystem. As a result to aid economic development, the need to possess geospatial algae knowledge is more concerning. UAVs facilitate these accurately by providing better resolution images of several algae kind. Furthermore, UAVs are also capable of incorporating engineering designs through several innovative approaches such as BwN (building with nature). Apart from this, technological implementations of UAVs (such as those equipped with LiDAR) assist sensitive monitoring of morphological changes in algal bodies. As water bodies act as an interface between algal body and its nutrient, they have also been resulting in significant economical advantages particularly when implemented along with engineering technologies. Furthermore, the economical advantage is also extended toward algal vegetation in potable as well as wastewater, such as fisheries, lakes thereby resulting in social as well as economical benefits. The abundance of algal communities owing to their climatic adaptation and its economical effects by agricultural industries, oil and plastic markets, etc., can certainly be expected to increase. For instance, algal production costs can already be seen translated through several technological implementations such as through fuel markets, biorefineries, and energy industries. Further, this need is also driven by stringent regulations imposed by governments over use of algal biomass replacing plastics and fossil fuels. This can be realized by a recent report (Transparency Market Research) which had projected from US\$ 608 million in 2016 to US\$ 1143 million till 2024 with a compound annual growth rate (CAGR) of 7.39% [http://www.transparencymarketresearch.com/algae-market.html].

Additionally, 42% and 53.7% of global algal biomass has been accounted to biofuel productions and pharmaceutical applications respectively. However, resulting expensive inputs into algal productions have been major hurdle in extensive commercialization and technological implementations. For example, the dependency of algal fuel production on floating costs of fossil fuels had impacted the algal biomass investments.

15.8 Climate change

CO_2 emissions have defined earths' temperature, ever since the industrial revolutions, these levels have substantially spiked. This spike is primarily attributed to the technological advances that have severely impacted the global CO_2 levels. This further influences the greenhouse effect, thereby resulting in climate change. Among many other factors, fossil fuel combustions required in powering transportation, energy, oil, and manufacturing industries are one of the major contributing factors responsible for climate change. Furthermore, the increase in world population since last century has also caused increased material demands, affecting overall economy, thereby resulting in environmental distress. For instance, melting ice sheets, glaciers are among many consequences of climate change. However, with recent advances in nanotechnology, implementing green methodologies, and sustainable approaches have attributed in improving the current scenario. Among many such technological implementations, remote sensing and geographical information systems have largely contributed to efficient control, monitoring, and protection of ecological resources. These technologies further de-stress environment by effective utilization, reuse, of natural and exploration of renewable resources through sustainable practices.

15.9 Conclusions

Cellulose, one of the most abundant natural polymers, has the potential to overcome challenges related to material biodegradability, renewability, cost, and energy. In recent decades, nanocellulose has shown good performance in numerous applications, including paper making, water, and wastewater treatment, biomedical engineering, and energy production. Compared with other nanocelluloses, CNCs have relatively green process, as the starting material is renewable, the acid can be recycled, and the degraded sugars may be separated for biofuel production. Also, their rigid structure, strength, amphiphilic nature, chemical purity, optical properties, ability to be completely redispersed from dried powder make them exceptional.

Various remote-sensing techniques have emerged to map the spatial coverage of *Cladophora* known for its use in the production of nanocellulose. The high spatial and temporal resolution of satellite datasets makes it possible for monitoring the marine phytoplanktons which exhibit high spatial and temporal variability. Furthermore, various multispectral sensors using visible bands have been used by scholars to detect HAB species. The remote sensing–based spectral indices (band ratio algorithms) have added to the monitoring and identification of phytoplankton blooms in a sustainable way. The algal bloom spectral indices viz. CAI, CI, GLI, NGRDI, NDVI, EXG, AI, etc. are some band ratio algorithms that assist in mapping and thereby making it possible for their management over large basin areas.

The advancement of UAVs further highlights the sustainable monitoring of algal blooms and HAB species in large spatial and temporal scales. The integration of hyperspectral sensors on UAVs is an added advantage to identify and map algal blooms at species level. However, the only key limiting factors in using UAVs for algal bloom monitoring are the difficulties in imaging water (more specifically, the difficulties in stitching together water images), flight issues due to weather conditions, expensive sensors, and a lack of standardization and interoperability in methodologies. As hyperspectral and LiDAR sensors diminish in price, and algorithms specific to freshwater and marine environments are established, UAV-based algal bloom research will continue to flourish. Indices that are modified from NDVI, such as the NGRDVI, GNDVI, and BNDVI, can be utilized to detect phytoplankton blooms using UAVs. Governments can benefit from this information as UAV-based algal bloom studies can prevent the destruction of aquaculture establishments, risk of consuming toxic shellfish, death of fish and other organisms, and illness of people who live near affected lakes, rivers, and tidal zones. Furthermore, identifying different colored algae with optical sensors on a UAV can help researchers understand environmental concerns such as insect emergence, nitrogen fixation and export, and carbon cycling in aquatic systems. As UAV technology continues to advance, airborne methodologies will allow data collectors to integrate their detection techniques into predictive algal bloom models, preventing public health outbreaks, and promoting adaptive strategies to combating deleterious effects of prolific and toxic algal blooms.

References

[1] M.M. Khin, A.S. Nair, V.J. Babu, R. Murugan, S. Ramakrishna, A review on nanomaterials for environmental remediation, Energy Environ. Sci. 5 (2012) 8075–8109, doi:10.1039/c2ee21818f.

[2] W.J. Ong, L.L. Tan, Y.H. Ng, S.T. Yong, S.P. Chai, Graphitic carbon nitride (g-C_3N_4)-based photocatalysts for artificial photosynthesis and environmental remediation: are we a step closer to achieving sustainability?, Chem. Rev. 116 (12) (2016) 7159–7329, doi:10.1021/acs.chemrev.6b00075.

[3] Y. Zeng, C. Liu, L. Wang, S. Zhang, Y. Ding, Y. Xu, Y. Liu, S. Luo, A three-dimensional graphitic carbon nitride belt network for enhanced visible light photocatalytic hydrogen evolution, J. Mater. Chem. A 48 (2016) 19003–19010.

[4] M.C. Cringoli, S. Kralj, M. Kurbasic, M. Urban, S. Marchesan, Luminescent supramolecular hydrogels from a tripeptide and nitrogen-doped carbon nanodots, Beilstein J. Nanotechnol. 8 (2017) 1553–1562, doi:10.3762/bjnano.8.157.

[5] P. Chen, N. Li, X. Chen, W.-J. Ong, X. Zhao, The rising star of 2D black phosphorus beyond graphene: synthesis, properties and electronic applications, 2D Mater 5 (2017) 014002, doi:10.1088/2053-1583/aa8d37.

[6] S. Kumar, A. Kumar, A. Bahuguna, V. Sharma, V. Krishnan, Two-dimensional carbon-based nanocomposites for photocatalytic energy generation and environmental remediation applications, Beilstein J. Nanotechnol. 8 (2017) 1571–1600, doi:10.3762/bjnano.8.159.

[7] T. Ma, Q. Fan, H. Tao. Z. Han, M. Jia, Y. Gao, W. Ma, Z. Sun, Heterogeneous electrochemical CO_2 reduction using nonmetallic carbon-based catalysts: current status and future challenges, Nanotechnology 28 (2017) 472001, doi:10.1088/1361-6528/aa8f6f.

[8] Y.L. Loow, E.K. New, G.H. Yang, L.Y. Ang, L.Y.W. Foo, T.Y. Wu, Potential use of deep eutectic solvents to facilitate lignocellulosic biomass utilization and conversion, Cellulose 24 (9) (2017) 3591–3618, doi:10.1007/s10570-017-1358-y.

[9] N.M. Park, J.B. Koo, J.Y. Oh, H.J. Kim, C.W. Park, S.D. Ahn, S.W. Jung, Electroluminescent nanocellulose paper, Mater. Lett. 196 (2017) 12–15, doi:10.1016/j.matlet.2017.03.003.

[10] Z. Karim, S. Claudpierre, M. Grahn, K. Oksman, A.P. Mathew, Nanocellulose based functional membranes for water cleaning: tailoring of mechanical properties, porosity and metal ion capture, J. Membr. Sci. 514 (2016) 418–428, doi:10.1016/j.memsci.2016.05.018.

[11] J.N. Putro, A. Kurniawan, S. Ismadji, Y-H. Ju, Nanocellulose based biosorbents for wastewater treatment: study of isotherm, kinetic, thermodynamic and reusability, Environ. Nanotechnol. Monit. Manage. 8 (2017) 134–149, doi:10.1016/j.enmm.2017.07.002.

[12] J. Liu, S. Willför, A. Mihranyan, On importance of impurities, potential leachables and extractables in algal nanocellulose for biomedical use, Carbohydr. Polym. 172 (2017) 11–19, doi:10.1016/j.carbpol.2017.05.002.

[13] X. Du, Z. Zhang, W. Liu, Y. Deng, Nanocellulose-based conductive materials and their emerging applications in energy devices—a review, Nano Energy 35 (2017) 299–320, doi:10.1016/j.nanoen.2017.04.001.

[14] A.F. Turbak, F.W. Snyder, K.R. Sandberg, Microfibrillated cellulose: a new cellulose product: properties, uses, and commercial potential, In Journal of Applied Polymer Science: Applied Polymer Symposium, Ninth International Cellulose Conference, 37, Wiley: NY, Syracuse, NY, 1983 May 24, 1982.

[15] D. Klemm, F. Kramer, S. Moritz, T. Lindström, M. Ankerfors, D. Gray, A. Dorris, Nanocelluloses: a new family of nature-based materials, Angew. Chemie. Int. Ed. 50 (24) (2011) 5438–5466. https://doi.org/10.1002/anie.201001273.

[16] G. Metreveli, L. Wågberg, E. Emmoth, S. Belák, M. Strømme, A. Mihranyan, A size-exclusion nanocellulose filter paper for virus removal, Adv. Healthcare Mater. 3 (2014) 1546–1550, doi:10.1002/adhm.201300641.

[17] C. Ruan, M. Strømme, J. Lindh, A green and simple method for preparation of an efficient palladium adsorbent based on cysteine functionalized 2,3-dialdehyde cellulose, Cellulose 23 (2016) 2627–2638, doi:10.1007/s10570-016-0976-0.

[18] S.A. El-Safty, M.A. Shenashen, M. Sakai, E. Elshehy, K. Halada, Detection and recovery of palladium, gold and cobalt metals from the urban mine using novel sensors/adsorbents designated with nanoscale wagon-wheel-shaped pores, J. Vis. Exp. (106) (2015) e53044, doi:10.3791/53044.

[19] Future Markets Inc, The Global Market For Nanocellulose 2017 to 2027. Research and Markets (2017).

[20] D. Klemm, E.D. Cranston, D. Fischer, M. Gama, S.A. Kedzior, D. Kralisch, F. Kramer, T. Kondo, T. Lindström, S. Nietzsche, K. Petzold-Welcke, F. Rauchfuß, Nanocellulose as a natural source for groundbreaking applications in materials science: today's state, Mater. Today 21 (7) (2018) 720–748.

[21] N. Mohammed, N. Grishkewich, R.M. Berry, K.C. Tam, Cellulose nanocrystal–alginate hydrogel beads as novel adsorbents for organic dyes in aqueous solutions, Cellulose 22 (2015) 3725–3738. https://doi.org/10.1007/s10570-015-0747-3.

[22] A. Mautner, K-Y. Lee, T. Tammelin, A.P. Matthew, A.J. Nedoma, K. Li, A. Bismarck, Cellulose nanopapers as tight aqueous ultra-filtration membranes, React. Funct. Polym. 86 (2015) 209–214. https://doi.org/10.1016/j.reactfunctpolym.2014.09.014.

[23] J-J. Wang, H-C. Yang, M-B. Wu, X. Zhang, Nanofiltration membranes with cellulose nanocrystals as an interlayer for unprecedented performance, J. Mater. Chem. A. 5 (31) (2017) 16289–16295. https://doi.org/10.1039/C7TA00501F.

[24] J. Lu, R.N. Jin, C. Liu, Y.F. Wang, X.K. Ouyang, Magnetic carboxylated cellulose nanocrystals as adsorbent for the removal of Pb(II) from aqueous solution, Int. J. Biol. Macromol. 93 (2016) 547–556, doi:10.1016/j.ijbiomac.2016.09.004.

[25] N. Mohammed, A. Baidya, V. Murugesan, A.A. kumar, M.A. Ganayee, J.S. Mohanty, K.C. Tam, T. Pradeep, Diffusion-controlled simultaneous sensing and scavenging of heavy metal ions in water using atomically precise cluster–cellulose nanocrystal composites, ACS Sustain. Chem. Eng. 4 (11) (2016) 6167–6176. https://doi.org/10.1021/acssuschemeng.6b01674.

[26] H. Zhu, X. Yang, E.D. Cranston, S. Zhu, Flexible and porous nanocellulose aerogels with high loadings of metal-organic-framework particles for separations applications, Adv Mater 28 (35) (2016) 7652–7657, doi:10.1002/adma.201601351.

[27] N. Bossa, A.W. Carpenter, N. Kumar, C-F. de Lannoy, M. Wiesner, Cellulose nanocrystal zero-valent iron nanocomposites for groundwater remediation, Environ. Sci. Nano. 6 (6) (2017) 1294–1303, doi:10.1039/c6en00572a.

[28] O. Schofield, R.A. Arnone, W.P. Bissett, T.D. Dickey, C.O. Davis, Z. Finkel, M. Oliver, M.A. Moline, Watercolors in the coastal zone: what can we see?, Oceanography 17 (2003) 24–31.

[29] K.H. Lee, R.S. Lunetta, Wetland detection methods, in: JG Lyon, J McCarthy (Eds.), Wetland and Environmental Application of GIS, Lewis Publishers, New York, 1996, pp. 249–284.

[30] M. Shaikh, D. Green, H. Cross, A remote sensing approach to determine environmental flow for wetlands of lower Darling River, New South Wales, Australia, Int. J. Remote Sens. 22 (2001) 1737–1751.

[31] A.M.B. May, J.E. Pinder, G.C. Kroh, A comparison of LANDSAT Thematic Mapper and SPOT multi-spectral imagery for the classification of shrub and meadow vegetation in Northern California, USA, Int. J. Remote Sens. 18 (1997) 3719–3728.

[32] E. Belluco, M. Camuffo, S. Ferrari, L. Modenese, S. Silvestri, A. Marani, M. Marani, Mapping salt-marsh vegetation by multispectral and hyperspectral remote sensing, Remote Sens. Environ 105 (2006) 54–67.

[33] P.H. Rosso, S.L. Ustin, A. Hastings, Mapping marshland vegetation of San Francisco Bay, California, using hyperspectral data, Int. J. Remote Sens. 26 (2005) 5169–5191.

[34] S. Martin, An Introduction to Remote Sensing, Cambridge University Press, Cambridge, England, 2004.

[35] S. Purkis, V. Klemas, Remote Sensing and Global Environmental Change, Wiley-Blackwell, Oxford, 2011.

[36] N.K. Hojerslev, Water color and its relation to primary production, Bound.-Layer Meteorol 18 (1980) 203–220.

[37] M.A. Montes-Hugo, K. Carder, R.J. Foy, J. Cannizzaro, E. Brown, S. Pegau, Estimating phytoplankton biomass in coastal waters of Alaska using airborne remote sensing, Remote Sens. Environ. 98 (2005) 481–493.

[38] M.T. Moita, P.B. Oliveria, J.C. Mendes, A.S. Palma, Distribution of chlorophyll-a and Gymnodinium catenatum associated with coastal upwelling plumes off central Portugal, Int. J. Ecol. 24 (2003) S125–S132.

[39] G.C. Pitcher, A.D. Boyd, D.A. Horstman, B.A. Mitchell-Innes, Subsurface dinoflagellate populations, frontal blooms and the formation of red tide in the southern Benguela upwelling system, Marine Ecol. Prog. Series 172 (1998) 253–264.

[40] J.J. Cullen, A.M. Ciotti, A.F. Davis, M.R. Lewis, Optical detection and assessment of algal blooms, Limnol. Oceanogr. 42 (5, part 2) (1997) 1223–1229.

[41] V. Klemas, Sensors and techniques for observing coastal ecosystems, in: X. Yang (Ed.), Remote Sensing and Geospatial Technologies for Coastal Ecosystem Assessment and Management, Springer-Verlag, Berlin, 2009, pp. 17–44.

[42] S.J. Lavender, S.B. Groom, The detection and mapping of algal blooms from space, Int. J. Remote Sens 22 (2001) 197–201.

[43] M.C. Tomlinson, R.P. Stumpf, V. Ransibrahmanakul, E.W. Truby, G.J. Kirkpatrick, B.A. Pederson, G.A. Vargo, C.A. Heil, Evaluation of the use of SeaWiFS imagery for detecting Karenia brevis harmful algal blooms in the eastern Gulf of Mexico, Remote Sens. Environ 91 (2004) 293–303.

[44] J.H. Landsberg, The effects of harmful algal blooms on aquatic organisms, Rev. Fish. Sci. 10 (2002) 113–390.

[45] R.P. Stumpf, M.C. Tomlinson, Remote Sensing of Harmful Algal Blooms, in: R.L. Miller, C.E. del Castillo, B.A. McKee (Eds.), Remote Sensing of Coastal Aquatic Environments, Springer, Dordrecht, The Netherlands, 2005, pp. 277–296. ISBN 1-4020-3099-1.

[46] A. Morel, L. Prieur, Analysis of variation in ocean color, Limnol. Oceanogr 22 (1997) 709–722.

[47] W. Philpot, Estimating atmospheric transmission and surface reflectance from a glint-contaminated spectral image, IEEE Trans. Geosci. Remote Sens. 45 (2007) 448–457.

[48] I.S. Robinson, Measuring the Ocean from Space: The Principles and Methods of Satellite Oceanography, Springer-Verlag, Berlin, 2004.

[49] Y. Wang, Remote Sensing of Coastal Environments, CRC Press, Boca Raton, Florida, 2010.

[50] A.A. Gitelson, Y.Z. Yacobi, J.F. Schalles, D.C. Rundquist, L. Han, R. Stark, D. Etzion, Remote estimation of phytoplankton density in productive waters, Arch. Hydrobiol., Special issue on Advanced Limnology 55 (2000) 121–136.

[51] V. Klemas, Remote sensing techniques for studying coastal ecosystems, J. Coast. Res 27 (2011) 1–16.

[52] K.-H. Szekielda, J.H. Bowles, D.B. Gillis, W. Snyder, W.D. Miller, Patch recognition of algal blooms and macroalgae, in: W. Hou, R.A. Arnone (Eds.), Ocean Sensing and Monitoring II, 7678, SPIE, 2010, pp. 1–9.

[53] D.J. Keith, Estimating chlorophyll conditions in southern New England coastal waters from hyperspectral remote sensing, in: Y. Wang (Ed.), Remote Sensing of Coastal Environments, CRC Press, Boca Raton, Florida, 2010, pp. 151–172.

[54] M. Wang, W. Shi, Satellite-observed algae blooms in China's Lake Taihu, AGU EOS Trans. 89 (2008) 201–202.

[55] M. Wang, Remote sensing of the ocean contributions from ultraviolet to near-infrared using the shortwave infrared bands: simulations, Appl. Optics 46 (2007) 1535–1547.

[56] J.R. Jensen, Remote Sensing of the Environment: An Earth Resource Perspective, Prentice Hall, Upper Saddle River, New Jersey, 2007.

[57] W.P. Bissett, R.A. Arnone, C.O. Davis, T.D. Dickey, D. Dye, D.D.R. Kohler, R.W. Gould Jr., From meters to kilometers: a look at ocean-color scales of variability, spatial coherence, and the need for fine-scale remote sensing in coastal ocean optics, Oceanography 17 (2) (2004) 32–43.

[58] J.F.R. Gower, SeaWiFS global composite images show significant features of Canadian waters for 1997–2001, Can. J. Remote Sens. 30 (2004) 26–35.

[59] J.F.R. Gower, L. Brown, G.A. Borstad, Observation of chlorophyll fluorescence in west coast waters of Canada using the MODIS satellite sensor, Can. J. Remote Sens. 30 (2004) 17–25.

[60] C. McClain, S. Hooker, G. Feldman, P. Bontempi, Satellite data for ocean biology, biogeochemistry, and climate research, AGU EOS Trans 87 (2006) 337–343.

[61] N. Kabbara, J. Benkhelil, M. Awad, V. Barale, Monitoring water quality in the coastal area of Tripoli (Lebanon) using high-resolution satellite data, ISPRS J. Photogramm. Remote Sens. 63 (2008) 488–495.

[62] C.M. Roelfsema, S.R. Phinn, W.C. Dennison, A.G. Decker, V.E. Brando, Monitoring toxic cyanobacteria Lyngbya majuscule (Gomont) in Moreton Bay, Australia by integrating satellite image data and field mapping, Harmful Algae 5 (2006) 45–56.

[63] B.A. Keafer, D.M. Anderson, Use of remotely-sensed sea surface temperatures in studies of Alexandrium tamarense bloom dynamics Toxic Phytoplankton Blooms in the Sea, in: T.M. Smayda, Y. Shimizu (Eds.), Proceedings of the 5th International Conference on Toxic Marine Phytoplankton, Elsevier, Amsterdam, 1993, pp. 763–768.

[64] P.A. Tester, R.P. Stumpf, F.M. Vukovich, P.K. Fowler, J.F. Turner, An expatriate red tide bloom: transport, distribution and persistence, Limnol. Oceanogr. 36 (1991) 1053–1061.

[65] P.I. Miller, J.D. Shutler, G.F. Moore, S.B. Groom, SeaWiFS discrimination of harmful algal bloom evolution, Int. J. Remote Sens 27 (2006) 2287–2301.

[66] R.E. Turner, N.N. Rabalais, Coastal eutrophication near the Mississippi river delta, Nature 368 (1994) 619–621.

[67] B. Dzwonkowski, X.H. Yan, Tracking of a Chesapeake Bay estuarine outflow plume with satellite-based ocean color data, Cont. Shelf Res. 25 (2005) 1942–1958.

[68] V. Klemas, Remote sensing of coastal plumes and ocean fronts: overview and case study, J. Coast. Res. 28 (1A) (2011) 1–7.

[69] V. Klemas, W. Philpot, Drift and dispersion studies of ocean-dumped waste using Landsat imagery and current drogues, Photogr. Eng. Remote Sens. 47 (1981) 533–542.

[70] A.C. Thomas, R.A. Weatherbee, Satellite-measured temporal variability of the Columbia River plume, Remote Sens. Environ 100 (2006) 167–178.

[71] C. Kislik, I. Dronova, M. Kelly, UAVs in support of algal bloom research: a review of current applications and future opportunities, Drones 2 (35) (2018) 1–14, doi:10.3390/drones2040035.

[72] R.M. Kudela, S.L. Palacios, D.C. Austerberry, E.K. Accorsi, L.S. Guild, J. Torres-Perez, Application of hyperspectral remote sensing to cyanobacterial blooms in inland waters, Remote Sens. Environ. 167 (2015) 196–205.

[73] S.J. Goldberg, J.T. Kirby, S.C. Licht, Applications of Aerial Multi-Spectral Imagery for Algal Bloom Monitoring in Rhode Island. SURFO Technical Report No. 16-01, University of Rhode Island: South Kingstown, RI, USA, 2016, pp. 28.

[74] F. Xu, Z. Gao, X. Jiang, W. Shang, J. Ning, D. Song, J. Ai, A UAV and S2A data-based estimation of the initial biomass of green algae in the South Yellow Sea, Mar. Pollut. Bull. 128 (2018) 408–414.

[75] D. Van der Merwe, K.P. Price, Harmful algal bloom characterization at ultra-high spatial and temporal resolution using small unmanned aircraft systems, Toxins 7 (2015) 1065–1078.

[76] S.W. Jang, H.J. Yoon, S.N. Kwak, B.Y. Sohn, S.G. Kim, D.H. Kim, Algal bloom monitoring using UAVs imagery, Adv. Sci. Technol. Lett. 138 (2016) 30–33.

[77] H.M. Kim, H.J. Yoon, S.W. Jang, S.N. Kwak, B.Y. Sohn, S.G. Kim, D.H. Kim, Application of unmanned aerial vehicle imagery for algal bloom monitoring in river basin, Int. J. Control Autom. 9 (2016) 203–220.

[78] R.J. Zomer, A. Trabucco, S.L. Ustin, Building spectral libraries for wetlands land cover classification and hyperspectral remote sensing, J. Environ. Manag. 90 (2009) 2170–2177. http://www.transparencymarketresearch.com/algae-market.html. Algae Market, By Application, By Cultivation Technology, and Geography - Global Industry Analysis, Size, Share, Growth, Trends, and Forecast - 2016-2024.

[79] A. García, A. Gandini, J. Labidi, N. Belgacem, J. Bras, Industrial and crop wastes: a new source for nanocellulose biorefinery, Ind. Crops Prod. 93 (2016) 26–38, doi:10.1016/j.indcrop.2016.06.004.

[80] C.S.T. Daughtry, E.R. Hunt Jr., IIIP.C. Doraiswamy, J.E. McMurtrey, Remote sensing the spatial distribution of crop residues, Agronomy J 97 (3) (2005) 864–871.

[81] G. Serbin, C.S.T. Daughtry, E.R. Hunt Jr, D.J. Brown, G.W. McCarty, Effect of soil spectral properties on remote sensing of crop residue cover, Soil Sci. Soc. Am. J 73 (5) (2009) 1545–1558.

[82] C. Daughtry, E.R. Hunt, C.L. Walthall, T.J. Gish, S. Liang, E.J. Kramer, Assessing the spatial distribution of plant litter, Proc. Tenth JPL Airborne Earth Science Workshop, 2001 105–114.

[83] C.S.T. Daughtry, Discriminating crop residues from soil by shortwave infrared reflectance, Agronomy J 93 (1) (2001) 125–131. https://doi.org/10.2134/agronj2001.931125x.

[84] P.L. Nagler, C.S.T. Daughtry, S.N. Goward, Plant litter and soil reflectance, Remote Sens. Environ. 71 (2) (2000) 207–215.

[85] N. Oppelt, W. Mauser, Hyperspectral monitoring of physiological parameters of wheat during a vegetation period using AVIS data, Int. J. Remote Sens 25 (1) (2004) 145–159, doi:10.1080/01431 16031000115300.

[86] K. Shang, L. Ke, Z. Feng, S. Karungaru, An Event-Driven Based Multiple Scenario Approach for Dynamic and Uncertain UAV Mission Planning. In: Y. Tan, Y. Shi, F. Buarque, A. Gelbukh, S. Das,

A. Engelbrecht (Eds.), Advances in Swarm and Computational Intelligence. ICSI 2015. Lecture Notes in Computer Science, vol. 9141 (2015). doi:10.1007/978-3-319-20472-7_33.

[87] E. Honkavaara, T. Hakala, J. Kirjasniemi, A. Lindfors, J. Mäkynen, K. Nurminen, P. Ruokokoski, H. Saari, L. Markelin, New light-weight stereosopic spectrometric airborne imaging technology for high-resolution environmental remote sensing case studies in water quality mapping, Int. Arch. Photogramm. Remote Sens. Spat. Inf. Sci. 1 (2013) W1.

[88] P. Lyu, Y. Malang, H.H.T. Liu, J. Lai, J. Liu, B. Jiang, M. Qu, S. Anderson, D.D. Lefebvre, Y. Wang, Autonomous cyanobacterial harmful algal blooms monitoring using multirotor UAS, Int. J. Remote Sens. 38 (2017) 2818–2843.

[89] B. Bollard-Breen, J.D. Brooks, M.R.L. Jones, J. Robertson, S. Betschart, O. Kung, S. Craig Cary, C.K. Lee, S.B. Pointing, Application of an unmanned aerial vehicle in spatial mapping of terrestrial biology and human disturbance in the McMurdo dry valleys, East Antarctica. Polar Biol. 38 (2015) 573–578.

[90] R. Aguirre-Gómez, O. Salmerón-García, G. Gómez-Rodríguez, A. Peralta-Higuera, Use of unmanned aerial vehicles and remote sensors in urban lakes studies in Mexico, Int. J. Remote Sens. 38 (2017) 2771–2779.

[91] S. Shang, Z. Lee, G. Lin, C. Hu, L. Shi, Y. Zhang, X. Li, J. Wu, J. Yan, Sensing an intense phytoplankton bloom in the western Taiwan Strait from radiometric measurements on a UAV, Remote Sens. Environ 198 (2017) 85–94.

[92] C. Koparan, A. Koc, C. Privette, C. Sawyer, In situ water quality measurements using an unmanned aerial vehicle (UAV) system, Water 10 (2018) 264.

[93] D. Blondeau-Patissier, J.F.R. Gower, A.G. Dekker, S.R. Phinn, V.E. Brando, A review of ocean color remote sensing methods and statistical techniques for the detection, mapping and analysis of phytoplankton blooms in coastal and open oceans, Prog. Oceanogr. 123 (2014) 123–144.

[94] T. Kutser, L. Metsamaa, N. Strömbeck, E. Vahtmäe, Monitoring cyanobacterial blooms by satellite remote sensing, Estuar. Coast. Shelf Sci. 67 (2006) 303–312.

[95] J.A.J. Berni, P.J. Zarco-Tejada, M.D. Suárez Barranco, E. FeresesCastiel, Thermal and narrow-band multispectral remote sensing for vegetation monitoring from an unmanned aerial vehicle, IEEE Trans. Geosci. Remote Sens. 47 (2009) 722–738.

[96] S. Liu, C. Zhang, Y. Zhang, T. Wang, A. Zhao, T. Zhou, X. Jia, Miniaturized spectral imaging for environment surveillance based on UAV platform, Proc. SPIE. 10461 (2017) 104611K.

[97] T.-C. Su, H.-T. Chou, Application of multispectral sensors carried on unmanned aerial vehicle (UAV) to trophic state mapping of small reservoirs: a case study of Tain-Pu Reservoir in Kinmen, Taiwan. Remote Sens. 7 (2015) 10078–10097.

[98] A. Zingone, H. Oksfeldt Enevoldsen, The diversity of harmful algal blooms: a challenge for science and management, Ocean Coast. Manag. 43 (2000) 725–748.

[99] A.E. Jochens, T.C. Malone, R.P. Stumpf, B.M. Hickey, M. Carter, R. Morrison, J. Dyble, B. Jones, V.L. Trainer, Integrated ocean observing system in support of forecasting harmful algal blooms, Technol. Soc. J. 44 (2010) 99–121.

Index

Printed in the United States
by Baker & Taylor Publisher Services